On the Way to Individuality: Methodological Issues in Behavioral Genetics

Contributors

Dorret I. Boomsma
University of Amsterdam and Free University
Amsterdam, The Netherlands

Thomas J. Bouchard, Jr.
University of Minnesota
Minnesota, USA

Remi J. Cadoret
University of Iowa College of Medicine
Iowa, USA

Alexei A. Chikanian
Yale University
Connecticut, USA
Lebedev Institute for Physics
Moscow, Russia

Hilary Coon
University of Utah School of Medicine
Utah, USA

Conor Dolan
University of Amsterdam and Free University
Amsterdam, The Netherlands

Lindon J. Eaves
Medical College of Virginia
Virginia Commonwealth University
Virginia, USA

Graham G. Giles
The Anti-Cancer Council of Victoria
Victoria, Australia

Elena L. Grigorenko
Yale University
Connecticut, USA
Moscow State University
Moscow, Russia

John K. Hewitt
University of Colorado, Boulder
Colorado, USA

John L. Hopper
The University of Melbourne
Victoria, Australia

Mark A. Jenkins
The University of Melbourne
Victoria, Australia

Michele C. LaBuda
Johns Hopkins Medical Institutions
Maryland, USA

Gregory T. Macaskill
The University of Melbourne
Victoria, Australia

Hermine H. Maes
Medical College of Virginia
Virginia Commonwealth University
Virginia, USA

Millenium Pharmaceuticals

Peter C. M. Molenaar
University of Amsterdam and Free University
Amsterdam, The Netherlands

Nancy Pedersen
Karolinska Institute
Stockholm, Sweden
Pennsylvania State University
Pennsylvania, USA

Andrew Pickles
Institute of Psychiatry
University of London
London, England

Richard Rende
Department of Psychology
Rutgers University
New Jersey, USA

Michael Rutter
Institute of Psychiatry
University of London
London, England

Judy L. Silberg
Medical College of Virginia
Virginia Commonwealth University
Virginia, USA

Emily Simonoff
Institute of Psychiatry
University of London
London, England

Kristian Tambs
National Institute of Public Health
Department of Epidemiology
Oslo, Norway

Edward P. Troughton
University of Iowa College of Medicine
Iowa, USA

George P. Vogler
Pennsylvania State University
Pennsylvania, USA

William R. Yates
University of Iowa College of Medicine
Iowa, USA

ON THE WAY TO INDIVIDUALITY: METHODOLOGICAL ISSUES IN BEHAVIORAL GENETICS

MICHELE C. LABUDA AND ELENA L. GRIGORENKO

Nova Science Publishers, Inc.
Commack, New York

Editorial Production: Susan Boriotti
Office Manager: Annette Hellinger
Graphics: Frank Grucci and John T'Lustachowski
Information Editor: Tatiana Shohov
Book Production: Donna Dennis, Patrick Davin, Christine Mathosian and Tammy Sauter
Circulation: Maryanne Schmidt
Marketing/Sales: Cathy DeGregory

Library of Congress Cataloging-in-Publication Data
available upon request

1-56072-427-7

6080 Jericho Turnpike, Suite 207
Commack, New York 11725
Tele. 516-499-3103 Fax 516-499-3146
e-mail: Novascience@earthlink.net
e-mail: Novascil@aol.com
Web Site: http://www.nexusworld.com/nova

Printed in the United States of America

CONTENTS

PREFACE

On the Way to Individuality: Current Methodological Issues in Behavioral Genetics is intended to provide a bridge between behavioral genetics, psychiatric genetics, developmental psychopathology and developmental psychology. In the past decade, there is a significant tendency for a blend of these areas. Many developmental psychologists enhance their study designs by incorporating behavioral genetic methodology; many geneticists are becoming aware of the importance of developmental, stage-specific manifestations of the studied disorder. The authors contributing to the current volume come from different professional fields and different methodological backgrounds yet the main objective of their research is to better understand genetic and environmental forces influencing the development of adaptive and maladaptive behaviors.

The scope of this area of research is immense and it is difficult to capture the variety of relevant approaches, methodologies, and applications. We believe, however, that the value of this volume is in presenting for the reader several specific areas currently important in the advancement of behavioral genetic contributions to the study of behavior. No longer is it sufficient to demonstrate that familial similarity for behavioral traits is due to a genetic component. Instead, the methodologies presented and illustrated in this book tackle difficult but necessary complexities such as the modeling of developmental processes, environmental specificities, and underlying genetic mechanisms. The accomplishment of these goals requires such studies to be of large magnitude which facilitates and even necessitates international collaboration and awareness. It is our hope that the chapters in this volume are representative of the many large-scale projects conducted in different countries.

The text is intended for anyone with an interest in behavioral genetics, not necessarily those with extensive expertise in the area. The book is designed to be comprehensible to students as well as professionals in the fields of psychology, psychiatry, and genetics. Some of the chapters are quite general, requiring no real genetic, psychological, or statistical expertise while others are more technically demanding. In order for these chapters to be accessible to a variety of readers, the authors have been sensitive to providing the broader context of their research.

The book is organized into 3 main sections. The first section deals with the integration of development into genetic modeling. The statistical methods of the three chapters in this section are not simply an exercise in designing increasingly more elaborate structural models but reflect our enhanced understanding of the true complexities of human behavioral development, orchestrated by the interplay of genes and environment. While the first chapter in this section discusses the current theoretical issues in the field, the two subsequent chapters introduce specific path-analytic methodological advancements aimed at incorporating developmental issues into genetic analyses.

In chapter 1, Rende outlines the general framework established by developmental psychopathology for understanding common roots in normal and abnormal behavioral development/expression. In doing so, he suggests to focus not on normal or pathological development but on the range of individual differences that is seen when one considers dimensions of problem behaviors in children and adolescents. Rende argues that it is more advantageous to focus on roots of variation rather than to cleave populations into "affected" and "unaffected" groups in order to understand the continuity or discontinuity in etiological factors of adaptive and maladaptive behaviors. The author illustrates his major points with varied examples including fragile X, autism, reading disability, attentional problems, and depressive symptomatology and provides the reader with a discussion not only of the importance of genetic and environmental factors but their intricate interplay in development.

In chapter 2, Meyer and colleagues expand some theoretical ideas introduced by Rende in the first chapter. In particular, the authors address how to incorporate developmental change in traditional behavioral genetic path analytic methodology. They begin with a review of the approaches currently used for developmental genetic analysis and suggest that these models can lead to erroneous conclusions about gene expression when individuals within the same age groups are at different developmental stages. The authors develop an alternative model to accommodate age and stage developmental differences and illustrate their model by applying it to data on body mass index in a prospective study of adolescent twin pairs.

In chapter 3, Molenaar, Boomsma, and Dolan also focus on developmental issues but approach longitudinal behavior genetic analysis with a specific goal to detect the presence of genotype-environment interaction The authors suggest that the presence of this interaction could be detected by the analysis of higher order moments of the joint distribution of genetic and environmental factor scores. These techniques are illustrated in a simulated dataset.

The second section, *Family, Twin, and Adoption Studies: Diversity of Approaches*, is designed to provide illustrations of various methods used in behavior and psychiatric genetic studies. The section begins with a variation of traditional twin design and extends into family design, highlighting various aspects of this methodology.

In chapter 4, Bouchard and Pedersen describe the methodology of the twin reared apart design. The data discussed in the chapter were obtained form two large-scale studies of twins reared apart, the Minnesota Study of Twins Reared Apart and the Swedish Adoption/Twin Study of Aging. The participants of these studies came from the United States and Sweden as well as United Kingdom, Australia, Canada, China, New Zealand and Germany. The authors study various aspects of personality and intelligence in twins reared apart and discuss the advantages and disadvantages of this naturally occurring experiment.

The authors of chapter 5, Yates, Cadoret, and Troughton, define the goal of their chapter as to provide evidence that adoption studies can play a significant role in correlating the clinical epidemiology of psychiatric conditions with the findings from human genome research. The authors present results from their 20-year experience of working with adoptees, discussing alternative interactive pathways leading to alcohol abuse and antisocial personality in the individuals adopted away. The models incorporate indices of genetic vulnerability, adoptive family environments, as well as personality characteristics of adoptees.

Tambs in chapter 6 takes on three specific methodological issues in familial analyses of behavioral data: non-random mating, effects of family environment, and age-specific genetic or environmental effects. The author presents a thorough discussion of each of these issues and provides empirical illustrations based upon self-reports of anxiety and depression in a large family sample that includes most of the adult population of one Norwegian county.

Hopper and colleagues (chapter 7) devote their paper to the statistical analysis of familial aggregation for a binary trait. The categorical nature of these yes-no traits assumes a particular way of data analysis, that have not been addressed in detail in any other chapters. The authors stress the advantage of these approaches in being capable of examining simultaneously possible genetic and environmental etiologies. The methodological issues are illustrated by application to data on smoking behavior in parents and their offspring as assessed in two Tasmanian health surveys. The distinct feature of these data is that the assessments were obtained in two generations approximately 25 years apart (i.e., parents and their offspring were assessed at approximately the same age).

The third section is entitled *Searching for Genes Involved in Behavioral Expression.* Its goal is to provide a background in linkage studies and introduce some current methodological developments in the field.

In chapter 8, Coon does a thorough job of introducing concepts and terminology of linkage analysis and bridging traditional behavior genetic designs and techniques of molecular genetics. She devotes special attention to parametric and nonparametric analyses of complex diseases and presents results from a family study of schizophrenia.

In chapter 9, Grigorenko and Chikanian explore the applicability of the method of allele-sharing by distantly affected relatives to whole-genome screening. The authors include several simulation studies which research the informativeness and power of this method as a prelude to extensive genotyping efforts.

Vogler in chapter 10 focuses on finding chromosomal regions involved in the etiology of quantitative traits. This methodology combines the partitioning of quantitative genetic variation due to loci on specific chromosomal regions with the classical twin study design. The application of the model is illustrated with simulated data.

SECTION I
INCORPORATING DEVELOPMENT IN THE SEARCH FOR ETHIOLOGICAL FACTORS IN HUMAN TRAITS

CHAPTER 1

ADAPTIVE AND MALADAPTIVE PATHWAYS IN DEVELOPMENT: A QUANTITATIVE GENETIC PERSPECTIVE

Richard Rende

Department of Psychology, Rutgers University, New Jersey, USA

INTRODUCTION

Over the past decade, the discipline of developmental psychopathology has emerged as a recognized field of inquiry. As is often the case with new areas of study, however, providing a definition of developmental psychopathology is not an easy task. Although theorists have begun to carve out the defining aspects of this discipline (e.g., Cicchetti, 1984; Cicchetti, 1990), there is still much discussion as to what makes developmental psychopathology unique.

Perhaps the best way to introduce developmental psychopathology is to emphasize that it attempts to integrate the study of development with the study of maladaptation (Lewis, 1992). For example, there was at one time a noticeable schism between developmental psychology and clinical child psychology (Cicchetti, 1984). Developmental psychology tended to focus on issues of normal development, especially in terms of defining age-appropriate behaviors and revealing developmental milestones. In contrast, clinical child psychology concentrated on problem behavior and maladaptation. Although it may have been recognized that the normative milestones revealed by developmental psychologists could be used as a framework for studying problem behavior, there was little emphasis on both theoretical and empirical approaches to integrating these different fields of study.

The integration of studying normal development and deviations from normality in a unifying perspective may thus be the most obvious contribution of developmental psychopathology. Rather than dichotomizing the study of normal and abnormal, pathways in development, and use the study of each to inform the other. As noted by Lewis (1992), "while developmental psychopathology seeks to understand normal development, it is the development of deviation, the definition of deviation, and individual differences that are the focal points" (p. 487).

The theme of this chapter is ways in which quantitative genetic research may be applied to understand the etiological bases of the range of individual differences that is seen when one considers dimensions of problem behaviors--including the far end of distributions which manifest as diagnosed disorders--in childhood and adolescence. Specifically, quantitative genetic approaches which may shed light on both the definition, and development, of deviation will be discussed. This focus on the roots of variation, rather than clearly cleaving development into affected and unaffected individuals, represents one important way in which quantitative genetic research contributes to understanding both adaptive and maladaptive behavior in an integrated fashion (e.g., Plomin, Rende, & Rutter, 1991; Rende & Plomin, 1990).

NORMATIVE VS. PATHOLOGICAL DEVELOPMENT

Behavioral genetic methods have been used historically to study both normal and pathological development. A major application of behavioral genetic methods has involved the examination of individual differences in quantitative, continuous traits studied by developmental psychologists. For example, there is a wealth of information concerning genetic and environmental contributions to individual differences in normal dimensions, such as cognitive abilities and personality traits (e.g., Plomin, DeFries, & McClearn, 1990). In addition, behavioral genetic methods have also been used to examine genetic and environmental contributions to pathological development, most often in the form of diagnosed conditions (e.g., Rutter, Macdonald, Le Couteur, Harrington, Bolton, & Bailey, 1990). However, to date there have been few attempts to relate behavioral genetic research on normal dimensions to work on pathological development. For example, most textbooks on behavioral genetic research review these domains separately.

Although there has been little integration of the study of normal and abnormal from a quantitative genetic perspective, research in genetics provides examples in which such an interplay has been useful. Research on mutations has historically been used to dissect the complexity of normal developmental processes. As an example, examinations of inborn errors of metabolism led to the "one-gene, one-enzyme" hypothesis that is the basic mechanism of normal gene action (Beadle & Tatum, 1941).

Behavioral genetic research on psychopathology can be viewed from a similar perspective: finding out what goes wrong in disorders is likely to reveal what normally goes right. An apt analogy is phenylketonuria (PKU). By focusing on this rare group of severely retarded individuals, the first single-gene mutation for retardation was found. Although the PKU gene causes retardation, by no means does this imply that the normal gene is responsible for normal cognitive development. Any one of many genes can disrupt normal cognitive development, but the normal range of cognitive development is orchestrated by a system of many genes as well as by many environmental influences. Nonetheless, the PKU gene provides one discreet, albeit tiny, window on normal development by implying that sufficient amounts of tyrosine are necessary for normal cognitive development.

Thus, a general theme which may be taken from genetics is that understanding maladaptation may magnify the processes involved in normal functioning. Hence, studying behavioral consequences of known genetic conditions provides one concrete example of forging links between the understanding of normal and pathological development. Consider as a second example a genetic condition which has received much attention by researchers in the past few years: the fragile X syndrome (see Rende & Plomin, 1994).

Fragile X refers to an observable break (or fragile site) in the structure of the X chromosome when cells are examined under specific culture conditions. Diverse phenotypic characteristics are associated with this syndrome. Fragile X is recognized as the second most frequent cause (surpassed only by Down's syndrome) of inherited mental retardation; it is found in 5 to 10% of retarded individuals. Retardation is usually a consequence for males with the fragile X marker, whereas the consequences for females are more variable. For males, additional forms of disturbance include hyperactivity, attentional deficits, and anxiety (Bregman, Leckman, & Ort, 1988); females have been reported to show increased schizotypal and affective symptomatology. Additionally, links between the fragile X marker and autistic behavior have been suggested for males, including deficits in social interaction, stereotyped behavior, atypical speech and language, and abnormal nonverbal communication (Bregman et al., 1988).

How is the study of the fragile X condition important for our understanding of genetic influences on normal behavioral development? One consideration is that fragile X demonstrates that a single genetic factor can influence a vast array of behavioral domains, including cognition and language as well as psychopathological characteristics such as atypical personality and affective symptoms. The diversity of domains affected reveals an important principle in genetics: localized genetic sites (such as the fragile site on the X chromosome) may affect multiple domains of functioning, a principle referred to in genetics as pleiotropy.

Another consideration is that the study of fragile X (and other single-gene disorders) holds out the promise of identifying the specific mechanisms by which genes have their diffuse effect on behavior. For fragile X, the recent discovery of the genetic site of the mutation will intensify this search (Oberle et al., 1991; Yu et al., 1991). Knowing the molecular basis of the fragile X syndrome may eventually reveal not only the specific gene product responsible for the mutation, but also the biological pathways which result in the behavioral symptoms associated with the fragile X site. Thus, the eventual combination of both molecular genetic techniques and other biological methods may reveal how this genetic condition affects neural development as well as pinpoint specific areas in the brain that underlie the behavioral consequences of fragile X. Again, in addition to understanding the cause of this condition, such research may also reveal genetic pathways involved in normal cognitive development, as was the case with research on PKU. It is this step--drawing implications about normal development from the study of pathological development--which reveals one way in which behavioral genetic approaches to psychopathology can inform the study of normal development.

Although studying known genetic conditions represents an important approach for understanding genetic influences on behavioral development--both pathological and normative--research in psychopathology most often takes on a different task: to begin with a behavioral condition of interest and to determine the relative contributions of genes and environment on the condition. Behavioral genetic research on psychopathology following this general framework may also make contributions to understanding normal development.

Consider as an example the syndrome of autism. Autism is one of the most disabling as well as baffling of childhood psychiatric disorders: to date the cause of the disorder is unknown. Recent twin studies have suggested a substantial genetic component to this disorder (see Rutter et al., 1990), although specific sites on the genome have not yet been identified. In terms of the theme of this chapter, what is especially interesting is that behavioral genetic approaches have not only suggested the potential importance of genetic factors in the etiology of the disorder, but have also helped to shape the way in which the disorder is characterized.

It has been known for a number of years that autistic individuals are impaired in multiple fundamental developmental domains, with core symptoms including aberrant social development, difficulties with language and communication, and bizarre motor problems. The peculiar features of autism have often led researchers to suggest that it represents a clear case of aberrant development that is qualitatively different from normal development. However, both family and twin studies have implied that this perspective may not be entirely accurate. Social and cognitive abnormalities have been found in the relatives of autistic individuals, suggesting that genetic influence on autism may operate more broadly through these areas of functioning rather than strictly through the specific clinical phenotype (see Rutter et al., 1990). The implication from this research is that the genetic contributions to autism may also lead to lesser forms of dysfunction which blend more easily into the normal range of variability. Hence, future research on autism may lead directly to further understanding of the genetic contributions to a wide range of social and cognitive domains necessary for normal development, and may help reveal in part the biological bases for core areas of human development.

Individual Differences in Development

The previous section focused primarily on research which contrasts pathological development from normative development. Much research in both developmental psychology and psychopathology has often applied such a framework. For example, social development can be studied in terms of developmental milestones which occur throughout the lifespan, including forming a relationship with caregivers in infancy, developing friendships in childhood, and embarking on romantic relationships in adolescence and adulthood. Given this developmental framework, the deficits in social development in autism can be characterized in part as failures to achieve each of these milestones. Presumably, then, research on the genetic basis of autism will reveal in part why social development goes awry in autistic individuals.

However, another prominent position in developmental research is to focus not on norms (and deviations from norms), but rather on variation across individuals in a given domain of development. From an individual differences perspective, rather than characterizing the typical or normative way in which infants develop a relationship with caregivers in infancy, the focus would be on variation in this process: for example, do some infants form "closer" relationships with their caregivers than others? Other similar questions could be asked: Are some infants more sociable than others? Do they remain more sociable when they reach middle childhood? Again, rather than define a norm, such research would emphasize the differences seen across individuals in a population.

An individual differences perspective on psychopathology would also focus more on variation in a population rather than consider individuals as either affected or unaffected. As was mentioned in the previous discussion on autism, lesser variants of pathological conditions may be seen, and these variants may blend more gradually into variations seen across individuals rather than clearly stand out as disorders. It is important to note that such an individual differences perspective does not imply that the "disordered" behavior is not problematic; rather it is seen as problematic against a background of variation instead of a background of a static norm.

Much behavioral genetic research is based on an individual differences perspective rather than a normative or "species typical" framework. The theory of quantitative genetics takes as its starting point variation in the population, and attempts to determine the extent to which genetic and environmental factors contribute to the observed variance (see Plomin et al., 1990). Although the utility of this approach has been well-recognized for studying normative dimensions of behavior such as cognitive abilities and personality traits, in recent years this general approach has begun to yield data on common problem behaviors in childhood and adolescence. Here the focus will be on the findings to emerge to date, as well as the implications for studying adaptive and maladaptive pathways in development.

COMMON BEHAVIOR PROBLEMS IN CHILDHOOD AND ADOLESCENCE

Common behavior problems offer an opportunity to apply behavioral genetic techniques aimed at individual differences because the defining symptoms are often present in varying degrees in individuals. Symptoms like oppositional behavior, aggressive behavior, and depressed mood may be experienced in varying degrees by children and adolescents. Empirical attempts to provide quantitative, continuous indices of common behavior problems have produced reliable and valid measures such as the Child Behavior Checklist (CBCL; Achenbach, 1991). Although there are empirical links between continuous dimensions of problem behavior and psychiatric diagnoses (e.g., Edelbrock & Costello, 1988), the overall focus of instruments such as the CBCL has been to characterize dimensions rather than generate diagnoses, and to examine variance in these dimensions across individuals.

The variance in problem behaviors which is observed among children and adolescents raises the question of the etiology of these differences. For example, from a quantitative genetic perspective, individual differences in symptoms of attention problems (ranging from no problems to some problems to many problems) could be due to genetic differences among individuals, environmental differences among individuals, or a combination of genetic and environmental effects (see Plomin, Rende, & Rutter, 1991). In terms of the theme of this chapter, by focusing on the range of symptoms in the population (from none to many), quantitative genetic approaches provide information on the etiology of both adaptive and maladaptive behavior.

Recent behavioral genetic studies have begun to provide a data base on the influences of genes and environment on common behavior problems in childhood and adolescence, and a few general trends have begun to appear. A number of studies have indicated that two forms of behavior problems in childhood and adolescence--aggressive behavior and attention problems--may be especially influenced by genetic differences across individuals. Two recent twin studies have revealed significant genetic influence on attention problems and related symptoms of hyperactivity in adolescence (Edelbrock, Rende, Plomin, & Thompson, 1995; Goodman & Stevenson, 1989). In addition, a sibling adoption study in middle childhood (focusing on comparisons between adoptive and nonadoptive sibling pairs) also suggested that the dimension of attention problems is highly heritable (Rende, Edelbrock, Plomin, Fulker, & DeFries, unpublished). In these three studies, it was also found that shared environmental factors were of little importance in explaining individual differences in this dimension. Because of the differences in methodology between twin and adoption designs (see Rutter, Bolton, Harrington, Le Couteur, Macdonald, & Simonoff, 1990), it is especially interesting when the different designs converge on the same conclusion, as is the case with the current evidence for a significant heritability of attention problems.

A similar finding has emerged for aggressive behavior, another core dimension of problem behavior assessed by instruments such as the Child Behavior Checklist. Both the Edelbrock et al (1995) twin study in adolescence and the Rende et al (unpublished) sibling adoption study in middle childhood converge on the conclusion that individual differences in aggressive behavior are also highly heritable. Again, shared environmental factors were negligible in both studies.

A different pattern of findings has emerged for delinquent behavior. Older twin studies of delinquency have suggested that genetic influences are not prominent, whereas shared environmental factors seem to be especially important (Rowe, 1983, 1986). The more recent twin study by Edelbrock et al (1995) also revealed significant effects of shared environment; however, genetic influence was also significant. In contrast to the twin studies, the recent sibling adoption study (Rende et al, unpublished) found evidence for neither genetic or shared environmental influences on delinquent behavior in early childhood. Hence, the conclusion reached about the relative influences of genes and environment on delinquent behavior is open to question at this time. Two considerations may be especially important. One is that as the twin studies suggest shared environmental influence, whereas the sibling adoption study did not, it may be that there are unique complementary influences between twins not observed with nontwin siblings. A second consideration is that the sibling adoption study focused on middle childhood, in contrast to the twin studies of adolescence. Hence, another possibility is that there may be a developmental effect in which the shared environmental influences become more prominent in adolescence. Importantly, the sibling adoption study, which was conducted as part of the longitudinal Colorado Adoption Project (CAP; Plomin & DeFries, 1985; Plomin, DeFries, & Fulker, 1988; DeFries, Plomin, & Fulker, 1994), is ongoing, and the children will be assessed again as adolescents. At that point, more direct comparisons between findings from twin vs. the sibling adoption design made be made without the confound of age.

A final area of interest with respect to common behavior problems is anxious and depressed mood. A recent study which combined a twin design with a step-family design revealed significant genetic influence on depressive symptoms in a sample of adolescents (Rende, Plomin, Reiss, & Hetherington, 1993). Similarly, the Edelbrock et al (1995) twin study also found genetic influence on symptoms of anxiety and depression. The only contrast to date arises from the sibling adoption study (Rende et al, unpublished), in which no evidence of genetic influence was found. Again, as noted above, the discrepancy may be due to either differences in methodology or differences in age. It should be mentioned, however, that the step-family component (i.e., comparisons of full, half, and unrelated siblings in step-families) used in the Rende et al (in press) study suggested genetic influence as well as the twin component (i.e., comparisons of MZ and DZ twins). As such, it may be that developmental differences in depressive symptoms are most salient, and that genetic influences on depression become especially prominent in adolescence.

IMPLICATIONS FOR STUDYING ADAPTIVE AND MALADAPTIVE DEVELOPMENT

The findings reviewed above may be seen as an initial attempt to construct a developmental model of genetic and environmental influences on diverse behavior problems through childhood and adolescence. It should be noted that the potential changes in heritability with age (i.e., no genetic influence seen in middle childhood but observed in adolescence) would be consistent with research demonstrating how genetic influences are not static over development, but rather often change across different points in the lifespan (Plomin, 1986). Returning to the theme of this chapter, what should be noted is that the focus is not on normative development or pathological development, but rather on the range of individual differences that is seen when one considers dimensions of problem behaviors in children and adolescents. Focusing on the roots of variation, rather than clearly cleaved affected and unaffected outcomes, represents a prototypical way in which quantitative genetic research contributes to understanding both adaptive and maladaptive behavior in an integrated fashion.

ETIOLOGY OF INDIVIDUAL DIFFERENCES AND EXTREME CASES

The previous two sections have contrasted two viewpoints on development. In the first viewpoint, pathological development was viewed as a deviation from normative, species-typical outcomes. In the second viewpoint, emphasis was placed on the wide range of variability that can be observed in terms of developmental outcome. In both cases, the ways in which quantitative genetic approaches can lead to greater understanding of both adaptive and maladaptive outcomes were discussed.

A third viewpoint is to merge the perspectives described above and generate empirical tests to determine the most appropriate models for specific forms of behavior. That is, rather than assume that pathological development is either quantitatively or qualitatively different from normative development, it would be important to develop methods to decide empirically the best model. As noted earlier, this perspective is perhaps best reflective of the theorizing which has gone into the emerging discipline of developmental psychopathology. Returning to a point made at the beginning of the chapter, Lewis (1992) reminds us that while developmental psychopathologists consider normal development, the focus is on variation around norms and especially the development, and definition, of deviance and its causes. It is the explicit focus on determining the continuity or discontinuity between adaptive and maladaptive development which is a hallmark of developmental psychopathology (e.g., Cicchetti, 1984).

In terms of quantitative genetics, the crucial question from this perspective is whether the genetic and environmental influences on disorders also contribute to normal variations in the general population (Rende & Plomin, 1990; Rende & Plomin, 1992). That is, the question asked is are there discrete etiological factors specific to a disorder, or is there a continuum of risk factors that contribute to the etiology of the disorder? Depending on the finding, the implications for normal development may vary.

Geneticists have long recognized that disorders that are defined categorically may be due to a continuum of genetic and environmental risk factors. Specifically, it has been suggested that a hypothetical distribution of risk or "liability" to categorical disorders may be assumed (Falconer, 1965). The normal distribution of liability may then have a "threshold" which determines if an individual is diagnosed as having the disorder. Heritability estimates of this construct of liability can be calculated from twin data which statistically separates the contributions of heredity and environment (see Smith, 1974).

The concept of liability is useful in research on diseases. For example, convulsive threshold is a dimension that underlies the diagnostic category of epilepsy, and it can be detected only by special tests, as it is not manifest in minor or infrequent seizures (see Plomin, Rende, & Rutter, 1991). However, a difficulty with the liability approach in research on psychopathology is that it assumes that a normal distribution of risk factors underlies a disorder, rather than assessing the risk factors empirically. It has been suggested that there can be no presuppositions that the extremes do, or do not, involve the same etiological mechanisms that operate in the normative range of functioning (Rutter, 1988); hence there must be empirical tests for similarities and dissimilarities between etiological mechanisms. That is, the challenge is to translate the hypothesized liability into something measurable, such as a behavioral dimension representing degrees of a categorical disorder or processes related to a disorder.

DF METHOD: LINKS AND BREAKS BETWEEN DIMENSIONS AND DISORDERS

Recently, DeFries and Fulker (1985, 1988) have developed a quantitative genetic method that assesses the extent to which dimensions are etiologically related to disorders. This technique, which has been called "DF" analysis (Plomin & Rende, 1991), makes it possible to assess the extent to which the magnitude of genetic and environmental influences on extreme scores (on a quantitative measure) differ from the magnitude of etiological influences on the entire range of individual differences. As such, the method represents a way of determining, in terms of etiology, the extent to which extreme symptomatology (e.g., characteristic of a psychiatric disorder) reflects a qualitative or quantitative departure from the normal range of variation (see Plomin, 1991; Plomin, Rende, & Rutter, 1991; Rende & Plomin, 1990).

UNDERLYING METHODOLOGY

Before describing the underlying methodology of DF analysis, some basic terms should be defined briefly. The term familiality will be used to describe resemblance between family members, specifically first-degree relatives. In quantitative genetic research, familiality is decomposed into two sources of resemblance: heritability (resemblance due to genetic similarity) and shared environment (resemblance due to environmental factors family members have in common). In the basic quantitative genetic model, the extent to which family members do not resemble each other is labeled nonshared environment (nonresemblance due to unique environmental factors which family members do not have in common as well as error of measurement). In general, quantitative genetics takes advantage of "natural" experiments such as differences between identical and fraternal twins in order to provide estimates of these latent constructs. For example, in the twin method, the similarity of identical twins (who share all genes) is compared to the similarity of fraternal twins (who, like siblings, share on average half of their genes) in order to provide estimates of heritability, shared environment, and nonshared environment (see Plomin et al., 1990).

The conceptual framework for DF analysis will be discussed first with respect to the more straightforward case of siblings (rather than twins). DF analysis introduces a new concept of familial resemblance (group familiality) which it contrasts with the traditional concept of individual familiality. The traditional concept of individual familiality is indexed by sibling resemblance for a quantitative trait. Specifically, the correlation between siblings indicates the extent to which individual differences--or variability--on the trait of interest may be due to familial factors (i.e., genetic or environmental factors operating to produce similarities within families), and hence is termed "individual familiality."

A novel index of familial resemblance derived from DF analysis estimates familial factors involved in the expression of extreme scores on the trait (as in the case of clinical populations). This estimate, called "group familiality," represents in general terms the extent to which family members--in this case siblings--deviate from normality on the trait of interest. For example, if extreme scores on a trait are due to familial factors, then the siblings of "probands" (i.e., individuals that are diagnosed as cases because their scores exceed a cut-off on the scale of interest) would also be expected to have scores which deviate from normality and perhaps begin to approach the clinical range.

Technically, group familiality is indexed by the regression toward the mean of sibling scores on the quantitative measure in order to provide a quantitative estimate of the degree to which extreme scores are due to familial factors. The basic quantitative approach is shown in Figure 1. For a quantitative measure relevant to a particular disorder, the mean of diagnosed probands will fall towards the extreme of the distribution (P in Figure 1). The mean of the siblings (S in Figure 1) of the probands will regress to the unselected population mean (mu) to the extent that familial factors are unimportant in the etiology of the disorder. In contrast, if familial factors are important, the mean of the siblings of the probands will be greater than the population mean.

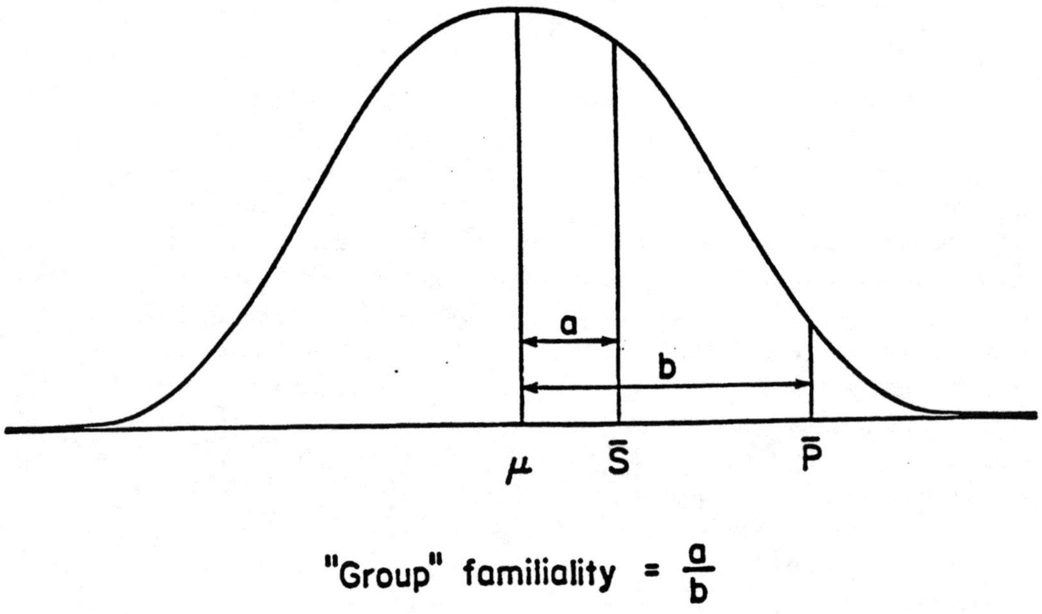

Figure 1. Hypothetical distribution of behavior variation of an unselected sample of siblings (mean = m), diagnosed probands (mean = P), and siblings of diagnosed probands (mean = S).

Continuing with Figure 1, *a* indicates the difference between the mean of the siblings (S) and the population mean (μ), and *b* indicates the difference between the mean of the probands (P) and the population mean (μ). The extent to which the mean difference between the siblings and the population approaches the mean difference between the probands and the population on a quantitative measure provides an estimate of the extent to which the mean difference between the probands and the population on the quantitative measure is due to familial factors. Referring to the figure, a large group familiality means that *a* is approaching *b*; in contrast, a small group familiality means that *a* is small relative to *b*.

One example of the utility of this approach is provided by research on mental retardation using IQ tests as a continuous measure (Nichols, 1984; see Plomin et al., 1991). Siblings of severely retarded children (IQs less than 50) had an average IQ of 103 and none was retarded. In contrast, the average IQ of siblings of mildly retarded children (IQs from 60 to 69) was 85 and one-fifth of the siblings were retarded. This research yielded two important findings: (1) severe retardation shows no familiality and is thus etiologically distinct from the rest of the IQ distribution; and (2) mild retardation is familial and is thus etiologically connected with the rest of the IQ distribution.

In the example of mental retardation, group familiality was not calculated and the mean IQ of probands was not reported. However, the calculation of individual and group familiality may provide more specific information about the magnitude of both types of indices of familial resemblance. As mentioned earlier, individual familiality is estimated by the correlation between siblings. In addition, the quantitative estimate of group familiality (*a/b* in Figure 1) may then be compared to the estimate of individual familiality. This comparison provides an empirical method for assessing whether the magnitude of familial influence on extreme scores differs from the magnitude of familial influence on the range of individual differences.

Twins or adoptees are needed to determine the extent to which familiality is due to heredity rather than environment shared by siblings. Group heritability (h_g^2) indicates the proportion of the difference between the probands and the unselected population that is due to genetic differences. It is based, for example, on the differential regression toward the mean for cotwins of MZ and DZ probands. For instance, h_g^2 is zero if MZ and DZ cotwins regress to the population mean to the same extent. In contrast, h_g^2 is 1.0 if MZ cotwins do not regress toward the mean and if DZ cotwins regress halfway back to the mean. Two important differences between the twin approach and the sibling approach should be emphasized. In the case of twins, note that two types of "siblings" are studied, and hence information is provided by comparing the difference between MZ and DZ cotwin means rather than using a single sibling mean. In addition, the sibling approach indicates the extent to which familiality is important, whereas the twin approach attempts to decompose familiality in genetic and environmental components.

Similar to individual familiality, individual heritability (h^2) is based on traditional estimates of heritability reviewed earlier, such as doubling the difference in correlations for MZ and DZ twins. In addition, estimates of both individual and group shared environment (c^2 and c_g^2, respectively) may also be determined in this approach. A multiple regression approach that estimates h2, h_g^2, c2 , and h_g^2 is presented by DeFries and Fulker (1985; 1988).

It should be emphasized that the reliance on quantitative measures in DF analysis does not preclude the use of clinically diagnosed individuals. Rather, the inclusion of diagnosed individuals as probands strengthens the approach. The key, however, is to utilize quantitative measures that are relevant to the symptomatology of interest, so that links between the normal and abnormal may be addressed empirically (see Rende & Plomin, 1990). Indeed, many behavioral disorders of interest--e.g., unipolar depression, hyperactivity, anxiety, and conduct problems--clearly are present in varying degrees in the population, and the etiological links or breaks between normative ranges of functioning and the extremes in these cases may be investigated directly.

Implications for Studying Maladaptive Behavior

The DF analysis was first applied to reading disability. Probands and twins were assessed using a continuous discriminant function score of reading-related tests (DeFries et al., 1987). Group heritability was found to be only about half the magnitude of individual heritability, suggesting that reading disability is etiologically different from the continuous dimension of reading ability. A multivariate extension of the DF analysis suggested that phonological coding ability (e.g. speed and accuracy in pronouncing nonwords such as "ter" and "tegwop") may be a key element in the genetics of reading disability (Olson et al., 1989).

Recently, DF analysis has been applied to the study of depressive symptomatology in adolescence (Rende et al., 1993). Data were collected from twins and siblings participating in the ongoing Nonshared Environment and Adolescent Development project. The sample consisted of 707 sibling pairs representing the following genetic relationships: MZ twins, DZ twins, and nontwin full siblings in intact families, and full siblings, half siblings, and biologically unrelated siblings in step-families. Depressive symptomatology was assessed using the Children's Depression Inventory (CDI; Kovacs, 1983); adolescent self-reports were obtained twice over a 1-month period.

An application of DF analysis to these data yielded the following results. There was moderate genetic influence on individual differences in depressive symptomatology, and no shared environmental influences. Moderate genetic influence also contributed to extreme scores on the CDI (i.e. adolescents who exceeded the suggested "cutoff" score of 13 which is used to index risk for clinical depression). However, shared environmental influences were pronounced for the extremes (accounting for over 40% of the mean difference between probands and the population), and were significantly greater than the shared environmental effects on individual differences. These results were interpreted using a diathesis-stress model. In this model, it is speculated that there is a continuum of genetic risk (or diathesis) for depression which results in a wide range of individual differences in basic symptoms of this disorder. It is also hypothesized that there may be specific familial psychosocial stressors (such as family discord or major life events) which lead to the expression of extreme depressive symptomatology in some adolescents. It must be stressed, however, that the adolescents in this study were not assessed clinically, and research with clinical samples is necessary before firm conclusions can be drawn.

Two recent twin studies have applied the DF methodology to assess genetic influence on diagnosed attention deficit disorder (ADHD). In both studies, group heritability was highly significant and substantial, providing strong evidence for genetic influence on this syndrome (Gillis, Gilger, Pennington, & DeFries, 1992; Stevenson, 1992). The findings in both studies are consistent with the results reported earlier for the dimension of attention problems (Edelbrock et al., 1995; Goodman & Stevenson, 1989; Rende et al., unpublished). However, as the focus of these studies was limited to diagnosed cases, formal statistical tests comparing group and individual heritability have not been conducted. The data from all these studies do indicate that genetic factors play a substantial role both in terms of the dimension of attention problems and symptoms characterizing diagnosed cases of the disorder.

IMPLICATIONS FOR STUDYING NORMAL DEVELOPMENT

This new quantitative genetic approach carries many implications for studying links and breaks between the normal and abnormal (see Rende & Plomin, 1990). Importantly, the implications are just as profound for investigating normal development as they are for abnormal development. Most notable is the attempt to consider both the normal and abnormal in an integrated fashion. Such work is a defining characteristic of the field of developmental psychopathology and may help to determine if the normal range of variability for different forms of symptomatology includes the extremes of the distribution.

The results of DF analysis are directly relevant to the study of normal development if the etiology of the extremes does not differ from the etiology of individual differences in the population. Such a finding would be consistent with the hypothesis that disorders are the extremes of the normal distribution. The payoff in terms of understanding normal development would be that the genetic and environmental contributions to affected cases may be the same as those contributing to normative or adaptive development. In this case, studying abnormal development may highlight the mechanisms that also contribute to positive outcomes. Such research may refine our study of the "normal" and "abnormal," and may lead to a more integrated approach that focuses on behaviors and processes along a continuum of adaptation/maladaptation as opposed to presence or absence of disorders. Note that this position was expressed earlier in the section of this chapter on individual differences; however the new point here is that rather than assume an etiological continuity across individual differences, such a continuity would be assessed empirically.

If, on the other hand, the etiology of the extremes differs from the etiology of individual differences in the population, the range in symptomatology that may be considered normative may be specified, as well as the range that indexes an etiologically different phenomenon. In this way, DF analysis may contribute to our definitions of both normal ranges of variability and the extremes which are considered as pathological. For example, one crucial piece of information would be the "breakpoint" in the distribution that best distinguishes the normal range of variability from extreme cases. Such a finding would suggest that specific molecular genetic techniques discussed later in this book, such as linkage analysis, may be especially appropriate for determining the mechanisms responsible for the disorder. In this way, DF analysis may be of great benefit in providing proper definitions of a phenotype which may be highly heritable prior to molecular genetic studies. Note, however, that in this case an empirical demonstration of etiological discontinuity would be available, rather than merely an assumption that this is the case.

In summary, then, the DF method represents an important way of studying both adaptive and maladaptive development from a quantitative genetic perspective. The most important contribution of this approach is to move away from arguments regarding the most appropriate etiological model, and replace rhetoric with empirical tests about links between variation in the population and the extreme cases which are assessed as disorders.

RESILIENCE IN DEVELOPMENT

The previous sections of this chapter have emphasized the search for etiological factors responsible for either variation in symptoms of psychopathology or psychopathological disorders. From this perspective, genetic influences are often conceptualized as risk factors for psychopathology: the genetic defects involved in PKU and fragile X syndrome were reviewed as risk factors for maladaptive pathways for development. In addition, evidence was reviewed suggesting that genetic factors may also play a role in the development of more common behavioral problems such as aggressive behavior and attention problems. From this perspective, the ways in which implications for adaptive pathways in development have been discussed.

Another way in which lessons about normal development can be gleaned from studies of abnormal development is by focusing on resilience. The concept of resilience--which is essentially individuals doing well in the face of well-established risk factors--is a central theme in developmental psychopathology (e.g., Rolf, Masten, Cicchetti, Nuechterlein, & Weintraub, 1990). Although quantitative genetic approaches to resilience are essentially still in a theoretical stage (see Rende & Plomin, 1993), the topic is an important one in the study of adaptive and maladaptive development, and a few general points will be made in this section of the chapter.

There are instances in which single genes are deterministic in leading to negative outcomes. For example, Huntington's Disease is due to a single gene; if an individual has the gene, they will eventually develop the disorder. Another general principle to consider, however, is that genetic factors may also be protective with respect to specific risk factors for disease. An especially interesting finding is that some genetic conditions which lead to diseases may nonetheless function as protective factors for other conditions. A notable example is that the gene which affects sickle-cell anemia also increases resistance to malaria (e.g., Childs, Moxon, & Winkelstein, 1992).

Although there are no comparable findings to date in research on psychopathology, the potential protective function of genes should be kept in mind when discussing the role of genetics in the development of psychiatric disorders. This point is important because genetic influences are often considered in psychopathology only as predisposing factors to maladaptation. One of the most prominent general frameworks for the development of psychopathology is a diathesis-stress model, in which a genetic predisposition to a disorder interacts with negative environmental factors to lead to psychopathology. In this framework, genetic influences are usually conceptualized as predisposing to psychopathology whereas their potential protective function are not considered (Rende & Plomin, 1992; Rende & Plomin, 1993). Hence, when examining the role of genes in the development of psychopathology, it may be important to not only consider genetic risk but also to consider heritable influences which may be protective. Considering genetic influences on resilience offers yet another way in which the study of adaptation and maladaptation are merged through the use of a quantitative genetic perspective.

ENVIRONMENTAL INFLUENCES AND ADAPTIVE AND MALADAPTIVE DEVELOPMENT

Behavioral genetics is often considered to be a fundamental way of investigating genetic influences on development. However, it is now appreciated that behavior genetics also provides many methods for studying the influence of the environment (e.g., Plomin & Rende, 1991). With few exceptions, quantitative genetic research converges on two conclusions concerning the environment. First, the environment is important in the sense that genetic factors only partially account for psychopathology. The second conclusion is more novel: The way in which the environment affects psychopathology may be very different from the way in which the environment was assumed to operate in traditional theories of environmental influence. Whatever the salient environmental factors might be, they appear to operate to make children in the same family no more similar than children reared in different families (Dunn & Plomin, 1990; Hetherington, Reiss, & Plomin, in press; Plomin & Daniels, 1987). Such environmental influences are called nonshared in that they are not shared by children in the same family, in contrast to environmental factors shared by siblings which would be expected to make siblings similar. In other words, nonshared environmental influences are experienced differently by children in the same family.

One example of how this conclusion has been reached involves identical twins. Differences within pairs of identical twins directly estimate nonshared environment (plus error of measurement). For schizophrenia, the concordance of identical twins is about 40%. This resemblance suggests strong genetic influence as compared to the population base rate of about 1% for schizophrenia. However, why are these genetically identical pairs of individuals more often than not discordant for schizophrenia? The answer is nonshared environment. There can be no genetic reason why identical twins are so often discordant for schizophrenia. Methods for estimating nonshared and shared environmental influence and an update of the evidence for the importance of nonshared environment is available elsewhere (Plomin, Chipuer, & Neiderhiser, 1995).

The message is not that family experiences are unimportant but rather that the relevant environmental influences are specific to each child, not general to an entire family. These findings suggest that instead of thinking about the environment on a family-by-family basis, the environment should be conceptualized on an individual-by-individual basis. With respect to understanding adaptive vs. maladaptive outcomes in development, an emphasis on differential effects of the environment may lead to a better understanding of the role of environmental factors in developmental pathways.

The study of nonshared environmental factors in the development of psychopathology is still in a very early stage. However, it has been suggested that investigations of nonshared factors related to psychopathology may help illuminate principles of normal development as well (Rende & Plomin, 1990). Because siblings often differ with regard to developmental outcome, the study of nonshared factors can indicate how environmental influences contribute to both positive and negative pathways in development.

IMPLICATIONS FOR STUDYING MALADAPTIVE BEHAVIOR

Consider as an example the negative consequences of parental behavior displayed by depressed adults, which have been documented by many researchers (see Rutter, 1990). To date, there have been no studies which have examined how a depressed parent interacts with multiple children within the family. One possibility is that because of the chronic nature of depression, severely depressed adults do not discriminate among children; their behavior during depressive episodes may be consistent across siblings in a family.

Another possibility, however, should be considered: a depressed parent may show differential behavior, perhaps based in part on differences among children. It has been shown that parent irritability, criticism, and hostility--behaviors related to depression--may be focused especially on children with specific temperamental characteristics (Rutter, 1986). Hence, although one child in a family may be especially vulnerable to environmental risk, a sibling may be protected from *exposure*. Documenting the extent to which siblings are exposed to negative experiences may eventually reveal in part why children in the same family may experience vastly different developmental outcomes.

As mentioned earlier, the strongest evidence for the importance of nonshared environment comes from studies of discordant identical twins. As mentioned earlier, discordant identical twins provide a unique opportunity to identify specific sources of nonshared environment. This is because their discordance must be due to nonshared environment, unlike fraternal twins or nontwin siblings, who might differ in outcome because of genetic differences. This approach has been used to explore nonshared environmental factors in schizophrenia, albeit with little success to date (Gottesman, 1991; Gottesman & Shields, 1982).

Although little progress has been made in specifying nonshared factors that contribute to discordance among identical twin pairs, the general approach has great potential for linking the environmental contributions to both adaptive and maladaptive development. That is, in addition to searching for negative influences that may lead to schizophrenic symptoms in one co-twin, it is also essential to investigate possible factors that protect the other co-twin from a negative outcome. That is, both risk and protective factors could be examined when assessing discordant identical twins. This approach is intriguing considering recent research on the offspring of twins discordant for schizophrenia. Such work indicates that the offspring of nonaffected identical twins have a risk for schizophrenia similar to that of the offspring of schizophrenic cotwins (Gottesman & Bertelesen, 1989). What this implies is that the same schizophrenia inducing genes are present in both the normal and schizophrenic twins, as well as in their offspring. Most importantly, this finding suggests that environmental factors determine whether or not schizophrenic symptomatology is manifested. Again, since only some of the twins and offspring development schizophrenia, such environmental factors must be nonshared, and it appears that these nonshared influences help determine whether an individual has an adaptive or maladaptive outcome.

Implications for Studying Normal Development

Previous discussion in this chapter suggested that there should be no presumption that the extremes in a distribution are due to either qualitative or quantitative variations in etiological factors. Although this point was made most strongly with respect to genetic influences, the principle is as applicable to environmental factors related to psychopathology. Indeed, the general implication that environmental factors play a significant role in the development of psychopathology suggests that attention be devoted to nongenetic, as well as genetic, etiological factors.

Similar to the argument made earlier on genetic factors, if the nonshared factors which influence the extremes in a distribution are similar to those which affect normal variation, then this finding would suggest that the study of nonshared environment should contribute to the understanding of both adaptive and maladaptive pathways in development. An important consideration is that the examination of siblings discordant for psychopathology may provide an especially powerful glimpse into nonshared processes, and this strategy may generate basic information on nonshared environmental factors applicable to normal development (Rende & Plomin, 1990; Rutter, Bolton, et al., 1990). In addition, the focus on discordant siblings may help uncover specific protective factors in the environment that contribute to the normative path of siblings in discordant pairs.

If, on the other hand, the extremes in terms of psychopathology are influenced by nonshared factors which are not normally experienced by individuals, then understanding the environmental roots of the extremes may highlight how normal development usually occurs. This point is especially important because the study of nonshared environmental factors is very much in a beginning stage in terms of uncovering specific sources of influence which are associated with developmental outcomes (e.g., Dunn & Plomin, 1990). Hence, by focusing on situations in which severe aberrations occur in the environment, the way in which nonshared environment generally works to promote adaptive outcomes may be magnified (as was the case of severe genetic conditions).

A final point to consider is that focusing on nonshared processes related to psychopathology may help clarify the ways in which both genetic and environmental influences contribute to pathways in development. An influential paper by Scarr and McCartney (1983) presented an overview of various developmental processes that involve the interplay of genetic and environmental influences, such as gene-environment interaction and correlation. It may be possible to use the discordant sibling method to highlight some of these models. As discussed earlier, two theories could be contrasted concerning the influence of the behavior of depressed parents on children. In one model, depressed parents would be expected to behavior similarly to multiple children in the family because of the chronic nature of depression. In a second model, it is hypothesized that depressed parents may especially target some children in the family, whereas others may not be exposed as strongly to maladaptive parenting. By using informative designs (such as the twin method, or comparisons of full, half, and unrelated siblings), it may be possible to not only test which model is correct, but also determine if there are indeed associations between children's heritable traits, their environment (i.e., parental behavior), and their outcomes. Placed within the framework of nonshared environment, such designs could help clarify etiological influences which lead siblings to discordant developmental pathways. Again, by focusing on discordant pairs, an integrated approach to both adaptive and maladaptive development would be achieved.

Concluding Remarks

This chapter has focused on ways in which quantitative genetic strategies can be integrated with the emerging discipline of developmental psychopathology. Although it is clearly recognized that quantitative genetic approaches offer powerful techniques for the study of maladaptive development (e.g., Rutter, MacDonald, et al., 1990), less attention has been given to the ways in which this approach also contributes to our understanding of normal development (see Rende & Plomin, 1990). However, since a defining feature of developmental psychopathology is an integrated approach to the normal and abnormal, it is useful to consider ways in which quantitative genetics can contribute to this overall goal.

The topics reviewed in this chapter present various ways in which the study of adaptive and maladaptive development may be integrated with a quantitative genetic framework. It must be emphasized that these approaches hold promise only to the extent that the do indeed provide novel information about the normal and abnormal. Again, the best historical examples come from the field of genetics, in which studies of normal functioning and dysfunction have been used to inform both. The hope is that current methods used in quantitative genetics be used in a similar manner, so that the thoughts expressed in this chapter may eventually be supplanted by data revealing more clearly the ways in which both genetic and environmental influences contribute to the study of adaptive and maladaptive development. Such data would carry profound implications not only for preventing maladaptation, but also for promoting positive developmental outcomes (Rende & Plomin, in press).

References

Achenbach, T.M. (1990). Conceptualization of developmental psychopathology. In M. Lewis & S. Miller (Eds.). *Handbook of developmental psychopathology* (pp. 3-14). New York: Plenum Press

Achenbach, T.M. (1991). *Manual for the child behavior checklist/4-18 and 1991 profile.* Burlington, VT: Author.

Beadle, G.W. & Tatum, E.L. (1941). Experimental control of developmental reaction. *American Naturalist, 75*, 107-116.

Bregman, J.D., Leckman, J.F., & Ort, S.I. (1988). Fragile X syndrome: genetic predisposition to psychopathology. *J Autism Dev Disord, 18*, 343-354.

Childs, B., Moxon, E.R. & Winkelstein, J.A. (1992). Genetics and infectious diseases. In R.A. King, J.I. Rotter & A.G. Motulsky (Eds.). *The genetic basis of common diseases* (pp. 71-91). New York: Oxford University Press

Cicchetti, D. (1984). The emergence of developmental psychopathology. *Child Dev, 55*, 1-7.

Cicchetti, D. (1990). An historical perspective on the discipline of developmental psychopathology. In J. Rolf, A. Masten, D. Cicchetti, K. Neuchterlein & S. Weintraub (Eds.). *Risk and protective factors in the development of psychopathology* (pp. 2-28). New York: Cambridge University Press

DeFries, J.C. & Fulker, D.W. (1985). Multiple regression analysis of twin data. *Behav Genet, 15*, 467-473.

DeFries, J.C. & Fulker, D.W. (1988). Multiple regression analysis of twin data: etiology of deviant scores versus individual differences. *Acta Genet Med Gemellol (Roma), 37*, 205-216.

DeFries, J.C., Fulker, D.W., & LaBuda, M.C. (1987). Evidence for a genetic aetiology in reading disability of twins. *Nature, 329*, 537-539.

DeFries, J.C., Plomin, R. & Fulker, D.W. (1994). *Nature and nurture in middle childhood.* New York: Blackwell.

Dunn, J. & Plomin, R. (1990). *Separate lives: Why siblings are so different.* New York: Basic.

Edelbrock, C. & Costello, A.J. (1988). Convergence between statistically derived behavior problem syndromes and child psychiatric diagnoses. *J Abnorm Child Psychol, 16*, 219-231.

Edelbrock, C., Rende, R., Plomin, R., & Thompson, L. (1995). A twin study of competence and behavioral problems in adolscence. *J Child Psychol Psychiatry*,36,775-785

Falconer, D.S. (1965). The inheritance of liability to certain diseases, estimated from the incidence among relatives. *Annals of Human Genetics, 29*, 51-76.

Gillis, J.J., Gilger, J.W., Pennington, B.F., & DeFries, J.C. (1992). Attention deficit disorder in reading-disabled twins: evidence for a genetic etiology. *J Abnorm Child Psychol, 20*, 303-315.

Goodman, R. & Stevenson, J. (1989). A twin study of hyperactivity--II. The aetiological role of genes, family relationships and perinatal adversity. *J Child Psychol Psychiatry, 30*, 691-709.

Gottesman, I.I. (1982). *Schizophrenia: The epigenetic puzzle.* Cambridge: Cambridge University Press.

Gottesman, I.I. (1991). *Schizophrenia genesis: The origins of madness*. New York: W. H. Freeman.

Gottesman, I.I. & Bertelson, A. (1989). Confirming unexpressed genotypes for schizophrenia. *Archives of General Psychiatry, 46*, 867-872.

Kovacs, M. (1983). *Children's Depression Inventory: A self-rated depression scale for school-aged youngsters*. University of Pittsburgh: Unpublished manuscript.

Lewis, M. (1992). Developing developmental psychopathology. *Journal of Applied Developmental Psychology, 13*, 483-488.

Nichols, P.L. (1984). Familial mental retardation. *Behav Genet, 14*, 161-170.

Oberle, I., Rousseau, F., Heitz, D., Kretz, C., Devys, D., Hanauer, A., Boue, J., Bertheas, M.F., & Mandel, J.L. (1991). Instability of a 550-base pair DNA segment and abnormal methylation in fragile X syndrome. *Science, 252*, 1097-1102.

Olson, R., Wise, B., Conners, F., Rack, J., & Fulker, D. (1989). Specific deficits in component reading and language skills: genetic and environmental influences. *J Learn Disabil, 22*, 339-348.

Plomin, R., Chipuer, H.M. & Neiderhiser, J.M. (1995). Behavioral genetic evidence for the importance of nonshared environment. In E.M. Hetherington, D. Reiss & R. Plomin (Eds.). *Separate social worlds of siblings: Impact of nonshared environment on development* Hillsdale, NJ: Lawrence Erlbaum Associates

Plomin, R. & Daniels, D. (1987). Why are children in the same family so different from each other? *Behavioral and Brain Sciences, 10*, 1-16.

Plomin, R. & DeFries, J.C. (1985). *The origins of individual differences in infancy: The Colorado Adoption Project*. New York: Academic Press.

Plomin, R., DeFries, J.C. & Fulker, D.W. (1988). *Nature and nurture in early childhood.* New York: Academic Press.

Plomin, R., DeFries, J.C. & McClearn, G. (1990). *Behavioral genetics : A primer*. New York: W. H. Freeman.

Plomin, R. & Rende, R. (1991). Human behavioral genetics. [Review]. *Annu Rev Psychol, 42*, 161-190.

Plomin, R., Rende, R. & Rutter, M. (1991). Quantitative genetics and developmental psychopathology. In D. Cicchetti & S. Toth (Eds.). *Internalizing and externalizing expressions of dysfunction: Rochester Symposium on Developmental Psychopathology (vol. 2)* (pp. 155-202). Hillsdale, NJ: Lawrence Erlbaum

Rende, R. & Plomin, R. (1990). Quantitative genetics and developmental psychopathology: Contributions to understanding normal development. *Development and Psychopathology, 4*, 393-407.

Rende, R. & Plomin, R. (1992). Diathesis-stress models of psychopathology: A quantitative genetic perspective. *Applied and Preventative Psychology, 1*, 177-182.

Rende, R. & Plomin, R. (1994). Genetic influences on behavioral development. In M. Rutter & D. Hay (Eds.). *Developmental principles and clinical issues in psychology and psychiatry* Oxford: Blackwell Scientific Publications

Rende, R. & Plomin, R. (1993). Families at risk for psychopathology: Who becomes affected and why? *Development and Psychopathology*,5,529-540

Rende, R., Plomin, R., Reiss, D., & Hetherington, E.M. (1993). Genetic and environmental influences on depression in adolescence: Etiology of individual differences and extreme scores. *J Child Psychol Psychiatry, 34*, 1387-1398.

Rolf, J., Masten, A.S., Cicchetti, D., Nuechterlein, K.H. & Weintraub, S. (1990). *Risk and protective factors in the development of psychopathology*. New York: Cambridge University Press.

Rowe, D. (1983). Biometric models of self-reported delinquent behavior: A twin study. *Behav Genet, 13*, 473-489.

Rowe, D. (1986). Genetic and environmental components of antisocial pairs: A study of 265 twin pairs. *Criminology, 24*, 513-532.

Rutter, M. (1986). Meyerian psychobiology, personality development, and the role of life experiences. [Review]. *Am J Psychiatry, 143*, 1077-1087.

Rutter, M. (1988). Epidemiological approaches to developmental psychopathology. *Archives of General Psychiatry, 45*, 486-495.

Rutter, M. (1990). Commentary: Some focus and process consideration re: the effects on children of parental depression. *Developmental Psychology, 26*, 60-63.

Rutter, M., Bolton, P., Harrington, R., Le Couteur, A., Macdonald, H., & Simonoff, E. (1990a). Genetic factors in child psychiatric disorders--I. A review of research strategies. [Review]. *J Child Psychol Psychiatry, 31*, 3-37.

Rutter, M., Macdonald, H., Le Couteur, A., Harrington, R., Bolton, P., & Bailey, A. (1990b). Genetic factors in child psychiatric disorders--II. Empirical findings [see comments]. [Review]. *J Child Psychol Psychiatry, 31*, 39-83.

Rutter, M. (1991). Nature, nurture, and psychopathology: A new look at an old topic. *Development and Psychopathology, 3*, 125-136.

Scarr, S. & McCartney, K. (1983). How people make their own environments: a theory of genotype greater than environment effects. *Child Dev, 54*, 424-435.

Smith, C. (1974). Concordance in twins: methods and interpretation. *Am J Hum Genet, 26*, 454-466.

Stevenson, J. (1992). Evidence for a genetic etiology in hyperactivity in children. *Behav Genet, 22*, 337-344.

Yu, S., Pritchard, M., Kremer, E., Lynch, M., Nancarrow, J., Baker, E., Holman, K., Mulley, J.C., Warren, S.T., Schlessinger, D., & et al, (1991). Fragile X genotype characterized by an unstable region of DNA. *Science, 252*, 1179-1181.

CHAPTER 2

VARIABLE AGE OF GENE EXPRESSION: IMPLICATIONS FOR DEVELOPMENTAL GENETIC MODELS

Joanne M. Meyer[1], *Judy L. Silberg*[1], *Lindon J. Eaves*[1], *Hermine H. Maes*[1], *Emily Simonoff*[2], *Andrew Pickles*[2] *Michael L. Rutter*[2], *And John K. Hewitt*[3].

[1]Department of Human Genetics, Medical College of Virginia-Virginia Commonwealth University, Richmond VA, U.S.A.
[2]MRC Child Psychiatry Unit, Institute of Psychiatry, London, England.
[3]Institute of Behavior Genetics, University of Colorado, Boulder, CO, U.S.A.

INTRODUCTION

In the spirit of adopting a comprehensive methodology for behavior genetic studies, path analytic models have been used to estimate genetic and environmental contributions to the developmental change of continuously distributed behavioral traits. The approaches currently used for these analyses are reviewed in the first part of this chapter. Following this review, we consider how the developmental models can lead to erroneous conclusions about gene expression when there are individual differences in developmental time. We illustrate an alternative model which accommodates these differences, and then apply the model to the development of the body-mass index in a prospective study of adolescent twin pairs.

PATH ANALYTIC MODELS FOR BEHAVIOR GENETIC STUDIES OF DEVELOPMENTAL CHANGE AND CONTINUITY

THE CHOLESKY FACTORIZATION

The use of the Cholesky or triangular factorization for the analysis of the genetic and environmental covariance structure of multiple phenotypes was introduced by Martin et al. (1982) in their twin study of finger ridge counts. The factorization remains popular in multivariate and longitudinal genetic analyses because it is a relatively simple way of obtaining positive definite estimates of genetic and environmental covariance matrices comprising the phenotypic structure. Additionally, for longitudinal data, the factorization can be interpreted from a developmental perspective.

The path diagram in Figure 1 illustrates a Cholesky decomposition of the genetic influences on a phenotype measured at three ages. This parameterization includes an initial additive genetic factor (G_1) influencing all occasions of measurement; a second factor (G_2) influencing the second and third occasion of measurement; and a final factor (G_3) influencing only the third occasion. The first factor represents those genetic influences which persist throughout development, while the second and third factors, uncorrelated with each other and with the first factor, index new genetic effects which are introduced during the period of follow-up. Typically, a significant impact of these subsequent factors on the phenotype is interpreted as the "switching on" of a new set of genes during development. As discussed later in the chapter, this interpretation may not always hold.

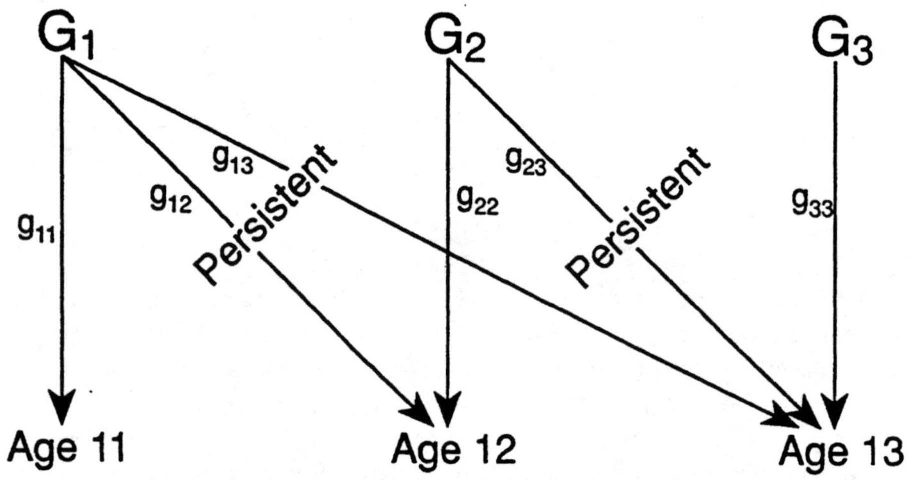

Figure 1. A Cholesky factorization of the additive genetic influences on a phenotype measured on three occasions. See text for a description of the factors and factor loadings. When this model is fit to genetically informative data, the genetic factors correlate across relatives according to their degree of genetic relationship.

In Figure 1, the impact of the genetic factors on the age-specific phenotypes is indexed by path coefficients g_{ij}, where i=1-3 and j=i-3. The genetic variance at each age is simply the sum of the squares of the path coefficients influencing that age (e.g., $\Sigma\ g^2_{i3}$, for occasion 3), while the genetic covariance of any two occasions is given by the sum of the products of path coefficients from each factor to the two occasions (e.g., $\Sigma\, g_{i2} * g_{i3}$, for occasions 2 and 3). Since equality constraints are not imposed on the path coefficients, the Cholesky factorization can yield different estimates of age-specific genetic variances and age-to-age genetic covariances.

Figure 2 illustrates a pattern of change in genetic variances over time which the flexible Cholesky model can accommodate. In this example, there is an increased expression of the initial genetic factor over time ($g_{11} < g_{12} < g_{13}$), coupled with the addition of new (persistent) genetic effects at ages 12 and 13 (g_{22}, g_{23}, g_{33}).

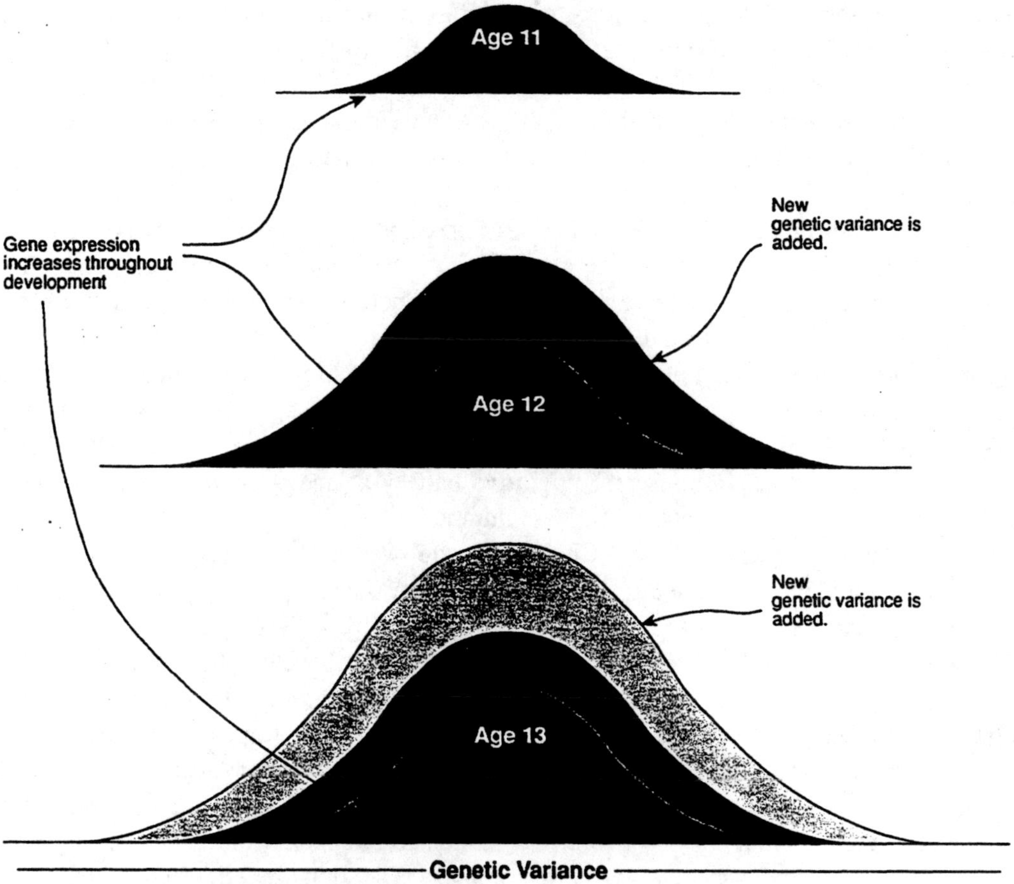

Figure 2. Changes in genetic variances over time which the Cholesky factorization can accommodate.

When the Cholesky factorization is applied to monozygotic (MZ) and dizygotic (DZ) twin data, the additive genetic factors are allowed to correlate 1.0 in MZ twins and 0.5 in DZ twins. Further, with reared-together twin data, the factorization and can be used to estimate the individual and family environmental covariance structures as well. The Cholesky model is considered to be a full parameterization of genetically informative multivariate data and will only fail to fit the data if there are significant differences in observed statistics which have the same expectation. For example, if the within-person, cross-occasion covariances for the first member of a twin pair are significantly different from those for the second member, the model would fail. The same would be true if there were MZ/DZ differences in the within-person covariance matrix. Generally, however, the model provides a suitable fit to the data (as assessed by a chi-square test) and is often used for exploratory analyses. Since the full Cholesky parameterization fits any set of twin covariance matrices which do not show the zygosity or twin order effects described above, the parameterization does not test a developmental hypothesis. That is to say, the goodness-of-fit of the full model cannot be used as a criterion for the acceptance or rejection of the Cholesky description of developmental change.

Applications of the Cholesky decomposition to genetically informative longitudinal data include: the analysis of the body mass index in juvenile adoptees and adult male veteran twins (Cardon and Fulker, 1992; Fabsitz et al., 1992); height in twins from birth to maturity (Phillips and Matheny, 1990); cognitive abilities in juvenile adoptees (Cardon and Fulker, 1991; Cardon et al., 1992a; Fulker et al., 1993); and alcohol consumption in Finnish and American adult twins (Carmelli et al., 1993; Kaprio et al., 1992).

The results of Fulker et al. (1993), in their investigation of general cognitive ability at ages 1, 2, 3, 4, 7 and 9, illustrate the usefulness of the Cholesky decomposition. A full Cholesky model with six additive genetic, six common environmental and six specific environmental factors was initially fit to cognitive data from biological siblings, adoptive siblings, and twins assessed at six ages (using the Barley Mental Development Index at ages 1 and 2 (Barley, 1969), the Stanford-Binet IQ test at ages 3 and 4 (Terman & Merrill, 1973), the Wechsler Intelligence Scale for Children-Revised at age 7 (Wechsler, 1974), and a telephone-administered cognitive battery at age 9 (Cardon et al., 1992b). Data were available from a total of 1,437 children.

A series of submodels of the full Cholesky model were fit to the cognitive data to test the significance of the genetic and environmental factor loadings. The parameter estimates under the best-fitting model (as determined by a likelihood ratio chi-square test) are provided in Table 1. What is most striking about these results is the differential factor structure for genetic, common environmental and unique environmental effects. The environmental patterns are simple: for the unique environment, all effects are occasion-specific; for the common environment, a single factor explains age-to-age covariances and variances. As a result, there is no continuity in unique environmental influences over time, but complete continuity of the common environmental effects.

The genetic factor structure is more complex. The first genetic factor influences cognitive ability over all measurement occasions, but its impact decreases over time. New genetic variance arises at ages 2 and 3, and continues to contribute to the phenotypic variance up to age 9. There is no evidence for new genetic variance at age 4, but by age 7, genetic effects uncorrelated with those expressed at previous occasions appear and persist through age 9. The authors have suggested that rapid changes in gene expression influencing cognitive ability during the first 3 years of life stabilize by age 4. However, as children enter school at age 7, new genetic variance arises, perhaps in response to the increased intellectual demands.

Table 1. Parameter estimates obtained for the best-fitting Choleksy decomposition fit to longitudinal data on cognitive acheivement at ages 1, 2, 3, 4, 7 and 9 (after Fulker et al., 1993).

	Genetic Factors						Common Environmental Factors						Unique Environmental Factors					
Age	1	2	3	4	5	6	1	2	3	4	5	6	1	2	3	4	5	6
1	0.73	--	--	--	--	--	0.33	--	--	--	--	--	0.60	--	--	--	--	--
2	0.41	0.71	--	--	--	--	0.32	--	--	--	--	--	--	0.46	--	--	--	--
3	0.24	0.52	0.45	--	--	--	0.32	--	--	--	--	--	--	--	0.60	--	--	--
4	0.20	0.46	0.57	--	--	--	0.33	--	--	--	--	--	--	--	--	0.57	--	--
7	0.21	0.35	0.28	--	0.43	--	0.31	--	--	--	--	--	--	--	--	--	0.69	--
9	0.04	0.34	0.13	--	0.78	--	0.32	--	--	--	--	--	--	--	--	--	--	39

A Common Factor Model with Transmission

In 1986, Eaves and colleagues outlined a alternative path model for the analysis of genetically informative longitudinal data. The primary importance of this work was the unification of genetical theory with psychometric (simplex) models for developmental continuity. Their model, shown in Figure 3 for MZ and DZ twin pairs, provides expectations (in terms of additive genetic (g) and individual environmental (e) parameters) for the covariance matrices of a trait measured on m occasions.

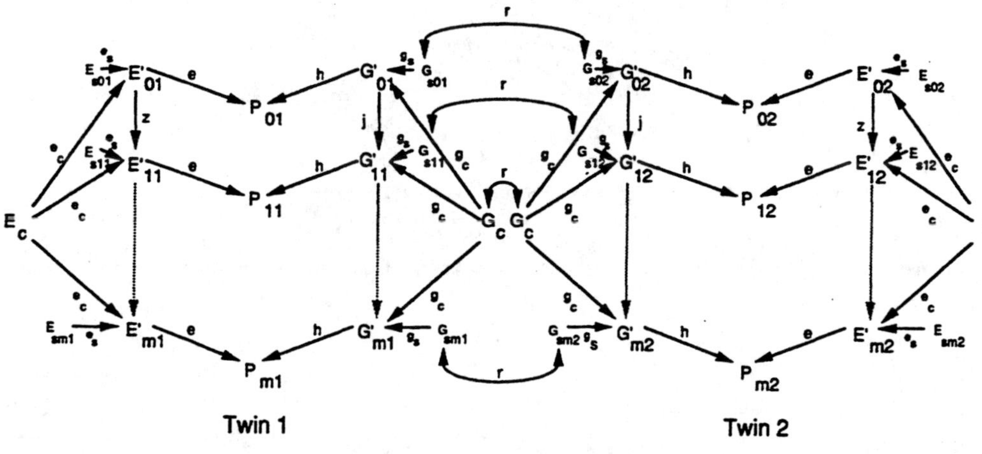

Figure 3. A model for the developmental change in the phenotypes (P) of twin pairs over m occasions of measurement. G_c and E_c indicate additive genetic and unique environmental factors common to all occasions of measurement, while G_s and E_s are factors specific to an occasion. The twin correlation of the additive genetic factors (r) is equal to 1.0 in MZ pairs and 0.5 in DZ pairs. Parameters j and z indicate transmission of genetic and environmental effects over time. (after Eaves et al., 1986.)

The mathematical expectations of this and similar models have been addressed fully by Eaves et al. (1986) and subsequent authors (Boomsma and Molenaar, 1987; Hewitt et al., 1988; Hewitt, 1990). Here, it is of primary importance to underscore the three mechanisms the model uses to explain consistency and change over time. These include: (1) **genetic** or **environmental innovations** at specific ages (factors E_s and G_s); (2) **a common set of genes or features of the environment** impacting the trait throughout time (factors E_c and G_c); and (3) **occasion to occasion transmission** of genetic or environmental variance (through paths z and j). The first mechanism contributes to age-specific variances and explains developmental change, while the latter two mechanisms result in age-to-age genetic and environmental covariances, and explain developmental continuity.

In the Eaves et al. (1986) model, the impact of the common factors on a trait may vary over time. This may reflect an up-regulation of the common set of genes or an increase in the importance of environmental differences on a trait. Figure 4 illustrates changes in genetic variance throughout puberty when a common genetic factor is up-regulated, and there are no age-specific genetic effects or an age-to-age transmission of effects. If the environmental variance remains stable, this example would produce an increase in trait heritability and trait variance over time. The occasion-to-occasion genetic correlations, however, would be one.

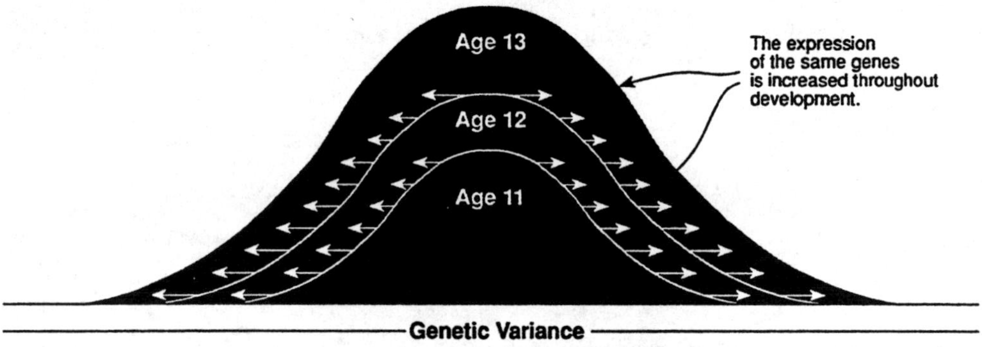

Figure 4. Genetic variance predictions under a common factor model in which factor loading increase over time and there are not occasion specific effects.

In contrast to the common factor model for developmental continuity, the transmission (or simplex) model postulates that the genetic or environmental variation present at an occasion is retained, to some degree, at the subsequent occasion of measurement. This is a model for learning since it allows genetic and environmental innovations occurring throughout development (paths g_s or e_s) to be passed on or "remembered" at the next occasion of measurement. The phenomenon of new gene expression, coupled with transmission of previous genetic effects, is diagrammed in Figure 5. Here, the genetic variance at each occasion persists throughout time as new genetic variance is added. If there were no environmental variance changes over time, then the transmission model, similar to the common factor model in Figure 4, would produce a developmental increase in trait heritability and variance. However, the age-to-age genetic correlations would not be unity due to innovations of new genetic variance. Additionally, the genetic correlations would be greatest for those occasions close in time and reduce in magnitude as the time between measurements increases.

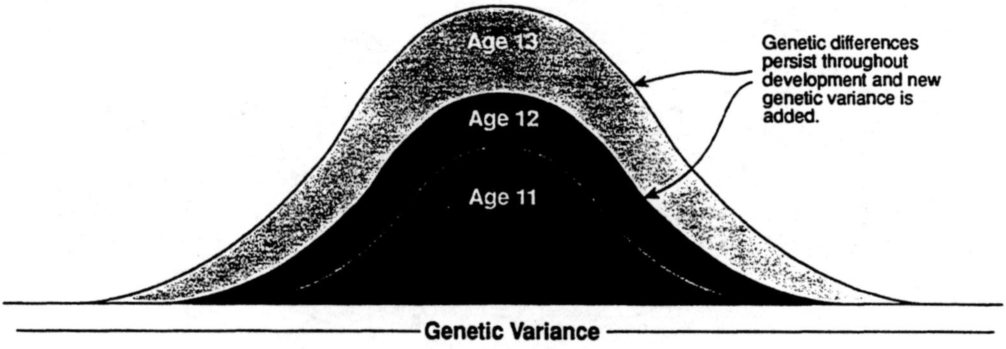

Figure 5. Genetic variance predictions under a transmission model for development which includes genetic innovations.

Since the common factor model and the transmission model for developmental change can lead to different predictions for occasion specific variances and age-to-age covariances, it is possible to discriminate between these two mechanisms for continuity through comparative model-fitting. However, it should be noted that five occasions of measurement are required to estimate the unconstrained parameters of the full model (with both a common factor and transmission) against which reduced models are compared. With fewer occasions of measurement, a restricted set of developmental hypotheses may be tested by placing constraints on parameters (Hewitt et al., 1988).

Example applications of the transmission model for developmental change (with and without common factors) include twin data on weight, the body mass index, blood pressure, psychomotor sensitivity to alcohol and cognitive ability (Fishbein et al., 1990; Dolan et al., 1991; Fabsitz et al., 1992; Hewitt et al., 1987; Boomsma et al, 1989; Eaves et al., 1986; Corey et al., 1986).

Boomsma et al. 's (1989) analysis of psychomotor activity following ingestion of alcohol illustrates the alternate application of a common factor and transmission model to longitudinal data. The authors analyzed data from 206 pairs of twins given a standard dose of alcohol (0.75 grams/kg body weight) and asked to complete a physical motor coordination test (the Vienna determination apparatus) once per hour for three hours following alcohol ingestion. Model-fitting results from two analyses are presented. In the first, estimates of genetic and environmental common factor and occasion-specific influences on motor coordination are obtained; in the second, the common factor structures for the environmental effects are kept, but the genetic structure is replaced with a transmission model. This latter model is referred to as the Hybrid Model. The parameter estimates under the best-fitting reduced models for these analyses are shown in Figure 6.

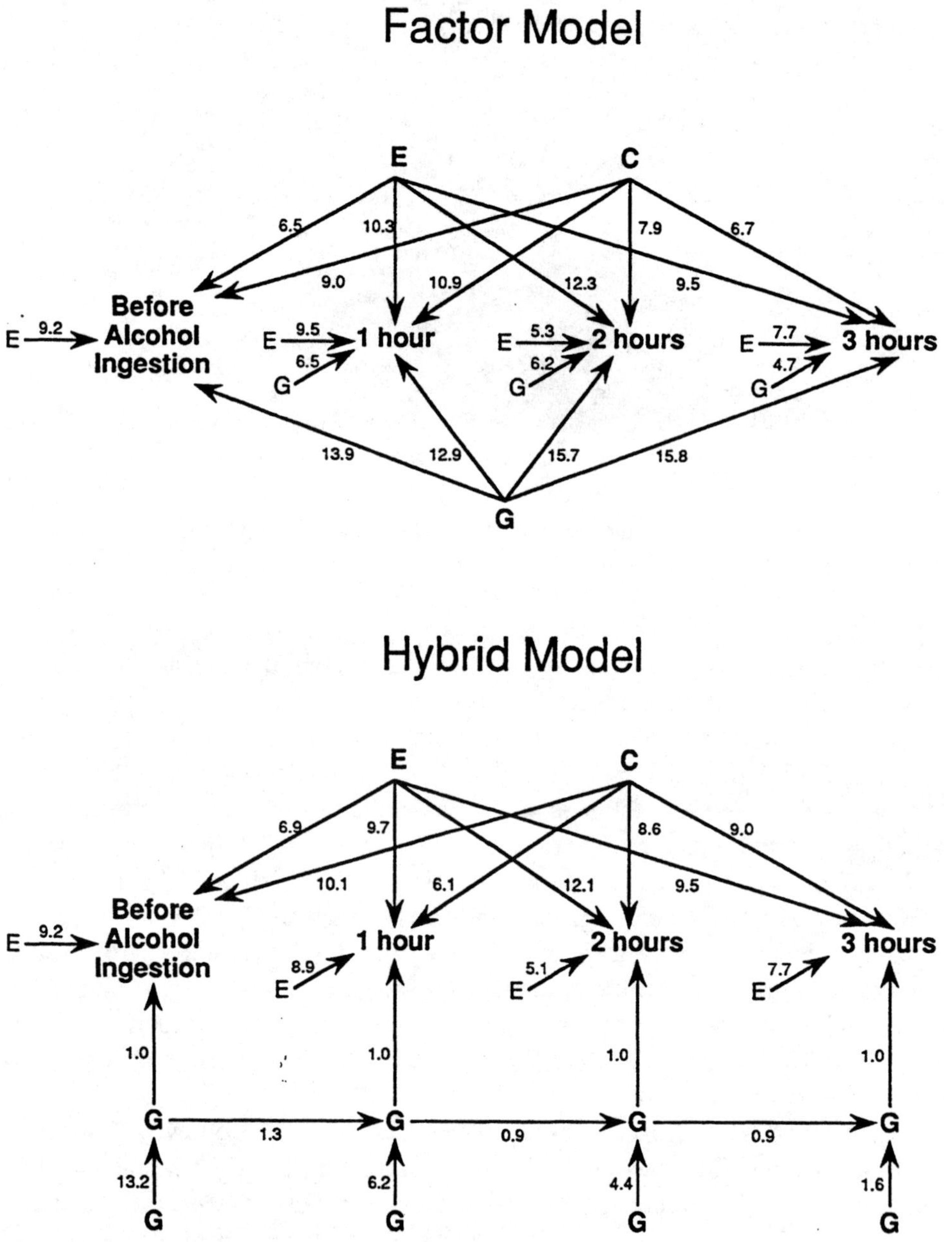

Figure 6. Model-fitting results described by Boomsma et al. (1989) in their analysis of motor coordination in MZ and DZ twin pairs at 1, 2 and 3 hours after ingesting of a standard dose of alcohol. Path coefficients are shown for the best-fitting common factor model (Factor Model) and common factor/transmission model (Hybrid Model).

The best-fitting Factor and Hybrid Models both fit the motor coordination data (Factor Model: $\chi^2 = 79.4$, d.f. = 81, p = 0.53; Hybrid Model: $\chi^2 = 77.11$, d.f. = 81, p = 0.60); but, the parameter values themselves provide different explanations of developmental continuity. Under the common factor model, there is a set of genes expressed at all occasions. Its impact diminishes somewhat during the first hour after ingestion, but rises at hours two and three. In addition, following ingestion, there are genetic influences on the phenotype which are occasion-specific. These explain 20, 13 and 8 percent of the total genetic variance at 1, 2 and 3 hours after ingestion. The occasion-specific influences result in occasion-to-occasion genetic correlations which are less than unity. Under the transmission model, the genetic variance present prior to ingestion is amplified at 1 hour after ingestion (since the transmission path coefficient is greater than one). Thus genetic differences which existed initially become more important soon after an individual is intoxicated. The genetic variance present at the first hour after ingestion is (partially) transmitted to the final two measurement occasions. Additional genetic innovations are added at each hour after ingestion, but, as with the occasion-specific genetic influences under the common factor model, they become increasingly less important.

Based on the small difference in the goodness-of-fit indices of these alternative models, it is inappropriate to suggest that the hybrid model provides a better explanation of the developmental trends than the factor model. With these parameter estimates, a larger sample is required to yield the power to reject the common factor model in favor of the hybrid model. If these data were available, then a more robust characterization of the mechanisms of underlying developmental continuity could be given.

LIMITATIONS OF CURRENT PATH ANALYTIC DEVELOPMENTAL MODELS

OVERVIEW

The path models described above are typically fit to the covariance matrices of a trait measured at defined ages in pairs of genetically informative relatives (e.g., MZ and DZ twins). Unfortunately, it is not always possible in longitudinal studies to assess individuals at the same chronological age, or to have enough subjects at any one given age for sufficient statistical power to test competing developmental hypotheses. Consequently, a range of ages within a defined age group is often used. It is assumed, however, that the phenotypes of the individuals in any age band are influenced by the same set of genetic and environmental factors. Thus, for instance, a genetic study of attention span in 8 to 10-year-old children makes the *a priori* assumption that the same set of genes (though variable in their gene products) influence the phenotypes of all children at this age. It is only by making this assumption that one can take a further step and assess whether an independent set of genes is expressed at a later age.

Here we ask whether it *is even reasonable* to assume that chronological time can be used to define developmental stages which are genetically homogenous. Is it the case, for example, that all 8 to 10-year-old children are at the same developmental stage of gene expression such that differences at *single* group of genetic loci explain all genetically determined differences in attention span? This empirical question cannot be fully answered until all genes which influence attention span are identified and their expression evaluated in 8-10 year-odds. However, some insight into the answer can be gained by rephrasing it into two questions. First, it may be asked whether there are individual differences in developmental time. The answer to this is certainly yes. There is a substantial body of evidence indicating that individuals differ in the speed of development and these differences are partially explained by genetic effects. For example, MZ twin similarity exceeds DZ twin similarity for the age at first tooth eruption, the age of voice development in males and breast development in females, the age at menarche, and the age at death (Ryman, 1975; Sharma, 1983; Gedda and Brenci, 1975; Meyer et al., 1991; Fishbein 1977; Kallman and Sander, 1948, 1949). Also found in the literature are individual differences in the rate of change of continuously distributed phenotypes over time. The best example of this are growth curves, which show genetically influenced variation in their slope and asymptote (Vandenberg and Falkner, 1965; Bock et al., 1973). Given that individual differences do exist in the timing of developmental milestones and the rate of change of continuous phenotypes, it must then be asked whether this timing has any impact on the phenotype of interest. For the example cited above, although it may be reasonable to assume that the age of first tooth eruption is not related to changes in attention span, it may not be correct to assume that the age at puberty is independent of such changes. Again, empirical assessments are required before determining how phenotypic "markers" of developmental time correlate with a phenotype of interest.

In light of the this discussion, how should the current path analytic models for developmental change be interpreted? The answer to this is not straight forward, but will depend upon the ages and phenotypes under study. At some ages (most likely those following the rapid period of adolescent development), it may be safe to assume homogeneity of causal mechanisms. Within adolescence, however, homogeneity of gene expression at a given age is to be doubted. If the timing of gene expression is related to changes in a phenotype of interest, then the modelling approaches outlined above will yield misleading results. The simulation study detailed below has been designed to illustrate this point.

Variable Age at Gene Expression: A Simulation Study

The Data

A simple scenario is considered in which the same set of genes influences a behavioral trait before and after puberty, but genetic variation is increased during puberty. Moreover, the assumption is made that the age at up-regulation varies and is heritable. Figure 7 diagrams this model, showing the genetic variance of a trait at age 11 (assumed here to be pre-pubertal); the variable onset of up-regulation of gene expression (with a mean age at up-regulation of 11.75 years); and the increased genetic variance at age 13. Note also that the up-regulation of gene expression is not immediate. Gene expression gradually increases over a relatively short period of time.

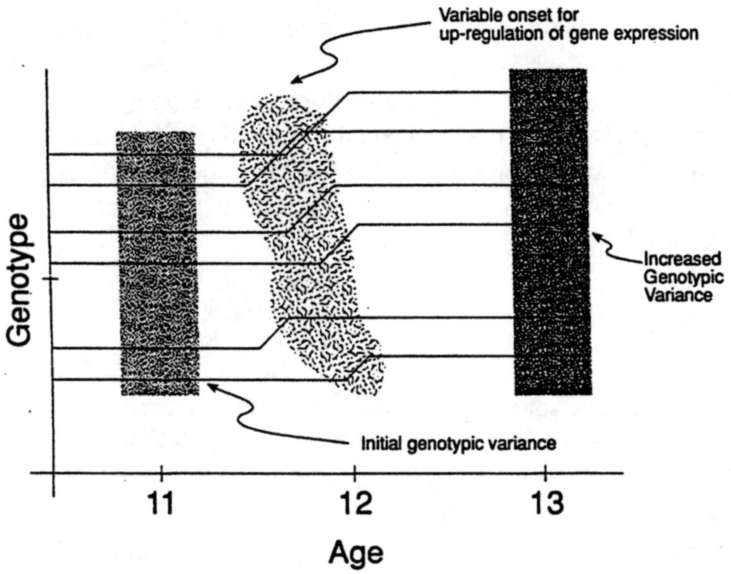

Figure 7. Changes in genetic variance through puberty, assuming an up-regulation of baseline genetic effects at a mean age of 11.75.

Twin data have been simulated under this model by making the following assumptions:

1. At pre-pubertal baseline, the additive genetic and unique environmental variance components (V_A and V_E) are equal to 0.5 and 0.5, respectively. There are no family environmental effects on the trait, thus the phenotypic variance (VP) is equal to $V_A + V_E$. Further, the monozygotic twin correlation (r_{mzbase}) = VA/VP = 0.50; while the dizygotic twin correlation (r_{dzbase})=1/2 VA/VP =0.25.

2. The up-regulation of gene expression (defining the onset of puberty) begins at a mean age of 11.75 (s.d. = 0.55). The onset of puberty is itself genetically influenced, with a heritability of 0.8 . Consequently, r_{mzage}=0.8 and r_{dzage}=0.4.

3. After six months, up-regulation is complete, with post-pubertal gene expression equal to four times the baseline expression.

4. There is no increase in the individual environmental variance component over time. Thus, the post-pubertal heritability is increased to 0.80 (given an additive genetic variance of 2.0 and a unique environmental variance of 0.5).

Under these conditions, data were simulated for 500 pairs of MZ twins and 500 pairs of DZ twins at **three ages of assessment**: 11, 12 and 13. These data were summarized in age-to-age phenotypic correlation matrices and age-specific variances for each zygosity group.

An alternative situation in which there was information available on the developmental stage of the twins was next considered. Using the same model outlined above, data were simulated for **three developmental stages**: early puberty (baseline), middle puberty (with gene expression equal to 2 times the baseline expression), and late puberty (with gene expression equal to 4 times the baseline expression). Again, the data were summarized in stage-to-stage phenotypic correlation matrices, and stage-specific variances for each zygosity group.

SUMMARY STATISTICS

In Table 2, the summary statistics for the age-to-age simulation study are presented.

There are three important features of these simulated data. The first is that trait variance rises at age 12, and then falls at age 13, but does not return to its original value. Compared to the phenotypic variance at age 11, the increased phenotypic variance at age 13 is predominately a consequence of the up-regulation of genes influencing the baseline trait. In contrast, the increase in phenotypic variance at age 12 reflects both the up-regulation of the baseline genetic effect **and** the variable age of onset of this up-regulation.

Table 2. Age-to-age phenotypic correlations for simulated data from MZ and DZ twin pairs.[a]

	T1 Age 11	T1 Age 12	T1 Age 13	T2 Age 11	T2 Age 12	T2 Age 13
T1 Age 11	-------	**0.35**	**0.39**	*0.22*	*0.20*	*0.21*
T1 Age 12	**0.35**	-------	**0.40**	*0.19*	*0.32*	*0.29*
T1 Age 13	**0.48**	**0.41**	-------	*0.14*	*0.20*	*0.36*
T2 Age 11	*0.62*	*0.37*	*0.40*	-------	**0.31**	**0.37**
T2 Age 12	*0.34*	*0.68*	*0.39*	**0.43**	-------	**0.40**
T2 Age 13	*0.46*	*0.40*	*0.78*	**0.57**	**0.43**	-------
Mean MZ	7.32	10.60	13.79	7.35	10.55	13.72
s.d. MZ	1.61	3.40	1.98	1.73	3.28	2.03
Mean DZ	7.24	10.83	13.92	7.28	10.75	13.69
s.d. DZ	1.41	3.44	1.80	1.50	3.30	2.02

[a]Correlations for MZ twins are shown in the lower triangle, while those for DZ twins are given the upper triangle. Within individual age-to-age correlations are bolded and cross-twin correlations are in italics.

The second feature of these data are the within-individual correlations. Note that these are smaller between ages 11 and 12 than between ages 11 and 13. This difference suggests that there is an additional source of variance at age 12 which does not influence the other occasions. From the simulations, we know that this source of variance is due to individual differences in the up-regulation of the baseline genetic factor. Without this prior knowledge, however, it becomes difficult to explain the patterns of correlations with a simple model for developmental continuity. Indeed, the transmission model for developmental continuity, for example, predicts a quite different correlation matrix, with age-to-age correlations decreasing as the time between measurements increases.

Also of note in the within-individual correlations is the variability in the estimates of the same relationship across individuals. For example, the age 11-13 correlations are 0.48, 0.57, 0.39 and 0.37 for first and second members of MZ and DZ twin pairs, respectively. This instability of the age-to-age correlations is also a consequence of individual differences in the timing of gene regulation.

Of final importance in Table 2 are the within-age MZ twin correlations. Since the data were simulated under a model of additive gene action and individual environmental variation, these correlations could be used as direct estimates of heritability. If so, the heritability estimates for the phenotype at ages 11, 12 and 13 are 0.62, 0.68 and 0.78, respectively. Curiously, the heritability estimate at age 11 exceeds the heritability of the baseline trait (0.5). It does so because twin correlations computed by age contain information about the genetic control of the baseline trait *and* the genetic control of age at onset. Consequently, these heritabilities are "compound" heritabilities, influenced by both genetic mechanisms.

Table 3. Stage-to-stage phenotypic correlations for simulated data from MZ and DZ twin pairs.[a]

	T1 Stage 1	T1 Stage 2	T1 Stage 3	T2 Stage 1	T2 Stage 2	T2 Stage 3
T1 Stage 1	——	**0.59**	**0.62**	*0.19*	*0.26*	*0.34*
T1 Stage 2	**0.60**	——	**0.73**	*0.23*	*0.31*	*0.33*
T1 Stage 3	**0.66**	**0.76**	——	*0.28*	*0.35*	*0.40*
T2 Stage 1	*0.52*	*0.64*	*0.65*	——	**0.57**	**0.62**
T2 Stage 2	*0.60*	*0.74*	*0.75*	**0.64**	——	**0.76**
T2 Stage 3	*0.65*	*0.78*	*0.78*	**0.65**	**0.75**	——
Mean MZ	7.05	10.57	14.07	7.02	10.52	14.03
s.d. MZ	1.07	1.31	1.58	1.00	1.31	1.67
Mean DZ	7.03	10.64	14.15	6.98	10.54	14.02
s.d. DZ	0.96	1.29	1.53	0.99	1.26	1.57

[a] Correlations for MZ twins are shown in the lower triangle, while those for DZ twins are given the upper triangle. Within individual age-to-age correlations are bolded and cross-twin correlations are in italics.

In Table 3, summary statistics for the stage-to-stage simulation study are shown. Compared to the age-specific phenotypic variances in Table 2, the stage-specific variances are both smaller and steadily increasing over time. These changes are to be expected, since the variances are no longer influenced by the variable age at onset of up-regulation. Instead, they solely reflect increased genetic variation of the baseline trait.

Also of note in Table 3 are the within-individual stage-to-stage correlations. These are not only larger and more stable than the age-to-age correlations, but follow a different pattern. Specifically, $r_{stage1\text{-}stage2} < r_{stage1\text{-}stage3} < r_{stage2\text{-}stage3}$, so an individual's phenotype at stage 1 correlates more highly with his/her phenotype at stage 3 than at stage 2. This is a direct reflection of the increase in the baseline genetic variance over time combined with the constant expression of occasion specific environmental effects. As a result, stage 3 shares more of its phenotypic variance with stage 1 than does stage 2.

Finally, it can been seen that the MZ twin correlations at stages 1 and 3 come quite close to the expected heritabilities of 0.5 and 0.8. They are no longer "compound" heritability estimates, affected by baseline genes and age at onset genes.

MODEL-FITTING

We next consider the consequences of fitting Cholesky models to the 6x6 age-to-age or stage-to-stage phenotypic covariance matrices of MZ and DZ twins. A full Cholesky model, containing decompositions of the genetic and individual environmental covariance structures, was initially fit to the data. Reduced models which eliminated a subset of parameters were also considered. To select the best-fitting model, likelihood ratio chi-square tests were used to compare the reduced models to the full models (c.f. Neale and Cardon, 1992). All model-fitting was done by maximum likelihood, implemented in the matrix-based model fitting program, *MX* (Neale, 1991).

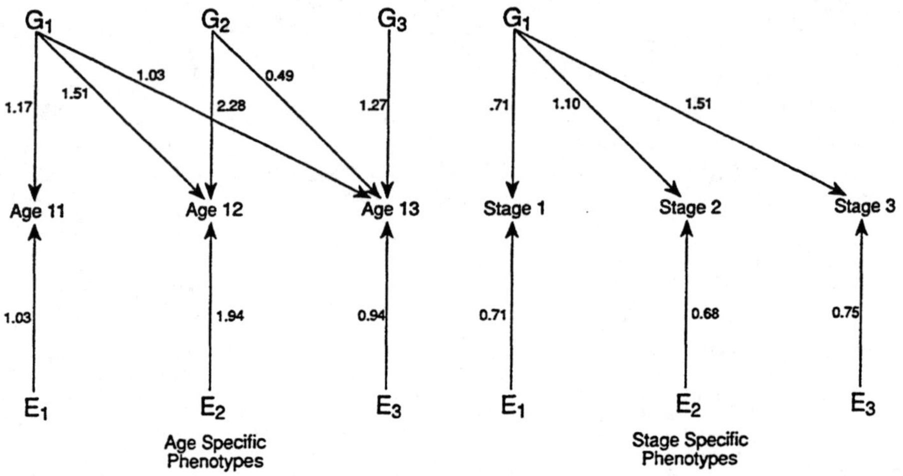

Figure 8. Cholesky model-fitting results for the simulation study, conditioning the data on age and pubertal stage. Results are shown for the best-fitting model in both cases.

In Figure 8, parameter estimates under the best-fitting Cholesky models are shown for the age and stage simulations. For the age simulations, the best fitting model was one which included all genetic parameters and occasion specific environmental parameters.

Under this model, $\chi^2_{33} = 65.33$, $p < 0.001$. For the stage simulations, only the first genetic factor was required, with loadings on all three developmental stages. The environmental structure was the same as that for the age simulations. Under this model, $\chi^2_{36}=40.78$, $p=0.27$. Genetic and environmental covariance and correlation matrices, computed under the best-fitting models are given in Table 4.

Table 4. Estimates of genetic and environmental covariance and correlation matrices from the Cholesky model fitting-results from the simulation study.[a]

	G1[b]	G2	G3	E1	E2	E3
Age Analysis	*1.37*			*1.05*		
	1.00			1.00		
	1.77	*7.48*		--	*3.76*	
	0.55	1.00			1.00	
	1.20	*2.67*	*2.93*	--	--	*0.89*
	0.60	0.15	1.00			1.00
Stage Analysis	*0.51*			*0.50*		
	1.00			1.00		
	0.78	*1.21*		--	*0.46*	
	1.00	1.00			1.00	
	1.00	*1.53*	*1.95*	--	--	0.55
	1.00	1.00	1.00			1.00
Age heritability	0.57	0.67	0.77			
Stage heritability	0.50	0.72	0.80			

[a] Covariance are in italics, correlations in plain type.
[b] Genetic and environmental factors.

These model-fitting results indicate the following:

1. The Cholesky model fails to fit the data summarized by age, but provides an acceptable fit to those data summarized by stage.
2. For the age analysis, the set of genes which influences the phenotype at age 11 persists in its effect throughout development. However, two new sets of genes are expressed at ages 12 and 13. For the stage analysis, one set of genes increases in expression over time. Consequently,
3. Age-to-age genetic correlations are significantly less than one, while stage-to-stage genetic correlations are unity.
4. For both age and stage analyses, cross-occasion environmental correlations are zero.

Now let us attempt to apply these findings to a situation encountered in practice. If there were no information available about the (genetic) stage of development, it would be only feasible to fit developmental models to age-specific phenotypes. Having done this, an investigator would note that the model fails to give a good description of the data; nonetheless, an attempt may be made to interpret the model-fitting results. The investigator would assert that new genetic effects, uncorrelated with those at age 11, influence the phenotype at each subsequent measurement occasion. **If the investigator were to ascribe this new genetic variance to the "switching-on" of two independent sets of genes which influence the phenotype, he or she would be incorrect.** At best, the investigator could speculate that the "new" genetic variance is solely the result of differences in the up-regulation in gene expression, and predict that the new variance would dissipate over time. In contrast, if data had been summarized by stage of development, the investigator would detect a single common genetic factor influencing the trait over time with increasing factor loadings. Remarkably, the simulation conditions would be recovered!

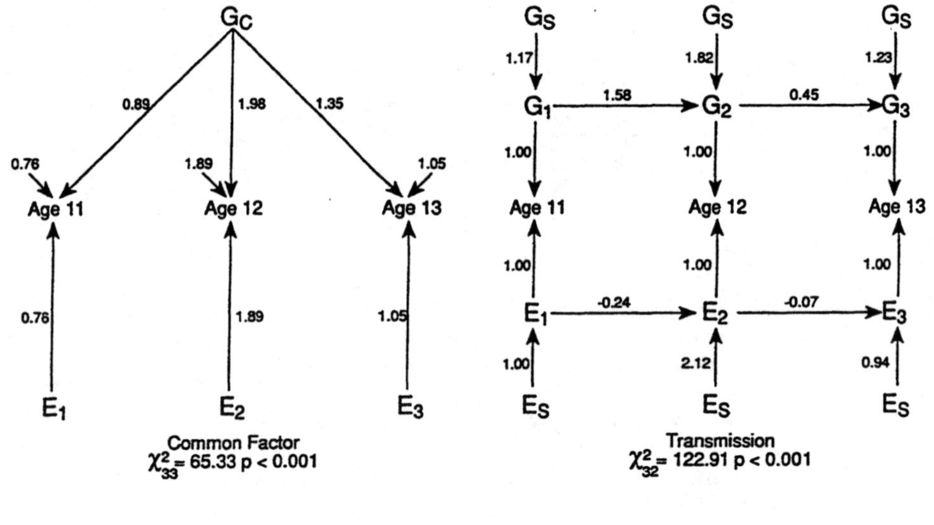

Figure 9. Model-fitting results for a common factor model and transmission model fit to the simulated longitudinal data summarized by age.

Given the unsatisfactory Cholesky model-fitting results for the age-to-age summary statistics, it is reasonable to ask whether other developmental models would provide a better description of the data. Unfortunately, for the models described previously, the answer is no. Model-fitting results for common factor and transmission models are shown in Figure 9. The common factor model gives the same fit to the data as the Cholesky model, and provides no additional insight into the mechanisms which contribute to continuity and change. The loadings on the single common genetic factor are significant, as are the occasion specific genetic influences. Thus, there is evidence for independent genetic effects influencing the phenotype throughout development. Similar to the Cholesky model-fitting results, there are no age-to-age environmental correlations under the common factor model. Figure 9 also shows that an unrestricted transmission model fits the data much worse than the common factor or Cholesky models. Further, transmission parameters are estimated which exceed unity and are negative. These values are not what would be expected in a model for learning; indeed they alert us to the fact that the transmission parameterization is not a useful way of describing these data.

DISCUSSION

The simulation exercise has demonstrated the pitfalls of applying developmental models to age-specific phenotypes when the timing of gene expression is variable and under genetic control. At the same time, it has highlighted the benefits of fitting models to measurements taken at various stages of development. Although the stage analysis appears to be an effective strategy, the reader is sure to ask how one can identify the stages which underlie developmental change in the phenotype of interest. For some behavioral traits, the "stages" may correspond to changes in gene regulation which are not indexed by a phenotypic marker (such as a developmental milestone). In these instances, stages of development cannot be identified without assaying for the presence of a particular gene product. In other cases, the use of developmental stages hypothesized to explain phenotypic change may suffice for exploratory analyses. For example, as illustrated in the next section, pubertal development is used as the conditioning variable for a genetic analysis of the body mass index. The usefulness of the conditioning variable is evaluated by comparing the parsimony and interpretation of developmental models fit to age-to-age and stage-to-stage data.

The limitations of the simulation exercise must also be emphasized. Several assumptions have been made which will not always be met in practice. Most importantly, it has been assumed that a **genetic** factor determines the timing of gene expression. It is the time dependence of this **single** factor which creates **multiple** age-specific genetic effects throughout puberty. If the individual differences in developmental time within an age category are not genetically determined, then age-specific genetic effects will not be detected. Certainly, it is reasonable to believe that environmental factors influence developmental time as well. The environmental factors could be unique to a individual or shared by members of a twin pair. For instance, changes in peer group norms which impact a behavioral trait may occur when children enter middle school. The timing of the change in norms will be age dependent (not genetically determined stage dependent) and perfectly correlated in twins. Thus, for this example, it would be useful to condition an analysis of developmental change on age rather than on a genetically determined stage.

Eventually, longitudinal analyses of genetically informative data may not be considered complete until genetic and environmental timing are individually addressed. For the present purpose, we recognize that differences in timing mechanisms may exist, but limit the discussion in this chapter to a single phenotypic developmental marker which is hypothesized to explain longitudinal change in a covarying phenotype.

A Developmental Genetic Analysis of the Body Mass Index of Adolescent Twins

In this example, prospective data on the body mass index of male adolescent twin pairs are summarized into three age categories and pubertal stages of development. Modified Cholesky path models are then fit to the data, and the age and stage model-fitting results compared.

The Sample

The Virginia Twin Study of Adolescent Behavioral Development (VTSABD) is a current longitudinal study of child psychopathology in a targeted twin sample of 1,894 caucasian pairs age 8-16. Since March 1990, 1,412 families (including the twins and their parents) have participated in home assessments of various domains of child psychopathology (full details are given in Hewitt et al., submitted). Further, 937 families with twins still within the targeted age range have been interviewed for a second time, approximately 15-months after their first home visit.

The targeted twin sample is part of a twin population of 5,021 like-sex and unlike-sex pairs born between 1968-1985 and identified by state-wide public and private schools in 1987. Parents of twin pairs were first contacted through a mailed questionnaire which included questions pertaining to zygosity diagnosis. Responses to the initial mailing were obtained from 85% of the families. The interview study is comprised of the responding families who met age and residence requirements at the time of home interview assignments (March 1990-December 1991). We continue to expend all efforts to include "soft-refusal" families into both waves of the home interview.

For like sex twins, zygosity has been diagnosed using three sources of information. The most definitive source is blood antigen typing or DNA typing. To date, the zygosity of 237 like-sex twin pairs have been diagnosed in this way. The remaining pairs have received zygosity diagnoses by considering parental responses to three questions about the similarity of the twins and the ratings of the twins' pictures by two experienced judges. The algorithm for deriving diagnoses from these two sources is described in full by Maes et al. (in preparation).

The current analysis is restricted to data on male twins participating in one or both waves of the home interview. This includes 292 MZ and 180 DZ pairs in Wave 1, and 172 MZ and 98 DZ pairs in Wave 2.

Measures

Although the primary focus of the VTSABD is to assess child psychopathology, additional data on height, weight and pubertal status are collected at the time of home interview. Height and weight measurements are taken by trained field interviewers who follow a standard protocol using portable scales and tape measures. For the analyses described below, the body mass index (BMI) has been computed from these measurements by the formula: weight (kg.)/height(m.)^2 . The BMI data were log-transformed to meet the assumption of normality in the genetic analyses.

Pubertal status was obtained through the child's self-report. Boys have been asked about skin changes, the presence of body hair, voice breaking and facial hair. All items other than skin changes are scored on a 4 point scale, with 0 indicating no development; 1, initial development; 2, moderate development and; 3, full development. For skin changes, a three point scale is used, indicating no change (0), some change (1); quite a bit of change (2). A summed score of the four items was used to define pubertal stage. Early puberty corresponded to a score of 0-3; middle puberty, 4-6; and late puberty, 7 and greater.

In Table 5 are shown the number of individual male twins in the three puberty groups at Waves 1 and 2. Also given are the number of individuals in the three age categories, defined as age 8-10, age 11-13 and age 14-16. For Wave 1, the age grouping divides the sample roughly in thirds, while the pubertal grouping places the majority of individuals (46%) in middle puberty. At Wave 2, the majority of twins (46%) are between the ages of 11 and 13 and in late puberty (47%).

Table 5. Male twins in each age and puberty group at Wave 1 and Wave 2 home visit. [a]

Wave 1	Wave 2					
		Age 8-10 **Early puberty**	Age 11-13 **Middle puberty**	Age 14-16 **Late puberty**	Wave 1 visit only	Total
	Age 8-10 **Early puberty**	92 **97**	90 **76**	-- **2**	131 **134**	313 (33.2%) **309 (32.7%)**
	Age 11-13 **Middle puberty**		89 **168**	103 **89**	156 **177**	348 (36.9%) **434 (46.0%)**
	Age 14-16 **Late puberty**			145 **99**	138 **102**	283 (30.0%) **201 (21.3%)**
	Total Wave 2	97 (18.3%) **92 (17.4%)**	244 (46.0%) **189 (35.7%)**	190 (35.8%) **248 (46.9%)**	–	

[a] Thrre number of individuals in each age group is shown in plain type, and in each puberty group, bold type.

DATA ANALYSIS

The aim of data analysis was to fit modified Cholesky models to the BMI data summarized by age or pubertal stage. Modified models must be fit to the data because individuals have only been observed at (the most) two sequential age or pubertal categories. Consequently, the standard Cholesky model for three occasions of measurement (which estimates the determinants of the phenotypic covariation of occasions 1 and 3) cannot be resolved with the current data set. However, the factors contributing to the cross-occasion correlation of sequential categories can be identified. This was done by using the path model diagrammed in Figure 10 to obtain estimates of h'_2, c'_2, and e'_2 (paths contributing to the covariation of the first two measurement occasions); h''_2, c''_2 and e''_2 (paths contributing to the covariation of the latter two measurement occasions), as well as h_1, c_1, e_1, h_2, c_2, e_2, h_3, c_3, and e_3, the occasion specific influences. The estimated path coefficients were then used to compute genetic and environmental cross-occasion correlations and occasion specific variances. It should be pointed out that this parameterization is equivalent to one which estimates the effect of one common factor on occasions 1 and 2, a second common factor on occasions 2 and 3, and specific influences on occasion 3. The parameterization in Figure 10 was chosen since it allows one to fix the specific influences on occasions 1 and 3 to zero without affecting cross-occasion correlations.

A modified Cholesky model fit to the BMI data summarized by age or pubertal stage. See the text for a description of parameter estimates.

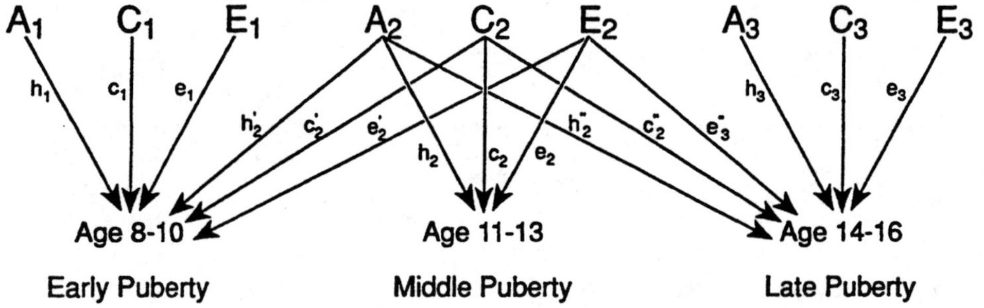

Figure 10. A modified Cholesky model fit to the BMI data summarized by age or pubertal stage. See the text for a description of parameter estimates.

Because the BMI data summarized by age and pubertal stage are not 'balanced' (i.e., the same number of individuals have not been assessed at each measurement occasion), the modified Cholesky model cannot be fit directly to cross-twin, cross-occasion phenotypic covariance matrices. Instead, the model has been fit to the raw (log-transformed) data using a maximum likelihood pedigree analysis approach (Lange et al., 1976) implemented in the matrix based model-fitting program *MX* (Neale, 1991). In pedigree analysis, occasion specific means are estimated along with the path coefficients of the developmental model. These means were constrained to be equal across zygosity and twin status. Thus, the full modified model for either the age or stage analysis includes 18 parameters: 3 means and 15 path coefficients.

After fitting the modified Cholesky model to the BMI data, two sub-models were initially considered. These were models with no genetic influences (13 parameters) and no common environmental influences (13 parameters). Both of the models were compared to the modified Cholesky model (termed the full model) using a likelihood ratio chi-square test. The best-fitting model among the three was then used as the starting point for additional sub-models. These models tested the significance of genetic and environmental cross-occasion correlations and occasion specific effects.

RESULTS

DATA SUMMARY

In Table 6 are shown mean and variance summary statistics for the (log-transformed) body-mass index assessed at three ages and three pubertal stages. Sample sizes reflect the number of individuals measured throughout the duration of the study (Waves 1 and 2). Repeat measurements within an age or pubertal category are not included. Thus, if an individual did not change age category or pubertal stage from Wave 1 to Wave 2, his Wave 2 measurement is dropped from the analysis. The statistics in Table 6 indicate that the BMI means for the three age categories and pubertal stages are fairly similar, with both increasing over time. In contrast, the BMI variances over ages and pubertal stages differ, decreasing more from middle puberty to late puberty than from ages 11-13 to 14-16.

Table 6. Summary statistics for the log-transformed body mass index of male adolescent twins at three puberty stages and three age categories.

	N	Mean	s.d.
Age 8-10	298	2.82	0.17
Age 11-13	424	2.92	0.18
Age 14-16	375	3.04	0.16
Early puberty	294	2.83	0.17
Middle puberty	490	2.93	0.19
Late puberty	284	3.05	0.14

A summary of the phenotypic correlations, computed from parameter estimates under the full Cholesky age and stage models, are shown in Table 7. The within-individual, cross-(MZ)twin and cross-(DZ)twin correlations are given for the three age groupings and pubertal stages. The within-occasion twin correlations indicate that MZ and DZ twin similarity for BMI is greater at the second and third occasions of measurement than at the first. This implies that over chronological time and throughout puberty, there is a reduction in the influence of the unique environment and/or an increase in the importance of familial factors on the body mass index. From the cross-occasion phenotypic correlations, information regarding the strength and nature of occasion-to-occasion continuity may be obtained. Within individuals, the cross-occasion correlations are consistently higher for the data summarized by age (0.86 and 0.88) than those summarized by stage (0.74 and 0.80). This indicates that factors influencing individual differences across age categories are more highly correlated than those influencing continuities across pubertal stage. This finding is actually the opposite of that illustrated in the simulation study. It could, however, arise if the division of the data into pubertal categories defines developmental stages which are more homogeneous (in terms of their genetic and environmental determinants) than the division of the data into age categories. Between co-twins, the cross-occasion correlations for ages 8-10 and 11-13 and those for stages 1 and 2 are similar. In contrast, the stage 2-3 cross-twin correlations are consistently lower than the corresponding age correlations. Taken together with the within-occasion twin correlations, this pattern indicates that familial factors are less highly correlated across the latter two pubertal stages than age categories.

Table 7. Estimated age-to-age and stage-to-stage phenotypic correlations for BMI data on MZ and DZ twin pairs.a

	Age 8-10	Age 11-13	Age 14-16
Age 8-10	**1.00**		
	0.71		
	0.36		
Age 11-13	**0.86**	**1.00**	
	0.70	*0.88*	
	0.35	0.51	
Age 14-16	-----	**0.88**	**1.00**
		0.81	*0.93*
		0.42	0.46
	Stage 1	Stage 2	Stage3
Stage 1	**1.00**		
	0.67		
	0.43		
Stage 2	**0.74**	**1.00**	
	0.72	*0.90*	
	0.38	0.46	
Stage 3	-----	**0.80**	**1.00**
		0.73	*0.86*
		0.33	0.55

a Within-individual correlations are given in bold, while MZ twin correlations are underlined, and DZ twin correlations are in plain type.

MODEL-FITTING

Before fitting sub-models to the age and stage data separately, a test of heterogeneity was conducted to determine whether the age and stage parameter estimates under the modified Cholesky model were significantly different from each other. The test of heterogeneity compared the log-likelihood of a model estimating one set of path coefficients for the age and stage data to a model which estimated two sets of path coefficients. The likelihood ratio χ^2 for the comparison of these models was 31.6, with 15 degrees of freedom, $p<0.05$. It was therefore concluded that the age and stage path coefficients were significantly heterogenous. Given this result, subsequent sub-models of the age and stage data were fit independently.

Table 8. Model-fitting results for the body mass index at three age categories.

Model	Parameters	vs. model	Chi-square (d.f.)	p
1. Full modified Cholesky model	18	--	--	--
2. No genes	13	1	134.4 (5)	<0.001
3. No common environment	**13**	**1**	**9.6 (5)**	**n.s.**
4. MODEL 3 with No unique environmental covariance occasions 1-2	12	3	21.7 (1)	<0.001
5. MODEL 3 with No unique environmental covariance occasions 2-3.	12	3	18.5 (1)	<0.001
6. MODEL 3 with No specific genes occasion 1.	12	3	7.4 (1)	<0.05
7. MODEL 3 with No specific genes occasion 3.	12	3	34.1 (1)	<0.001

Table 9. Model-fitting results for the body mass index at three pubertal stages.

Model	Parameters	vs. model	Likelihood Ratio Chi-square (d.f.)	p
1. Full modified Cholesky model	18	--	--	--
2. No genes	13	1	98.38 (5)	<0.001
3. No common environment	13	1	3.02 (5)	n.s.
4. MODEL 3 with No unique environmental covariance occasions 1-2	12	3	0.19 (1)	n.s.
5. MODEL 3 with No unique environmental covariance occasions 2-3	12	3	5.75 (1)	<0.05
6. MODEL 4 with NO specific genes occasion 1	**11**	**4**	**1.44 (1)**	**n.s.**
7. MODEL 4 with No specific genes occasion 3	11	4	15.91 (1)	<0.001
8. MODEL 6 with No genetic covariation occasions 2-3	10	6	57.51 (1)	<0.001

Summaries of the age and stage model-fitting results are provided in Tables 8 and 9, respectively. Initially, it was determined whether either additive genetic effects (Model 2) or common environmental effects (Model 3) could be dropped from the models. For both the age and stage analyses, models without genetic influences were significantly worse fitting than the fuller model. In contrast, models without common environmental influences were not significantly worse. Consequently, all subsequent models fit to the data did not include the common environmental parameters.

For the age analysis, Model 3 was selected as the best description of the data since models without age-to-age unique environmental covariances (Models 4 and 5) and age-specific genetic effects (Models 6 and 7) fit the data significantly worse than Model 3. For the stage analyses, a model without a unique environmental covariance between occasions 1 and 2 (Model 4) did not fit the data worse than Model 3. In contrast, when the unique environmental covariance between occasions 2 and 3 was fixed to zero (Model 5), the resulting model was significant worse fitting than Model 3. In Model 6, the unique environmental covariance between occasions 1 and 2 remains fixed at zero, while the specific genetic influences on occasion 1 are removed. This model does not fit the data worse than Model 3. However, when the specific genes on occasion 3 are removed (Model 7) or the genetic covariation between occasions 2 and 3 is fixed to zero, the fit of the model becomes significantly worse. It was therefore concluded that Model 6 is the best description of the data conditioned on pubertal stage.

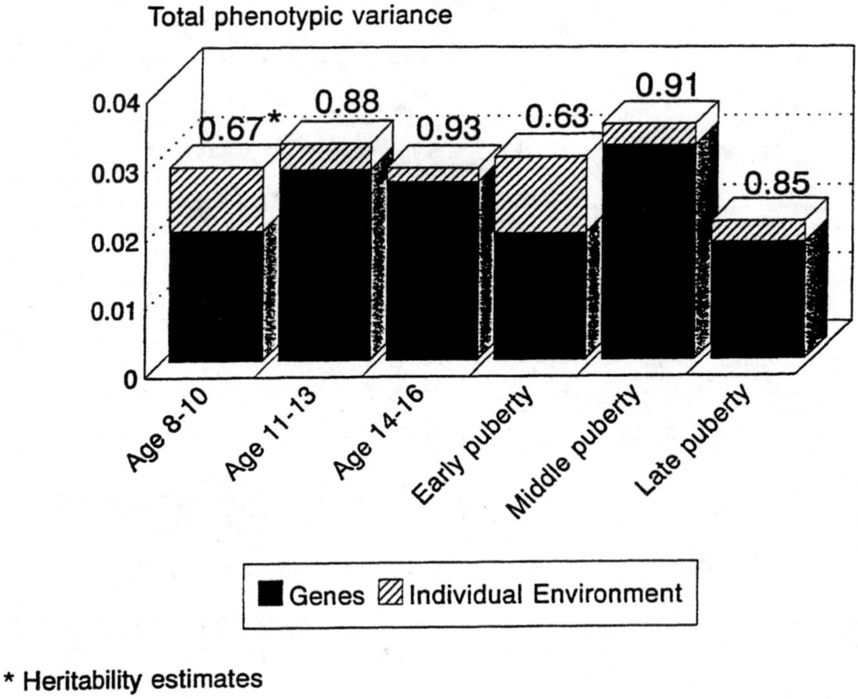

Figure 11. Age and stage-specific genetic and environmental variance components for BMI. The parameter estimates under the best-fitting Cholesky model have been used to compute the variances

Using the parameter estimates under the best-fitting age and stage models, within-occasion genetic and unique environmental variances, and cross-occasion correlations were computed. The variance estimates and the corresponding heritability estimates are shown in Figure 11, while the cross-occasion correlations are presented in Table 10. In Figure 11, noteworthy findings include: (1) an increase in trait heritability in the latter two age and stage categories; (2) a marked increase in genetic variation in the second age category and pubertal stage; and (3) a greater decline in genetic variation at late puberty than at ages 14-16.

Table 10. Estimated BMI age-to-age and stage-to-stage genetic and individual environmental correlations under best-fitting Cholesky models.a

	Age 8-10	Age 11-13
Age 11-13	**0.91** 0.24	-----
Age 14-16	-----	**0.89** 0.72
	Stage 1	Stage 2
Stage 2	**1.00** 0.00	-----
Stage 3	-----	**0.81** 0.65

a Genetic correlations are given in bold, while environmental correlations are in plain type.

The changes in heritability estimates over developmental and chronological time reflect both *decreases* in unique environmental variance components and *increases* in genetic variance components. It is the pattern of change in the environmental influences, however, which is most consistent in the age and stage results. In both cases, the size of the variance component is roughly halved after the first occasion of measurement. From these data alone, it is not possible to determine the cause of this decreased variance. It should be noted, though, that the cross-occasion environmental correlations between ages 8-10 and 11-13 and between stages 1-2 are very low (0.24 and 0.0, respectively, from Table 10.) This indicates that the factors contributing to the large environmental variance at occasion 1 are primarily occasion specific.

The marked increase in genetic variance at the second age and stage of assessment could be explained by the expression of new genetic effects which add to the 'old' genetic variation and/or an enhancement of existing genetic differences between individuals. If the first explanation accounts for the data, then the genetic correlations between occasions 1 and 2 would necessarily be less than one. This would not be the case if the second explanation alone accounted for the increase in genetic variation. From the correlations in Table 10, it is clear that new genetic effects **are not** being expressed during middle puberty. In contrast, there is evidence that a small proportion of the genetic influence on ages 11-13 is specific to that occasion, since the cross-occasion genetic correlation of 0.91 is slightly but significantly less than one. Nevertheless, the genetic variation at ages 11-13 which is not correlated with the previous occasion (estimated as 0.005) is not large enough to fully explain the increase in genetic variation within these ages. As with pubertal stage, the increase must involve existing genetic differences among individuals. Two mechanisms discussed earlier in this chapter, the transmission of genetic influences over time (coupled with the constant expression of a common genetic factor) and the up-regulation of gene expression, could account for the increase. A third possible explanation is that within the second occasion of assessment, individual genotypes differ in their rate of BMI development in a random fashion (e.g, the timing of change within the occasion is **not** under independent genetic control). Unfortunately, the data from just two occasions of measurement do not allow us to determine which of these three alternative hypotheses best explains the data.

The final feature of these data is the marked decrease in genetic variation between stages 2 and 3 compared to that between ages 11-13 and 14-16. Mechanisms contributing to this decline are similar to those yielding an increase in genetic variation: either the existing genetic differences between individuals become less important or there is a change in the genetic loci which influence the trait. Again, the cross-occasion correlations provide some insight into these possibilities. For both the age and stage analyses, the occasion 2-3 genetic correlations are significantly less than one, indicating that some of the genetic influences on the third occasions of assessment are specific. However, because a large percentage of the genetic variation at occasion 3 is shared with the previous occasion, its small magnitude cannot be totally explained by the switching on of "new" less variable genetic effects. Indeed, the low genetic variation at occasion 3 indicates that the variation identified at occasion 2 is decreasing over time. This effect is more apparent at the third pubertal stage than the third age category. One possible explanation for the decline in genetic variance at occasion 3 is the dissipation of genotypic differences in BMI development which arose in the previous occasion. The varying age and stage results suggest that individuals age 14-16 may be more developmental heterogenous than those in late puberty.

DISCUSSION

This exercise has demonstrated the utility of approaching a longitudinal genetic analysis from both chronological and developmental perspectives. The heterogeneity of the model-fitting results for the age and stage analyses indicates that different inferences can be made from the same data summarized in alternative ways. For BMI, the most important difference is the change in the genetic variance component over time. An analysis based on age alone would lead an investigator to state that genetic differences among older individuals (14-16) are elevated in comparison to younger individuals (8-10). The investigator could not predict that these differences would dissipate over time. In contrast, the pubertal analysis defines changes in genetic variation which correlate with developmental time. The result from this analysis allows one to predict that the genetic variance component for the body mass index will return to its baseline (early pubertal) value after development is complete.

The value of the pubertal analyses should in no way diminish the importance of the age analysis. Indeed, quantifying the extent of genetic and environmental variation in any defined age group only helps to underscore the differences which exist among individuals who are often given the same environmental treatment (e.g., schooling) based on age. However, as illustrated in the simulation study, one must proceed with caution in interpreting the pattern of age-to-age genetic covariances and age-specific genetic effects. For the BMI analysis, the pattern was comparable for the age and stage data, but this won't always be the case. Clearly, the results from both analyses are helpful in describing longitudinal change.

Several limitations of the current approach to developmental analyses must be emphasized. First and most important is the selection of the putative phenotypic marker of pubertal development. The summed score of the self-report puberty items is at best a fair indicator of biological change. A more complete assessment of pubertal development would, for example, include hormonal assays and bone density measurements. These measures could then be used to define pubertal stages more rigorously, or, perhaps even be used as continuous indices of development in a model which did not require the division of the data into developmental stages.

The use of continuous indices is especially appealing since it would eliminate the necessity of placing some individuals in the same pubertal stage over time.

Other limitations of the BMI analysis are related to the data themselves. The body mass index has become popular in genetic epidemiological studies primarily because the collection of height and weight data is simple and cost efficient. Unfortunately, the BMI provides no information about the percentages of individual fat mass and lean mass and therefore it cannot be used to define a clinically meaningful obesity phenotype. For this reason, genetic analyses of the BMI should be viewed as exploratory. Hypotheses generated from the analyses will eventually be tested using more refined measures of obesity. A final limitation of the BMI data is that the adolescent twins have been assessed at just two sequential ages or pubertal stages. Consequently, only a limited number of hypotheses regarding occasion-specific effects and occasion-to-occasion correlations could be tested. As the Virginia Twin Study of Adolescent Behavioral Development continues, a more informative analyses of the body mass index, and its pubertal correlates will become possible.

REFERENCES

Bayley, N. (1969). *Manual for the Bayley Scales of Infant Development*. New York: Psychological Corporation.

Bock, R.D., Wainer, H., Petersen, A., Thissen, D., Murray, J., & Roche, A. (1973). A parameterization for individual human growth curves. *Human Biology*, *45*, 63-80.

Boomsma, D.I., Martin, N.G., & Molenaar, P.C. (1989). Factor and simplex models for repeated measures: application to two psychomotor measures of alcohol sensitivity in twins. *Behavior Genetics, 19,* 79-96.

Boomsma, D.I. & Molenaar, P.C. (1987). The genetic analysis of repeated measures. I. Simplex models. *Behavior Genetics, 17,* 111-123.

Cardon, L.R., Corley, R.P., DeFries, J.C., Plomin, R., & Fulker, D.W. (1992). Factorial validation of a telephone battery of specific cognitive abilities. *Personality and Individual Differences, 13,* 1047-1050.

Cardon, L.R. & Fulker, D.W. (1991). Sources of continuity and change in infant predictors of adult IQ. *Intelligence, 15,* 279-293.

Cardon, L.R. & Fulker, D.W. (1995). Genetic influences on body fat from birth to age 9. *Genetic Epidemiology,*

Cardon, L.R., Fulker, D.W., DeFries, J.C., & Plomin, R. (1992). Continuity and changes in general cognitive ability from 1 to 7 years. *Developmental Psychology, 28,* 64-73.

Carmelli, D., Heath, A.C., & Robinette, D. (1993). Genetic analysis of drinking behavior in World War II veteran twins. *Genetic Epidemiology, 10,* 201-213.

Corey, L.A., Eaves, L.J., Mellen, B.G., & Nance, W.E. (1986). Testing for developmental changes in gene expression on resemblance for quantitative traits in kinships of twins: application to height, weight, and blood pressure. *Genetic Epidemiology, 3,* 73-83.

Dolan, C.V., Molenaar, P.C., & Boomsma, D.I. (1991). Simultaneous genetic analysis of longitudinal means and covariance structure in the simplex model using twin data. *Behavior Genetics, 21,* 49-65.

Eaves, L.J., Long, J., & Heath, A.C. (1986). A theory of developmental change in quantitative phenotypes applied to cognitive development. *Behavior Genetics, 16,* 143-162.

Fabsitz, R.R., Carmelli, D., & Hewitt, J.K. (1992). Evidence for independent genetic influences on obesity in middle age. *Int J Obes Relat Metab Disord, 16,* 657-666.

Fischbein, S. (1977). Onset of puberty in MZ and DZ twins. *Acta Geneticae Medicae Gemellologia, 26,* 151-158.

Fischbein, S., Molenaar, P.C., & Boomsma, D.I. (1990). Simultaneous genetic analysis of longitudinal means and covariance structure using the simplex model: application to repeatedly measured weight in a sample of 164 female twins. *Acta Geneticae Medicae Gemellologia, 39,* 165-172.

Fulker, D.W. & Cardon, L.R. (1993). What can twin studies tell us about the structure and correlates of cognitive abilities? In T.J Bouchard,Jr. & P. Propping (Eds.). *Twins as a Tool of Behavioral Genetics* New York: John Wiley and Sons, Ltd.

Fulker, D.W., Cherney, S.S. & Cardon, L.R. (1993). Continuity and change in cognitive development. In R. Plomin & G.E. McClearn (Eds.). *Nature, Nurture, and Psychology* Washington, D.C.: American Psychological Association

Gedda, L. & Brenci, G. (1975). Twins as a natural test of chronogenetics. *Acta Geneticae Medicae Gemellologia, 24*, 15-30.

Hewitt, J.K. (1990). Changes in genetic control during learning, development, and aging. In M.E. Hanh, J.K. Hewitt, N.D. Henderson & R.H. Benno (Eds.). *Developmental Behavior Genetics: Neural, Biometrical and Evolutionary Approaches* New York: Oxford University Press

Hewitt, J.K., Eaves, L.J., Silberg, J.L., Rutter, M., Simonoff, E., Meyer, J.M., Lober, R., Neale, M.C., Erickson, M., Kendler, K.S., Heath, A.C., Pickles, A., Truett, K.R. & Maes, H. (1995). *Genetics and developmental psychopathology: The Virginia Twin Study of Adolescent Behavioral Development.* (un pub)

Hewitt, J.K., Carroll, D., Sims, J., & Eaves, L.J. (1987). A developmental hypothesis for adult blood pressure. *Acta Geneticae Medicae Gemellologia, 36*, 475-483.

Hewitt, J.K., Eaves, L.J., Neale, M.C., & Meyer, J.M. (1988). Resolving causes of developmental continuity or "tracking." I. Longitudinal twin studies during growth. *Behavior Genetics, 18*, 133-151.

Kallman, F.J. & Sander, G. (1948). Twin studies on longevity and aging. *J Heredity, 39*, 349-357.

Kallman, F.J. & Sander, G. (1949). Twin studies on senescence. *American Journal of Psychiatry, 106*, 29-36.

Kaprio, J., Viken, R., Koskenvuo, M., Romanov, K., & Rose, R.J. (1992). Consistency and change in patterns of social drinking: a 6-year follow-up of the Finnish Twin Cohort. *Alcohol Clin Exp Res, 16*, 234-240.

Lange, K., Westlake, J., & Spence, M.A. (1976). Extensions to pedigree analysis. III. Variance components by the scoring method. *Ann Hum Genet, 39*, 485-491.

Maes, H., Meyer, J.M., Silberg, J.L., Eaves, L.J. & Hewitt, J.K. (1995). *Zygosity determination in the Virginia Twin Study of Adolescent Behavioral Development.* (un pub)

Martin, N.G., Eaves, L.J., & Loesch, D.Z. (1982). A genetical analysis of covariation between finger ridge counts. *Ann Hum Biol, 9*, 539-552.

Meyer, J.M., Eaves, L.J., Heath, A.C., & Martin, N.G. (1991). Estimating genetic influences on the age-at-menarche: a survival analysis approach. *Am J Med Genet, 39*, 148-154.

Neale, M.C. (1991). *Mx: Statistical Modelling.* Box 3 MCV Station, Richmond, VA 23298: Department of Human Genetics.

Neale, M.C. & Cardon, L.R. (1992). *Methodology for Genetic Studies of Twins and Families.* Dordrecht: Kluwer academic Publishers.

Phillips, K. & Matheny, A.P.,Jr. (1990). Quantitative genetic analysis of longitudinal trends in height: preliminary results from the Louisville Twin Study. *Acta Geneticae Medicae Gemellologia, 39*, 143-163.

Ryman, N., Lindsten, J., Leikrans, S., Filipsson, R., Hall, K., Hirschfeld, J., & Swan, T. (1975). A genetic analysis of the normal body-height growth and dental development in man. *Ann Hum Genet*, *39*, 163-171.

Sharma, J.C. (1983). The genetic contribution to pubertal growth and development studied by longitudinal growth data on twins. *Ann Hum Biol*, *10*, 163-171.

Terman, L.M. & Merill, M.A. (1973). *Stanford-Binet Intelligence Scale: 1982 norms edition*. Boston: Houghton-Mifflin.

Vandenberg, S.G. & Falkner, F. (1965). Hereditary factors in human growth. *Human Biology*, *37*, 357-365.

Wechsler, D. (1974). *Manual for the Wechsler Intelligence Scale for Children-Revised*. New York: Psychological Corporation.

This work was supported by NIH grants MH-48604 and MH-45268 and a grant from the Carman Trust.

CHAPTER 3

THE DETECTION OF GENOTYPE-ENVIRONMENT INTERACTION IN LONGITUDINAL GENETIC MODELS

Peter C.M. Molenaar, Dorret I. Boomsma and Conor V. Dolan

University of Amsterdam and Free University, Amsterdam, The Netherlands

INTRODUCTION

Recently a number of models have been suggested for the analysis of longitudinal twin data (see Loehlin, 1991 for a brief overview). These models can be used to study the genetic and environmental contributions to the variances and covariances of phenotypic measures at a single measurement occasion and to the stability and change of individual differences over time. McArdle (1986) applied the latent growth curve model to longitudinal twin data, Boomsma and Molenaar (1987) proposed modeling development by means of autoregressive or simplex models and Eaves, Long and Heath (1986) suggested a model combining both autoregressive and confirmatory factor analysis models (see also Boomsma, Martin and Molenaar, 1989; Hewitt, Eaves, Neale and Meyer, 1988; Loehlin, Horn and Willerman, 1989; Molenaar, Boomsma and Dolan, 1991). Although these models incorporate different developmental hypotheses, they share the basic assumption that the phenotypic deviation score at each measurement occasion is related to latent genetic and environmental deviation scores according to a simple linear (or additive) model:

$$P = hG + eE + cC \qquad (1)$$

where P stands for the phenotypic deviation score of an individual (subject subscript is discarded) and G, E and C stand for the genetic, unshared and shared (common to family members) environmental deviation scores. The coefficients h, e and c are standardized regression coefficients (factor loadings). All variables are expressed as deviations from the mean so that their expected values: E[P] = E[E] = E[G] = E[C] = 0. If the unobserved or latent genetic and environmental factors are standardized to have unit variance then the variance of the phenotype is equal to :

$$V_P = h^2 + e^2 + c^2 \tag{2}$$

In this linear measurement model, possible non-linear effects arising through the interaction among any combination of G, E and C are assumed to be absent. The absence of genotype-environment interaction implies that an environmental effect (or 'treatment') has the same effect regardless of the genotype of the individual upon whom it is imposed (Neale and Cardon, 1992, page 22). Plomin, DeFries and McClearn (1990) define genotype-environment (GxE) interaction as follows: 'Genotype-environment interaction denotes an interaction in the statistical sense of a conditional relationship: The effect of environmental factors depend on the genotype' (page 250).

Statistical analysis of genotype-environment interaction can be conducted by means of various approaches, including analysis of variance and regression analysis, in combination with direct or indirect measures of the environment (Neale and Cardon, 1992, Chapter 11; Eaves, 1984; Freeman, 1973) or of the genotype (Martin, Eaves and Heath, 1987) or both (Plomin, 1986, Chapter 5). If measures of either environment or genotype are not available -i.e. in the majority of the quantitative genetic studies of metric human phenotypes- the detection of genotype-environment interaction is more difficult. One test for genotype by environment suggested by Jinks and Fulker (1970) involves examing the association between the means and standard deviations of MZ twin pairs. For MZ twins reared together, the difference between members of a twin pair reflects the magnitude of environmental differences within families and the sum of their scores reflects genetic (or environmental) differences between families. For MZ twins reared together, this test thus detects interactions between genotype and individual-specific environmental factors. For MZ twins reared apart, interactions with all postnatal environmental effects are included in the test.

Another approach to test for genotype-environment interaction when measures of the environment or the genotype are not available, has recently been suggested and is based on the analysis of the higher-order moments of genetic and environmental factor scores. Molenaar and Boomsma (1987) and Molenaar, Boomsma, Neeleman and Dolan (1990) have shown that the effects of certain types of interaction cannot be detected at the level of second-order moments (i.e. variances and covariances), but do lead to specific values of the third- and fourth-order moments (i.e. skewness and kurtosis) of genetic and environmental factor scores. These methods do not require measurements of the environment or the genotype, but require multiple indicators of the phenotype for the calculation of factor scores (Boomsma, Molenaar and Orlebeke, 1990) and the estimation of the higher-order moments of these factor scores.

The object of this paper is to study the effects of genotype-environment interaction in the context of longitudinal data using the genetic simplex model (Boomsma and Molenaar, 1987). To explore the effects of interaction in the standard genetic model based on second-order statistics we assume that measurements are available at three time points on three congeneric tests (i.e. tests that are, except for errors in measurement, perfectly correlated). At each occasion a common additive genetic and a common unshared environmental factor account for the covariance between the observations. The variance specific to each variable at each occasion is error variance. The simplex part of the model consists of the covariance between factors across time being attributable to the first-order autoregressions of the common additive genetic and unshared environmental factors. In a first-order autoregressive process latent factors are only influenced by the latent factors directly preceeding them, so that the partial correlation between factors at time points i and k $r_{ik.j} = 0$, whenever $i<j<k$. We choose this somewhat simple model because our main objective is a theoretical exploration of the consequences of genotype-environment interaction in developmental data. In the illustrative simulations we will look at the possibility that part of the developmental process consists of genotype x unshared environment (G x E) interaction at each time-point. This simple scenario gives rise to several interesting and unexpected results of genotype-environment interaction. In order to arrive at a somewhat self-contained chapter, we first present a summary of results from simulation studies on the estimation of individual factor scores in genetic covariance structure models and on the detection of different types of interaction using these factor scores. Next we introduce the analysis of longitudinal twin data by means of the genetic simplex model and discuss the estimation of factor scores in a longitudinal twin design. In the last part, we consider the detection of genotype-environment interaction in this longitudinal model.

Summary of Results Relating to the Calculation of Individual Genetic and Environmental Factor Scores and Statistical Tests of Genotype-Environment Interaction

We consider the multivariate version of Equation 1 (Martin and Eaves, 1977) in matrix notation (discarding the subject index):

$$P = hG + eE + cC + \varepsilon \qquad (3)$$

where $P = (P_1,...,P_p)'$ denotes a random p-dimensional vector of zero means phenotypes and ' denotes transposition. The vectors G, E, and C represent common (i.e. to the components of P) genetic, within family (unshared) and between family (shared) environmental factors with p-dimensional loadings h, e, and c. The unique part in each phenotype, e, is a random p-dimensional vector composed of influences unique to each phenotype P_j, j=1,...,p. The common factors G, E, and C are taken to be mutually uncorrelated normally distributed variables with zero mean and unit variance. On an individual basis G, E and C represent individual factor scores, i.e. an individual's genetic, unshared environmental and shared environmental deviation scores. Let Ψ denote the correlation matrix of the 3 common factors G, E, and C, and let Θ denote the covariance matrix of the unique components ε. The (p x p) covariance matrix, Σ_P, of P is then:

$$\Sigma_P = \Lambda\Psi\Lambda' + \Theta \qquad (4)$$

Within individuals, the 3 x 3 matrix Ψ is an identity matrix and the p x 3 matrix Λ contains the factor loadings: Λ = [h, e, c]. Assuming that all parameters are known (i.e. the elements of Λ and Θ), we may calculate an individuals deviation scores on the common factors in a number of distinct ways (see McDonald and Burr, 1967; Lawley and Maxwell, 1971; Saris, De Pijper and Mulder, 1978). We limit the discussion to the regression method for estimating factor scores which is investigated extensively in Boomsma, Molenaar and Orlebeke (1990) for the genetic common factor model and in Boomsma, Molenaar and Dolan (1991) for the genetic simplex model. Factors scores of individual i are calculated according to the regression method by multiplying the observations P_i for individual i by a weight matrix W. W is constructed in such a way that the sum of squares of the difference between estimated and true factor scores is minimized:

$$\eta_i = W\,P_i \tag{5}$$

$$W = \Psi\Lambda'\,\Sigma_P^{-1} \tag{6}$$

where $\eta_i' = [G_i, E_i, C_i]'$. With data from genetically related individuals the weight matrix W can be extended to include the observations from family members in the construction of the factor scores. Standard errors of these scores can also be calculated so that confidence intervals can be constructed around the individual genetic and environmental deviation scores (Boomsma, Molenaar and Orlebeke, 1990). As an example of this technique, Table 1 shows correlations between simulated factor scores and factor scores estimated by the regression method for MZ and DZ twins. The simulated data consisted of a 5-variate factor model with low unique variance for all 5 variables. From simulations such as these, it is clear that individual factor scores may be estimated reliably, given that a good-fitting multivariate model has been obtained on the original observations that supplies the parameter estimates of Λ and Θ needed to construct the weight matrix for the computation of factor scores.

Table 1. Correlations of simulated and estimated factor scores for MZ and DZ twins (decimal point omitted) using the regression method. Simulations were based on 100 MZ and 100 DZ twin pairs, using a 5-variate common factor model for G, E, and C. Unique variances were between 5 and 11% for each variable.

	MZ			DZ		
	E(G)	E(E)	E(C)	E(G)	E(E)	E(C)
G	910*	081	211	884*	213	100
E	020	879*	096	295*	773*	369*
C	124	100	936*	045	341*	897*

* p < .001

If each of the zero-mean unit-variance common factors G, C or E were replaced by interaction terms, the presence of such interactions would not show up in the second-order moments of the data. However, estimates of the higher-order moments of the distribution of factor scores would reveal genotype-environment interactions in the data even when both the genotype and the environment are not directly observed. Molenaar and Boomsma (1987) considered the case in which interactions give rise to additional factors in the standard genetic covariance model. For instance, replacement of a second genetic factor G by G*=GxE or a second shared environmental factor C by C*=CxE gives rise to a model with two common unshared environmental factors. Replacement of C by C*=CxG gives rise to a second common genetic factor. Molenaar and Boomsma used a factor rotation method devised by McDonald (1967) to distinguish between a true second environmental (or genetic) common factor and a second common factor attributable to the mentioned forms of interaction. The test is based on a special rotation of the multiple within family environmental (or genetic) factors that maximizes the third-order moments of factor scores in order to determine whether the second factor that behaves as an additional E (or G) factor really is an interaction factor.

Table 2. Characteristics of second and fourth-order moments of latent interaction factors

Model	interaction	characteristics of 2nd-order statistics	characteristics of 4th-order moments
G*,E,C	G*=GxC	var(G*)=1 cor(G*,E)=0 cor(G*,C)=0 cor(G*,G*)mz = 1 cor(G*,G*)dz=0.5 G* behaves like G	$E[G^4]=9$ and $E[G^2C^2]=3$ when G*=GxC, whereas $E[G^4]=3$ and $E[G^2C^2]=1$ when G* is not an interaction factor
G,E*,C	E*=ExG	var(E*)=1 cor(G,E*)=0 cor(C,E*)=0 cor(E*E*)mz = 0 cor(E*E*)dz = 0 E* behaves like E	$E[E^4]=9$ and $E[E^2G^2]=3$ when E*=ExG, whereas $E[E^4]=3$ and $E[G^2E^2]=1$ when E* is not an interaction factor
G,E*,C	E*=ExC	var(E*)=1 cor(G,E*)=0 cor(C,E*)=0 cor(E*E*)mz = 0 cor(E*E*)dz = 0 E* behaves like E	$E[E^4]=9$ and $E[E^2C^2]=3$ when E*=ExC, whereas $E[E^4]=3$ and $E[C^2E^2]=1$ when E* is not an interaction factor

Molenaar, Boomsma, Neeleman and Dolan (1990) presented a more general approach to the test of genotype-environment interactions underlying multivariate observations. This test can be applied to covariance structure models in which only one common genetic, one common within-family and one common between-family environmental factor (or a subset of these factors) is present and in which the factors that make up the interaction term are not present as separate factors in the model. Detection of interaction in this case requires a test of fourth-order moments of factor scores. Table 2 contains a summary of the characteristics of the second- and fourth-order moments such interaction factors. The interactions described in Table 2 cannot be detected at the level of second-order statistics, but the fourth-order moment expressions can serve as simple tests for the presence of of various forms of interaction. Simulation studies suggest that application of these expectations to estimated factor scores makes it possible to detect genotype-environment interaction even with realistic sample sizes. The test can also be generalized to the case where genes that control sensitivity to the environment are different from genes that control average response over all environments (Mather and Jinks, 1982; Eaves, 1984; Martin, Eaves and Heath, 1987).

SPECIFICATION OF THE GENETIC SIMPLEX MODEL

In this section we introduce the genetic simplex model and discuss the estimation of longitudinal individual genetic and environmental profiles. The stage will then be set for a consideration of interactions of the type shown in Table 2 for developmental data. The basic model that we employ is shown in Figure 1. We assume that MZ and DZ scores are available consisting of three indicators of a phenotype at three measurement occasions. At each time point, individual differences are determined by an additive genetic and an unshared environmental factor. The longitudinal part of the model is represented by autoregressions of the latent variables on earlier latent variables. Let P denote the 6 dimensional vector of phenotypic congeneric deviation scores of twin pair i at occasion t. The phenotypic vector is related to the common genetic and unshared environmental factors through a linear measurement model:

$$P_{ti} = \Lambda_t \, \eta_{ti} + \varepsilon_{ti} \tag{7}$$

where

$$P'_{ti} = [P_{t11}\ P_{t12}\ P_{t13}\ P_{t21}\ P_{t22}\ P_{t23}]_i$$

$$\eta'_{ti} = [G_{t1}\ E_{ti}\ G_{t2}\ E_{t2}]_i$$

$$\varepsilon'_{ti} = [\varepsilon_{t11}\ \varepsilon_{t12}\ \varepsilon_{t13}\ \varepsilon_{t21}\ \varepsilon_{t22}\ \varepsilon_{t23}]_i$$

where i is the twin pair index (i; i=1,...N). The subscripts of the phenotypic variables (P_{tij}),the common latent factors (G_{ti}, E_{ti}) and specific error terms (ε_{tij}) indicate measurement occasion (t; t=1,2,3) and phenotypic variable (j=1,2,3). The matrix of factor loadings at each occasion equals:

$$\Lambda_t = \begin{bmatrix} \Lambda_{gt} & \Lambda_{et} & 0 & 0 \\ 0 & 0 & \Lambda_{gt} & \Lambda_{et} \end{bmatrix}$$

where Λ_{gt} and Λ_{et} are (3 x 1) matrices of genetic and environmental factor loadings. The longitudinal models for G and E are specified as first-order autoregressions:

$$G_{t+1} = \beta_{gt+1,t}\, G_t + \zeta_{gt+1} \tag{8}$$

$$E_{t+1} = \beta_{et+1,t}\, E_t + \zeta_{et+1} \tag{9}$$

where β_g and β_e represent the regressions of latent factors G and E on the previous genetic and environmental factors and ζ represents a random input term or innovation.

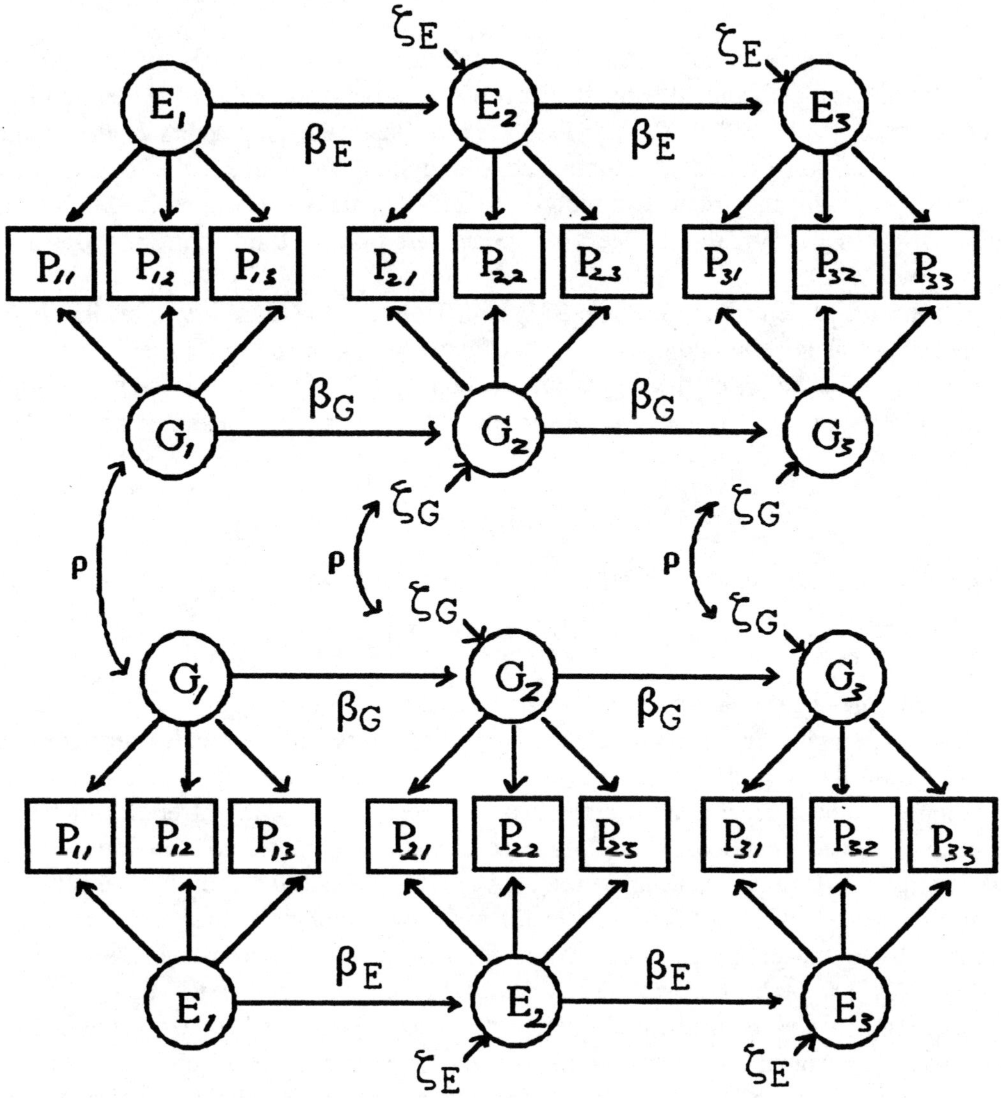

Figure 1. Genetic simplex model: 3 observations (squares) on 3 time-points in MZ and DZ twins pairs. The parameter ρ represents the additive genetic correlation (ρ=1 in MZ and 0.5 in DZ twin pairs). Latent variables G(enotype) and E(nvironment) are represented by circles, the b's represent the influence of a latent variable at an earlier time-point on a latent variable at a later time-point.

The implied covariance structure of the genetic factors is (exactly the same expectations obtain for the environmental factors):

$$\mathrm{var}(G_1) = \mathrm{var}(\zeta_{g1})$$

$$\mathrm{cov}(G_{t+1}, G_t) = \beta_{gt+1,t}\, \mathrm{var}(G_t)$$

$$\mathrm{var}(G_{t+1}) = \beta^2_{gt+1,t}\, \mathrm{var}(G_t) + \mathrm{var}(\zeta_{gt+1})$$

By fitting this model to MZ and DZ covariance matrices, estimates of the parameters in the model can be obtained with standard software packages such as LISREL (Jöreskog and Sörbom, 1988) or Mx (Neale, 1991). To estimate the parameters in the model a number of loss functions can be minimized. Throughout we minimize the likelihood ratio function to obtain maximum likelihood estimates (Neale and Cardon, 1992). Different ways of parameterizing this problem in LISREL are discussed in Boomsma, Martin and Molenaar (1989).

Boomsma, Molenaar and Dolan (1991) have looked at the feasibility of estimating the time-dependent genetic and environmental factor scores in the genetic simplex model. Using a generalization of the regression method described above (Priestley and Subba Rao, 1975; Brown, 1983), factor scores for the i-th twin pair are calculated as: $\eta_i = W\, P_i$, where:

$$W = E[\eta\eta']\, \Lambda'\, \Sigma^{-1}$$

for MZ $$W = [(I-B)^{-1}\, \Psi_{mz}\, (I-B')^{-1}]\, \Lambda'\, \Sigma^{-1}_{mz} \qquad (10)$$

for DZ $$W = [(I-B)^{-1}\, \Psi_{dz}\, (I-B')^{-1}]\, \Lambda'\, \Sigma^{-1}_{dz} \qquad (11)$$

where η'_i contains the latent genetic and environmental trajectories of the i-th MZ twin pair. For MZ twins, the genetic trajectories will of course be identical, whereas for DZ twins they will be correlated 0.5 on average at each time-point.

Applying these equations to simulated time-series data with different numbers of indicators for the phenotype at each time point and different genetic and environmental autoregressive parameters, a correlation between the true factor scores and the calculated factor scores of above 0.9 for MZ and DZ twins was obtained when three congeneric indicators for the phenotype were available (Boomsma, Molenaar and Dolan, 1991). When only one indicator was measured for the phenotype at each time point, the correlation between the true and the estimated factor scores was between 0.7 and 0.8 for DZ twins and around 0.8 for MZ twins. However, decomposition of univariate, and to a lesser extent bivariate, time-series yielded estimates of independent G_t and E_t scores that were intercorrelated. These intercorrelations depended somewhat on the difference in size between the genetic and environmental autoregressions, but to obtain independent estimates of individual genetic and non-genetic time-series, at least three measured indicators are needed at each time-point.

Interactions and Fourth-Order Moments in the Genetic Simplex Model

The model for the latent genetic and unshared environmental trajectories is an autoregression as in equations 8 and 9 ($\eta_t = b_{t,t-1}\,\eta_{t-1} + \zeta_t$). Suppose that both η_{t-1} and ζ_t are the outcome of a multiplicative interaction factor of two standard normal variables where the variables contributing to the innovation η_{t-1} are uncorrelated with those contributing to ζ_t. To test whether η_t is the outcome of a sum of two such interaction terms, we have to derive the fourth-order moment of η_t. We first consider the fourth-order statistics at an arbitrary time point t, where t>1. Expressing the contributions of η_{t-1} and ζ_t to the total variance of η_t as proportions we write:

$$\eta_t = a\eta_{t-1} + b\zeta_t$$

where $a = \sqrt{\ }(\beta^2/(1+\beta^2))$ and $b = \sqrt{\ }(1-a^2)$, as $var(\eta_t) = \beta^2_{t,t-1}\,var(\eta_{t-1}) + var(\zeta_t)$ and $var(\eta_{t-1}) = var(\zeta_t) = 1$. The fourth-order moment of η_t then equals:

$$E[\eta_t^4] = 9 - 12a^2 + 12a^4 \quad (12)$$

So we find that $E[\eta_t^4]$ can vary between 9 (a = 1, that is when there is no innovation, or a = 0, that is when b = 0, i.e. no transmission) and 6 (a = 0.7). This implies a dependence of the expected fourth-order moment of the interaction of factor scores $E[\eta_t^4]$ on the value of the innovation ζ_t and the transmission parameter $\beta_{t,t-1}$. This dependency between $E[\eta_t^4]$ and a is pictured in Figure 2.

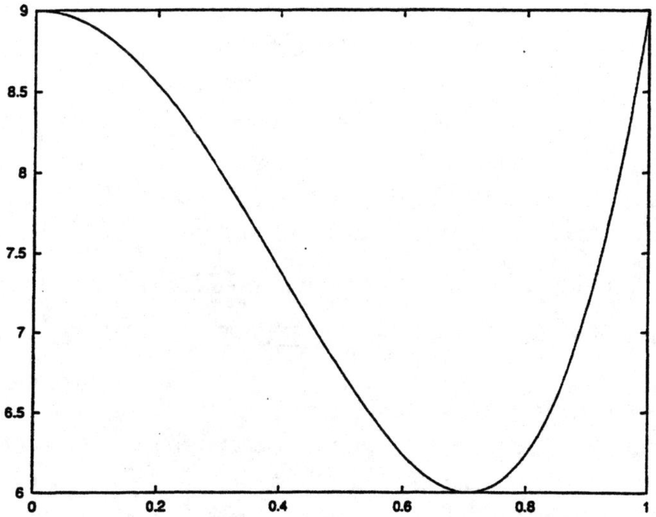

Figure 2. Fourth-order moment of factor scores $E[\eta_t^4]$ as a function of a, where $a = \sqrt{\ }(\beta^2/(1+\beta^2))$, is the proportion of variance which is transmitted from t-1 to t and where η is an interaction factor.

Even more importantly, we may expect the fourth-order moment to decrease as the number of interaction components contributing to η_t increases. A situation then arises where the Central Limit theorem obtains: the distribution of η_t will tend towards normality. According to the Central Limit theorem the distribution of the sum of N independent random variables converges to the normal distribution if N approaches infinity. In fact there are several variants of the Central Limit theorem that apply under still weaker conditions (e.g. sums of autocorrelated random variables). We will only need the standard version of the Central Limit theorem, however, applying to sums of independently identically distributed (i.i.d.) random variables. The i.i.d. variables concerned are the ζ_t innovations in equations 8 and 9. These equations can be rewritten in such a way that the latent factor η_{t+1} is expressed as an infinite weighted sum of ζ_{t+1}, ζ_t, ... Application of the Central Limit theorem to this so-called moving-average of infinite order shows that the distribution of η always will converge to the normal distribution, irrespective of the distribution of ζ. In the case presently considered the distribution of ζ, where ζ is a pure interaction innovation process, is rather complex. Yet the distribution of the latent genetic and environmental factors still will converge to a normal distribution as time proceeds.

At present there are two interaction components contributing to η_t, viz. η_{t-1} and ζ_t. But if the measurement at t=1 is recorded late in the developmental process, η_{t-1} will consist of an accumulation of interaction terms built up during development prior to t=1 so that the Central Limit theorem applies and η_t tends to normality. Surprisingly, this implies that the presence of fourth-order moments that equal the expected value under normality does not mean that the developmental process is not the outcome of an accumulation of interaction components.

Given the presence of an additive genetic and unshared environmental series, we may consider the fourth-order moment $[G^2_t\, E^2_t]$. As above the model for the latent variables is given by equations 8 and 9 for the genetic and environmental autoregressions. We assume that E_t is attributable to an interaction between G_{t-1} and E_{t-1} (both standard normal variables) and that ζ_{et} is the outcome of an interaction between two standard normal variables at time t. As usual, G_t and ζ_{gt} are defined as random zero-mean unit-variance variables.

For E_t we may write (as above): $E_t = a\, E_{t-1} + b\, \zeta_{et}$

and similarly for G_t: $G_t = c\, G_{t-1} + d\, \zeta_{gt}$

Their crossproduct $E[G^2_t\, E^2_t]$ then equals: $E[G^2_t\, E^2_t] = 3 - 2a^2 - 2c^2 + 4a^2c^2$ (13)

where a and b are as defined above (and thus depend on the value of the innovations and the transmission parameters) and c and d depend on the genetic autoregression parameters and innovations. The dependency of $[G^2_t\, E^2_t]$ on a and c is illustrated in Figure 3. The crossproduct $[G^2_t\, E^2_t]$ can vary between 3 (a=0 and c=0 or a=1 and c=1) and 1 (a=0 and c=1 or a=1 and c=0). We now have a fourth-order moment that is dependent on the autoregressive coefficients b_g.and b_e. In this case the Central Limit theorem also applies.

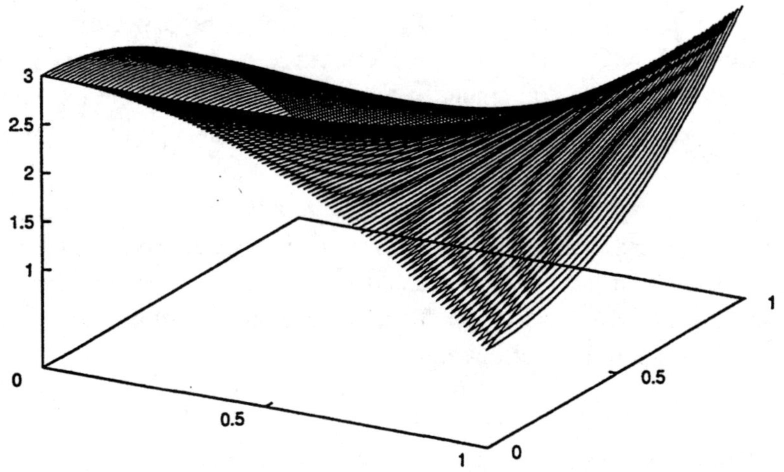

Figure 3. The crossproduct $[G_t^2E_t^2]$ as a function of a and c, where $a = \sqrt{(\beta_e^2/(1+\beta_e^2))}$ and $c = \sqrt{(\beta_g^2/(1+\beta_g^2))}$, are the proportions of respectively G x E variance and genetic variance that are transmitted from t-1 to t.

Illustration Using Simulated Data

To illustrate the application of fourth-order moments in the detection of genotype-environment interaction as outlined above, we provide some results obtained by analyzing simulated data. Data sets were simulated with IMSL subroutine FTGEN (IMSL, 1979) for 500 monozygotic and 500 dizygotic twin pairs at three occasions. The latent factors influencing the phenotype at each occasion were a common additive genetic and a common unshared environmental factor. The environmental and additive genetic factors were orthogonal. The longitudinal model for the genetic and environmental factors was a first-order autoregression. The phenotype consisted of three congeneric phenotypic tests (see Figure 1).

The factor loadings of the phenotypic variables on the latent variables were constant over time and equaled 0.9, 0.7, and 0.5 for the additive genetic factor and 0.5, 0.7, and 0.9 for the unshared environmental factor. Unique variance (error variance) was 1 for each phenotypic variable at each measurement occasion. Series of five data sets were simulated according to a G,E model without genotype-environment interaction and according to an interaction model where the E factor was made up of G x E interaction. Each of the five data sets were generated with the following values of the autoregressive coefficients, β_g.and β_e: 1, 1.5, 2, 2.5, and 3. The initial latent variances var(G_1) and var(E_1) and all innovation variances [var(ζ_{g2}), var(ζ_{g3}), var(ζ_{e2}) and var(ζ_{e3})] were all set to equal 1.

The G x E interaction was introduced at the first time point by multiplying the environmental deviation scores with the additive genetic deviation scores ($E^*_1 = E_1 \times G_1$). This interaction factor was transmitted to subsequent time points by the autoregressive process. At time point t = 2 and t = 3, the innovation terms also consisted of G x E interaction ($\zeta^*_{e2} = \zeta_{e2} \times \zeta_{g2}$ and $\zeta^*_{e3} = \zeta_{e3} \times \zeta_{g3}$). The expected values fourth-order moments of G_1, G_2 and G_3 are 3. At t = 1, the expected fourth-order moments $[E_1^4]$ and $[E_1^2G_1^2]$ are given in Table 2 as 9 and 3, respectively. At t = 2, the expected values for $[E_2^4]$ and $[E_2^2G_2^2]$ depend on the autoregressive coefficients (β_g and β_e) and are given in equations 12 and 13. For example, for $\beta_g = \beta_e = 3$, the expected fourth-order moment for $[E_2^4]$ is 7.9 and the expected value for $[E_2^2G_2^2]$ equals 2.64. For $\beta_g = \beta_e = 2$, these expected values are 7.1 and 2.36, and for $\beta_g = \beta_e = 1$, they are respectively 6 and 2. The expected values of the fourth-order moments, $[E_3^4]$ and $[E_3^2G_3^2]$ can be derived in the same manner as those for $[E_2^4]$ and $[E_2^2G_2^2]$ (these are not derived in the present paper, but will be considered in a future publication).

Table 3A. Fourth-order moments of genetic and environmental factor scores from simulated longitudinal data on 500 MZ and 500 DZ twin pairs, without G x E interaction. Estimated moments are given separately for twin 1 and twin 2, and are pooled over zygosities. Expected values for G_t^4 and E_t^4 equal 3, and expected values for the crossproduct $[E_t^2G_t^2]$ equal 1. Results are presented from five simulations with different values for the autoregressive coefficients β_g and β_e (3,2.5,2,1.5, and 1).

	Twin 1				Twin 2	
	I $\chi^2(333) = 348.0$ (p=.27)			$\beta_g = 3.014$ (3)	$\beta_e = 3.014$ (3)	
t	$E[G_t^4]$	$E[E_t^4]$	$E[E_t^2G_t^2]$	$E[G_t^4]$	$E[E_t^4]$	$E[E_t^2G_t^2]$
1	2.80	3.08	1.08	2.94	2.97	0.94
2	2.79	3.06	1.08	2.87	2.94	0.95
3	2.78	3.05	1.08	2.87	2.93	0.94
	II $\chi^2(333) = 349.8$ (p=.25)			$\beta_g = 2.469$ (2.5)	$\beta_e = 2.506$ (2.5)	
t	$E[G_t^4]$	$E[E_t^4]$	$E[E_t^2G_t^2]$	$E[G_t^4]$	$E[E_t^4]$	$E[E_t^2G_t^2]$
1	2.75	2.98	1.06	2.83	2.94	0.93
2	2.76	3.00	1.06	2.88	2.93	0.93
3	2.77	3.00	1.06	2.89	2.91	0.93
	III $\chi^2(333) = 381.9$ (p=.03)			$\beta_g = 2.007$ (2)	$\beta_e = 2.019$ (2)	
t	$E[G_t^4]$	$E[E_t^4]$	$E[E_t^2G_t^2]$	$E[G_t^4]$	$E[E_t^4]$	$E[E_t^2G_t^2]$
1	2.99	2.98	1.03	2.89	3.08	1.12
2	3.00	2.97	1.03	2.95	3.04	1.06
3	3.04	2.97	1.03	2.97	3.05	1.07
	IV $\chi^2(333) = 368.8$ (p=.09)			$\beta_g = 1.491$ (1.5)	$\beta_e = 1.492$ (1.5)	
t	$E[G_t^4]$	$E[E_t^4]$	$E[E_t^2G_t^2]$	$E[G_t^4]$	$E[E_t^4]$	$E[E_t^2G_t^2]$
1	3.05	2.88	1.08	2.96	2.78	1.02
2	2.95	2.83	1.11	2.97	2.77	1.02
3	2.93	2.81	1.13	2.97	2.78	0.97
	V $\chi^2(333) = 362.0$ (p=.13)			$\beta_g = .997$ (1)	$\beta_e = 0.976$ (1)	
t	$E[G_t^4]$	$E[E_t^4]$	$E[E_t^2G_t^2]$	$E[G_t^4]$	$E[E_t^4]$	$E[E_t^2G_t^2]$
1	3.18	3.27	1.27	3.25	2.81	1.30
2	3.16	3.27	1.19	3.23	2.83	1.15
3	2.94	2.98	1.04	3.06	2.86	1.23

Table 3B. Fourth-order moments of genetic and environmental factor scores from simulated longitudinal data on 500 MZ and 500 DZ twin pairs for a simulation model including genotype-environment interaction, where E* = G x E. Estimated moments are given separately for twin 1 and twin 2, and are pooled over zygosities. Results are shown for well-fitting models.

	Twin 1			Twin 2		
	I $\chi^2(333) = 364.9$ (p=.11)			$\beta_g = 2.962$ (3)	$\beta_e = 3.016$ (3)	
t	$E[G_t^4]$	$E[E_t^4]$	$E[E_t^2G_t^2]$	$E[G_t^4]$	$E[E_t^4]$	$E[E_t^2G_t^2]$
1	3.18	7.04	2.60	3.17	8.26	2.92
2	3.17	6.95	2.58	3.12	8.05	2.68
3	3.19	6.94	2.57	3.13	8.01	2.66
	II $\chi^2(333) = 336.3$ (p=.44)			$\beta_g = 2.499$ (2.5)	$\beta_e = 2.482$ (2.5)	
t	$E[G_t^4]$	$E[E_t^4]$	$E[E_t^2G_t^2]$	$E[G_t^4]$	$E[E_t^4]$	$E[E_t^2G_t^2]$
1	2.69	5.72	2.17	2.87	6.10	2.10
2	2.65	5.65	2.11	2.81	6.06	2.06
3	2.67	5.60	2.10	2.82	6.11	2.08
	III $\chi^2(333) = 349.0$ (p=.26)			$\beta_g = 1.993$ (2)	$\beta_e = 1.997$ (2)	
t	$E[G_t^4]$	$E[E_t^4]$	$E[E_t^2G_t^2]$	$E[G_t^4]$	$E[E_t^4]$	$E[E_t^2G_t^2]$
1	3.55	6.83	2.39	3.66	5.72	2.47
2	3.29	6.59	2.16	3.42	5.24	2.18
3	3.38	6.51	2.16	3.49	5.18	2.14
	IV $\chi^2(333) = 353.0$ (p=.21)			$\beta_g = 1.479$ (1.5)	$\beta_e = 1.525$ (1.5)	
t	$E[G_t^4]$	$E[E_t^4]$	$E[E_t^2G_t^2]$	$E[G_t^4]$	$E[E_t^4]$	$E[E_t^2G_t^2]$
1	3.35	6.55	2.64	3.42	6.75	2.32
2	2.89	5.69	1.80	2.88	5.38	1.96
3	2.88	5.43	1.73	2.83	5.08	1.92
	V $\chi^2(333) = 348.3$ (p=.27)			$\beta_g = .986$ (1)	$\beta_e = 1.006$ (1)	
t	$E[G_t^4]$	$E[E_t^4]$	$E[E_t^2G_t^2]$	$E[G_t^4]$	$E[E_t^4]$	$E[E_t^2G_t^2]$
1	3.79	5.70	3.07	3.98	5.90	2.97
2	3.23	5.29	2.37	3.44	5.40	2.40
3	2.90	4.29	1.80	2.97	4.56	1.87

Table 3A contains the results obtained by analyzing the simulated data without G x E interaction. For each of the five data sets we first give the χ^2 and associated probability and the parameter estimates for β_g.and β_e obtained from fitting the true model. Next, the the fourth-order moments of the estimated factor scores for G_t and E_t are shown for each data set (whose expected values are 3 in data without interaction) and the values obtained for the crossproduct $[E_t^2G_t^2]$ (which equals 1 in data without interaction). It can be seen that for all values of b the fourth-order moments of factor scores are close to their expected values and would lead to the correct conclusion of no G x E interaction.

Tables 3B and 3C contain the fourth-order moments obtained by analyzing simulated data containing a latent factor that behaves in the analysis of second-order moments (the covariance analysis) as an unshared environmental factor E, but is made up of G x E interaction terms according to the simulation model outlined above. If we compare the estimated moments for $[E_t^4]$ and for $[E_t^2G_t^2]$ with their expected values in the absence of

G x E interaction (respectively 3 and 1), it is clear that the values reported in Tables 3B and 3C indicate that E should be regarded as an interaction factor, instead of pure environmental one. Although these values for the fourth-order moments are in some instances not as high as their expectations, it clear that they are substantially higher than the values in Table 3A for the case of no G x E interaction. Comparing Tables 3B and 3C it may be seen that there is 'trade-off' between χ^2 goodness-of-fit and the value of the fourth-order moments: these are closer to their expected values under interaction when χ^2 is higher (Table 3C). This probably is related the amount of kurtosis in the raw data, where a higher kurtosis makes it easier to detect G x E interaction, but also leads to a higher χ^2. In contrast, the observed fourth-order moments [G_t^4] are close to their expected value of 3 in all simulations, indicating that G is a pure genetic factor.

Table 3C. Fourth-order moments of genetic and environmental factor scores from simulated longitudinal data on 500 MZ and 500 DZ twin pairs for a simulation model including genotype-environment interaction, where E* = G x E. Estimated moments are given separately for twin 1 and twin 2, and are pooled over zygosities. Results are shown for the case where the true model does not show a good fit (as indicated by a significant χ^2) to the data.

	Twin 1				Twin 2	
I	$\chi^2(333) = 415.1$ (p=.001)			$\beta_g = 3.024$ (3)	$\beta_e = 2.981$ (3)	
t	$E[G_t^4]$	$E[E_t^4]$	$E[E_t^2G_t^2]$	$E[G_t^4]$	$E[E_t^4]$	$E[E_t^2G_t^2]$
1	2.79	9.03	2.63	2.79	8.50	2.61
2	2.75	8.61	2.46	2.72	8.29	2.50
3	2.75	8.52	2.43	2.71	8.26	2.48
II	$\chi^2(333) = 418.9$ (p=.001)			$\beta_g = 2.473$ (2.5)	$\beta_e = 2.515$ (2.5)	
t	$E[G_t^4]$	$E[E_t^4]$	$E[E_t^2G_t^2]$	$E[G_t^4]$	$E[E_t^4]$	$E[E_t^2G_t^2]$
1	2.95	7.17	2.61	2.98	7.35	2.62
2	2.93	6.87	2.42	2.95	7.43	2.61
3	2.94	6.80	2.39	2.95	7.41	2.56
III	$\chi^2(333) = 422.8$ (p=.001)			$\beta_g = 1.976$ (2)	$\beta_e = 1.992$ (2)	
t	$E[G_t^4]$	$E[E_t^4]$	$E[E_t^2G_t^2]$	$E[G_t^4]$	$E[E_t^4]$	$E[E_t^2G_t^2]$
1	3.00	5.26	1.97	3.34	8.98	3.16
2	3.05	4.84	1.89	3.26	8.17	2.85
3	3.11	4.71	1.79	3.22	7.86	2.69
IV	$\chi^2(333) = 408.8$ (p=.003)			$\beta_g = 1.463$ (1.5)	$\beta_e = 1.498$ (1.5)	
t	$E[G_t^4]$	$E[E_t^4]$	$E[E_t^2G_t^2]$	$E[G_t^4]$	$E[E_t^4]$	$E[E_t^2G_t^2]$
1	3.75	5.87	2.74	3.60	5.24	2.27
2	3.46	5.17	2.16	3.16	4.78	1.91
3	3.45	5.06	2.05	3.21	4.59	1.72
V	$\chi^2(333) = 391.5$ (p=.015)			$\beta_g = 1.001$ (1)	$\beta_e = .958$ (1)	
t	$E[G_t^4]$	$E[E_t^4]$	$E[E_t^2G_t^2]$	$E[G_t^4]$	$E[E_t^4]$	$E[E_t^2G_t^2]$
1	3.86	5.37	3.66	3.98	5.77	2.99
2	3.87	5.45	2.78	3.44	5.51	2.03
3	2.46	4.84	2.07	2.97	4.33	1.65

DISCUSSION

It was shown that the change between two consecutive time points t-1 and t in the fourth-order moments of the factor scores associated with an interaction factor series depends upon the proportion of variance which is transmitted from t-1 to t. In the genetic simplex, which involves first-order autoregressions describing the latent genetic and environmental factor series, the total amount of transmitted variance between two consecutive time points is a simple function of the autoregressive beta-coefficients: $\beta^2 \text{var}(\eta_{t-1})$. Hence the proportion of transmitted variance of an interaction factor series at time t depends upon the relative magnitude of the autoregressive β-coefficient in comparison with the standard-deviation of the innovation term at t. For this rather simple scheme, explicit expressions for the change in fourth-order moments were derived. In particular it was found that this change is absent only if the variance of the innovation term is zero. In that special case the fourth-order moments keep their initial values (9 and 3, respectively) at all time points. Notice that if the innovation variance of a latent autoregression in our simplex model becomes zero, then this part of the simplex model reduces to a common factor model as described by Eaves, Long and Heath (1986).

Only the change in the fourth-order moments of an interaction factor between two consecutive time points t-1 and t were derived. The change between t-1 and t+1 (spanning two lags) then follows immediately by a recursive application of our derivation. In fact, the change between an arbitrary number of lags can thus be derived. Moreover, the same principles can be used to determine the changes concerned for interaction factor series obeying more complex time-series models such as higher-order autoregressions and moving-averages.

The simulations of G x E interaction in longitudinal models that were presented in this paper are not exhaustive, but serve to illustrate the possibility of detecting such interactions without measures of the environment or the genotype. We showed that estimates of fourth-order moments of factor scores are close to their expected values of 3 and 1 in data without interaction. In simulated longitudinal data with interaction the value of these fourth-order moments are larger indeed indicate the presence of interaction.

Our approach to the detection of interaction factors hinges upon the estimation of fourth-order moments. Unfortunately, the sampling variability of these estimates is very high and therefore one will need a large sample of phenotypical values in order to secure the reliability of the detection tests. In our simulation studies (where it is certain that the generated phenotypical values constitute a homogeneous sample) it was found that estimates of fourth-order moments strongly depend upon the extreme phenotypical values in a sample and that removal of these extreme observations (interpreted as outliers) could lead to severe bias. In future explorations of the approach we intend to consider alternative L-statistics characterizing the kurtosis of latent factor series (Hosking, 1990). It appears that, compared with the conventional fourth-order moments, L-kurtosis is less subject to bias in estimation, approximates its asymptotic distribution more closely in finite samples, and is more robust to the presence of genuine outliers in the phenotypical data.

Perhaps the most surprising conclusion of the present study, at least to us, is that the interactive nature of a developmental process becomes invisible after a sufficient amount of time, even under the most favorable circumstances. Started as a pure interaction factor (with fourth-order moments of 9 and 3, respectively), the repetitive addition of pure interaction innovations in time combined with non-zero transmission over time leads to a developmental process whose interactive nature becomes impossible to detect (fourth-order moments of 3 and 1, respectively). It could be that such processes account for the absence of a shared environmental component in many twin and family studies since a G x C interaction factor will look like a genetic factor. Also, this might be part of the explanation for the high contributions of unshared environmental factors to many characteristics (Plomin and Daniels, 1987) since all interactions with E look like E. We believe that this phenomenon can be best understood by an appeal to the Central Limit theorem: because of the repeated addition of independently and identically distributed interaction innovations the random process approaches normality as time proceeds. Notice that the applicability of the Central Limit theorem does not depend upon the details of our interaction detection procedure, but pertains to any developmental process involving nonzero transmission in combination with innovations. That is, the presence of interactive causes underlying such processes will become invisible as development proceeds, even if the causal interaction is enduring, stable, and effective during the entire life span. As far as we know this is the first time that this result, which may have far-reaching implications transcending the field of behavior genetics proper, has been noticed in the published literature.

REFERENCES

Boomsma, D.I., Martin, N.G., & Molenaar, P.C. (1989). Factor and simplex models for repeated measures: application to two psychomotor measures of alcohol sensitivity in twins. *Behav Genet*, *19*, 79-96.

Boomsma, D.I. & Molenaar, P.C. (1987). The genetic analysis of repeated measures. I. Simplex models. *Behav Genet*, *17*, 111-123.

Boomsma, D.I., Molenaar, P.C., & Dolan, C.V. (1991). Estimation of individual genetic and environmental profiles in longitudinal designs. *Behav Genet*, *21*, 243-255.

Boomsma, D.I., Molenaar, P.C.M., & Orlebeke, J.F. (1990). Estimation of individual genetic and environmental factor scores. *Genet Epidemiol*, *7*, 83-91.

Brown, R.G. (1983). *Introduction to random signal analysis and Kalman filtering*. New York: John Wiley & Sons.

Eaves, L.J. (1984). The resolution of genotype x environment interaction in segregation analysis of nuclear families. *Genet Epidemiol*, *1*, 215-228.

Eaves, L.J., Long, J., & Heath, A.C. (1986). A theory of developmental change in quantitative phenotypes applied to cognitive development. *Behav Genet*, *16*, 143-162.

Freeman, G.H. (1973). Statistical methods for the analysis of genotype-environment interactions. *Heredity*, *31*, 339-354.

Hewitt, J.K., Eaves, L.J., Neale, M.C., & Meyer, J.M. (1988). Resolving causes of developmental continuity or "tracking." I. Longitudinal twin studies during growth. *Behav Genet*, *18*, 133-151.

Hosking, J.R.M. (1990). L-moments: Analysis and estimation of distributions using linear combinations of order statistics. *Journal of the Royal Statistical Society B*, *52*, 105-124.

IMSL, Inc. (1979). *IMSL library reference manual, edition 7*. Houston, TX: IMSL Inc..

Jinks, J.L. & Fulker, D.W. (1970). Comparison of the biometrical genetical, MAVA, and classical approaches to the analysis of human behavior. *Psychol Bull*, *73*, 311-349.

Joreskog, K.G. & Sorbom, D. (1988). *LISREL VII: A guide to the program and applications*. Chicago: Spss Inc..

Lawley, D.N. & Maxwell, A.E. (1971). *Factor analysis as a statistical method*. London: Butterworths.

Loehlin, J. (1991). Behavior genetic studies of change. In L.M. Collins & J.L. Horn (Eds.). *Best methods for the analysis of change. Recent advances, unanswered questions, future directions* Washington, D.C.: American Psychological Association

Loehlin, J.C., Horn, J.M., & Willerman, L. (1989). Modeling IQ change: evidence from the Texas Adoption Project. *Child Dev*, *60*, 993-1004.

Martin, N.G. & Eaves, L.J. (1977). The genetical analysis of covariance structure. *Heredity*, *38*, 79-95.

Martin, N.G., Eaves, L.J., & Heath, A.C. (1987). Prospects for detecting genotype X environment interactions in twins with breast cancer. *Acta Genet Med Gemellol (Roma)*, *36*, 5-20.

Mather, K. & Jinks, J.L. (1982). *Biometrical Genetics: The study of continuous variation*. London: Chapman and Hall.

McArdle, J.J. (1986). Latent variable growth within behavior genetic models. *Behav Genet*, *16*, 163-200.

McDonald, R.P. (1967). Factor interaction in nonlinear factor analysis. *Br J Math Stat Psychol*, *20*, 205-215.

McDonald, R.P. & Burr, E.J. (1967). A comparison of four methods of constructing factor scores. *Psychometrika*, *32*, 381-401.

Molenaar, P.C. & Boomsma, D.I. (1987). Application of nonlinear factor analysis to genotype-environment interaction. *Behav Genet*, *17*, 71-80.

Molenaar, P.C., Boomsma, D.I., Neeleman, D., & Dolan, C.V. (1990). Using factor scores to detect G X E interactive origin of "pure" genetic or environmental factors obtained in genetic covariance structure analysis. *Genet Epidemiol*, *7*, 93-100.

Molenaar, P.C.M., Boomsma, D.I. & Dolan, C.V. (1991). Genetic and environmental factors in a developmental perspective. In D. Magnusson, L.R. Bergman, G. Rudinger & B. Thorestad (Eds.). *Problems and methods in longitudinal research: Stability and change* (pp. 250-273). Cambridge: Cambridge University Press

Neale, M.C. (1991). *Statistical modeling.* Box 3, MCV, Richmond, VA 23298: Department of Human Genetics.

Neale, M.C. & Cardon, L.R. (1992). *Methodology for genetic studies of twins and families (NATO ASI Series D: Behavioral and Social Sciences-vol 67).* Dordrecht, The Netherlands: Kluwer Academic Publishers B. V..

Plomin, R. (1986). *Development, genetics and psychology.* Hillsdale: Lawrence Erlbaum Associates.

Plomin, R. & Daniels, D. (1987). Why are children in the same family so different from one another? *Behavioral and Brain Sciences, 10,* 1-60.

Plomin, R., DeFries, J.C. & McClearn, G.E. (1990). *Behavior Genetics: A primer.* San Francisco: W.H.Freeman and Company.

Priestly, M.B. & Subba Rao, T. (1975). The estimation of factor scores and Kalman filtering for discrete parameter stationary processes. *International Jounal of Control, 21,* 971-975.

Saris, W.E., dePijper, M., & Mulder, J. (1978). Optimal procedures for estimation of factor scores. *Sociological Methods & Research, 7,* 85-106.

CHAPTER 4

TWINS REARED APART: NATURE'S DOUBLE EXPERIMENT

Thomas J. Bouchard, Jr.
University of Minnesota, Minneapolis

Nancy Pedersen
Karolinska Institute
Pennsylvania State University

THE ORIGIN OF THE TWINS REARED APART METHOD

Galton is most often credited with the founding of human behavioral genetics and the discovery of the twin method (Pearson, 1924). This priority has been challenged by Rende, Plomin and Vandenberg (1990) who prefer to credit Curtis Merriman (1924) and Herman Siemens (1924). In contrast Bouchard and Propping (1993b) have argued that there are many twin methods and that while Merriman and Siemens deserve credit for putting some pieces of the puzzle together, their claim to fame is minor in comparison to that of Galton. A similar argument might be made regarding the adoption method. Galton introduced it systematically in his book Hereditary Genius (Galton, 1869) but as with the twin data his arguments were qualitative rather than quantitative.

With regard to the twin reared apart (TRA) method there is no controversy as to who introduced the method as apparently no one knows. Given his scientific imagination it is a puzzle that Galton never mentioned the study of twins reared apart as a method which utilizes both twins and adoptees simultaneously. In the expectation that perhaps some comment about the method might be found in his papers, the first author has read all of the archival material in the Galton archives at the University of London dealing with twins. While no mention of the twin reared apart method could be found, it is of interest that one of the letters to Galton regarding the similarity between twins mentions a pair of reared apart twins. The letter states: "I have tried to learn for you something of a case in which twin brothers separated through life rejoined one another at about 50 [writing ambiguous], but I find real evidence unattainable. The brothers were raised [writing ambiguous] merchants I think later in life, and were great-uncles of mine. I have heard their story all my life, but I can [writing ambiguous] who can give first hand testimony" (Townsend, 1874-75).

The first article on monozygotic twins reared apart that we have been able to locate is by Popenoe (Popenoe, 1922). The article is largely discursive. Fortunately this pair of twins was followed up systematically by Herman J. Muller (1925). While published in the Journal of Heredity, the paper focused almost entirely on mental traits. Neither paper discusses the origin of the method. A number of case studies of monozygotic twins reared apart followed. Saudek (1934) reports on the first known case traced in the British Isles, again there is no mention of the origin of the method. He was aware that Newman, Freeman, and Holzinger were in the process of analyzing a series of cases. The Newman Freeman and Holzinger study of twins reared together had begun in 1926, and at that time they had no intention of studying MZA twins. During the course of the study it occured to them how useful such twins would be, and Newman wrote a short article about the twin work they were doing and made an urgent appeal for information about cases of identical twins reared apart. In 1937, Newman, Freeman and Holzinger (1937) published their book on this series of 19 pairs of MZA twins and 50 pairs of MZT and 50 pairs of DZT twins. Occasional case studies continued to appear (Burks, 1942; Burks & Roe, 1949; Stephens & Thompson, 1943; Yates & Brash, 1941) including the 20th and last case in the NFH series (Gardner & Newman, 1940). In 1962, Shields (Shields, 1962) reported on 38 pairs of MZA twins and a few cases of DZA twins most of whom were located in the British Isles. Finally in 1965 Juel-Nielsen (Juel-Nielsen, 1965) reported on 12 pairs of twins located in Denmark.

Two recent studies of twins reared apart, the Minnesota Study of Twins Reared Apart (MISTRA) and the Swedish Adoption/Twin Study of Aging (SATSA) will provide the basis for much of the information in this chapter. MISTRA is a longitudinal study of monozygotic and dizygotic twins reared apart begun in 1979. The assessment of the twins is 50% medical/psychiatric and 50% psychological and extends over one week (50 hours). Twins are recruited from a wide variety of sources (Bouchard, Lykken, McGue, Segal, & Tellegen, 1990). Most participants have come from the United States and the United Kingdom but others have come from Australia, Canada, China, New Zealand, Sweden and West Germany. The average age of the twins in MISTRA is about 42 years old. Twins in MISTRA are followed-up after ten years and thirty-four pairs have been re-assessed as of mid 1995. Comparison data for the MISTRA reared apart twins is gathered by mail from twins in the Minnesota Twin Family Registry (Lykken, Bouchard, McGue, & Tellegen, 1990). SATSA is a longitudinal study of all twin pairs in the population-based Swedish Twin Regsitry who indicated that they had been separated before the age of 10 and reared apart, and a control sample of conventionally reared twins, matched on the basis of sex, country and date of birth. Both members of 351 pairs of twins reared apart responded to a questionnaire in 1984 and have been followed at three year intervals. The average age of the SATSA twins in 1984 was 59. The fourth wave of in-person testing is currently underway. A subsample of these pairs (i.e. those who are 50 or older) have participated in personal interview/health assessments/cognitive evaluation (Pedersen, McClearn, Plomin, Nesselroade, Berg, & DeFaire, 1991).

The studies reported prior to 1970, including a number of odd cases not mentioned here, have been discussed in some detail by Farber (1981). Her book, however, has serious flaws and has been severely criticized by many reviewers (Bouchard 1982a; Heston, 1982; Loehlin, 1981; Rose, 1981). Consider the prescient quote from Loehlin:

> "A second aspect to the book is an elaborate statistical treatment of the IQ data from the separated MZ twin studies. Some interesting analyses are provided, but the readers are hereby cautioned to watch out for the graphs and summaries in Chapter 7. These suggest that the amount of contact between separated MZ twins accounts for some 20-30% of the IQ variance. Perhaps, but only if one assumes that the mechanisms involved work in opposite directions in males and females (see Appendix E, page 350). For the sexes combined, the amount of contact between the twins does not predict their resemblance in IQ" (p. 297).

Consistent with Loehlin's comments neither the Minnesota Study of Twins Reared Apart (MISTRA) nor the Swedish Adoption Twin Study of Aging (SATSA) has been able to demonstrate a relationship between amount of contact and IQ in their TRA samples (Bouchard, et al., 1990; Pedersen, Plomin, Nesselroade, & McClearn, 1992). Loehlin's characterization of the data analysis as elaborate is a euphemism. Bouchard (1982a) labeled it 'pseudoanalysis' and has shown that many treatments of the archival TRA data are very seriously flawed (Bouchard, 1982b; Bouchard, 1983; Bouchard, 1993a).

For completeness we mention here two studies for which only limited information has been collected by questionnaire. The first was carried out in Japan (Hayakawa, 1987; Hayakawa & Shimizu, 1987), the second in Finland (Kaprio, Koskenvuo, & Rose, 1990; Langinvainio, Kaprio, Koskenvuo, & Lonnquist, 1984; Langinvainio, Koskenvuo, Kaprio, Lonnquist, & Tarkkonen, 1981).

It would be remiss of us to fail to mention the controversial work of Sir Cyril Burt who purported to have studied a large group of young MZA twins reared apart (Burt, 1955; Burt, 1958; Burt, 1966; Burt, 1972). Burt's work was severely criticized by Kamin (Kamin, 1974), who showed among other things, that Burt continually reported a correlation of .771 even as his sample continually got larger and larger (Fancher, 1985, p. 208). Because of Kamin's criticism and the poor documentation of his work there is agreement that Burt's data should not be included in any summaries of kinship correlations of the IQ literature (Jensen, 1974; Rowe & Plomin, 1978). Whether Burt was the fraud he is purported to have been (Hernshaw, 1979) has become a contentious issue and many of the initial accusations made against him have been shown to be invalid (Fletcher, 1991; Joynson, 1990). There is a consensus that he was both a brilliant man and a rogue. It is remarkable, however, that the summary figure for the IQ correlation for all twins reported in the literature up until MISTRA and SATSA is .771 (Bouchard, 1982a). In addition MISTRA reported two of three IQ correlations at .78 (the other one was .69) and SATSA reported it's IQ correlation at .78 (Bouchard, et al., 1990; Pedersen, et al., 1992).

PATH MODELS FOR THE TWIN REARED APART METHOD

The quantitative nature of the twin reared apart method can be easily described in terms of path analysis, a useful scheme for characterizing "causal models" (Neale & Cardon, 1992). Before discussing twins, however, let us discuss a pervasive problem in psychology—the psychometric problem of reliability. A common method of determining the reliability of a test is to give alternate forms of the test to a group of individuals, call them test A and test B. The correlation between the two tests is r_{AB}. This correlation reflects the variance in A and B caused by the influence of the true score (a latent hypothetical trait). The correlation is a measure of what is called the "true score variance". This correlation is often computed via analysis of variance (Hayes, 1973, p. 535) and a correlation computed in this fashion is called an intraclass correlation. The important fact to note here is that the correlation between the two tests is a direct measure of the true score variance. The correlation is not squared (Jensen, 1971). This idea is shown in the path diagram in Figure 1a below:

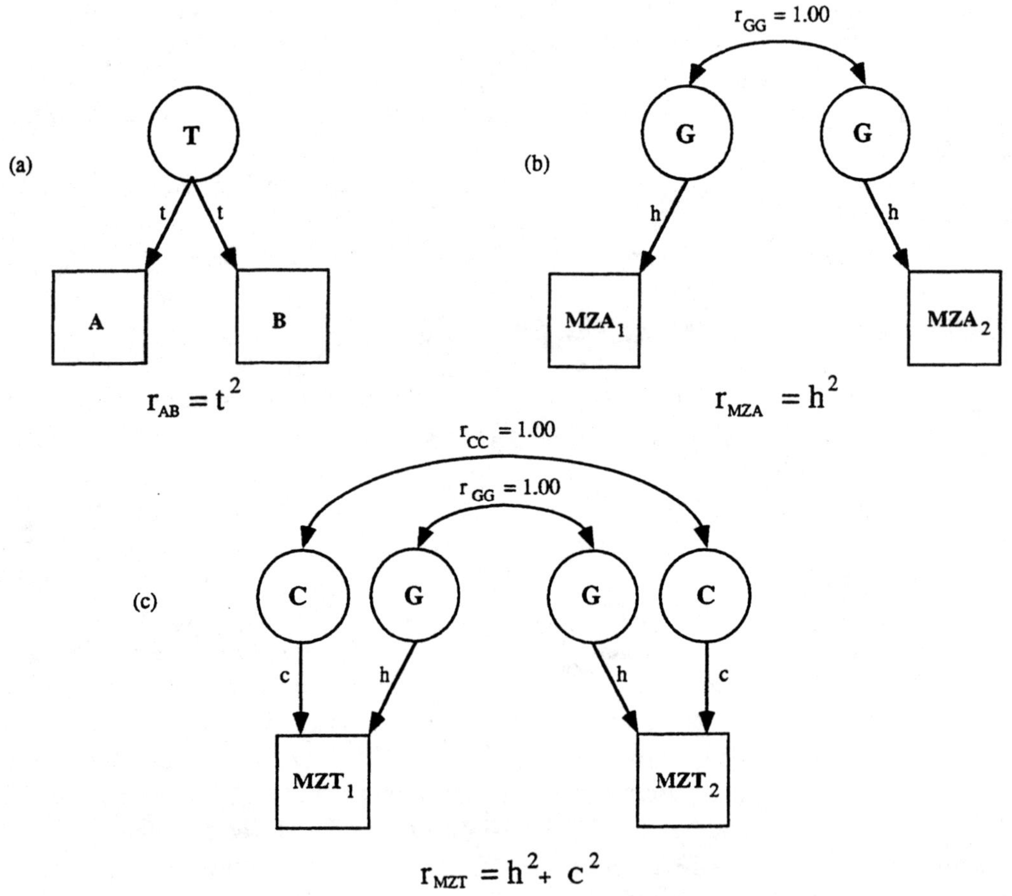

Figure 1. Path diagrams for reliability (a), monozygotic twins reared apart (b) and monzygotic twins reared together (d).

The boxes represent variables, the circles represent latent constructs, the directional arrows represent causal influences (if the arrow pointed in the other direction this would imply that the test could be a cause of the true score, an implausible situation). In this diagram, T is the influence of a true score (in substantive terms the true score might be a trait of some sort--i.e., the hypothetical construct intelligence), Y and Z are sources of error variance and A and B are test scores (perhaps IQ test scores). The lower case "t" is a path coefficient, and it shows the degree of influence of T on each test. Via the rules of path analysis the correlation between the tests is the cross product of the path coefficient linking the two tests. Thus $r_{AB} = t * t$ or t^2. Thus $t = \sqrt{r_{AB}}$. This is routine psychometrics for psychologists, but what many of them do not realize is that it is also elementary modeling in genetics.

If the scores are generated by identical twins reared apart we replace T in the figure with two G's that are correlated 1.00 (linked by two headed arrows) to show that the twins are identical. The "t's" become "h's" and the correlation would read $r_{MZA} = h * h$ or h^2. This model is shown in Figure 1b. The term h^2 is called the heritability. If the correlation between identical twins reared apart were zero, h^2 would be zero and we would conclude that genetic effects had no influence on individual differences in the trait being measured.

In Figure 1c we generalize the model to identical twins reared together. In this figure we have added a latent trait C (for common environmental influence) and lower case "c's" for the path coefficients. The correlation for identical twins reared together is therefore $r_{MZT} = h^2 + c^2$. The path diagrams and the equations conform to most people's intuitive belief that identical twins reared together are alike because they share many of the same (common) environmental influences and perhaps some genetic influence. For dizygotic twins reared apart (DZA) and reared together (DZT) the path models simply contain r_{GG}=.5 instead of r_{GG}=1.00 in the case of MZAs and MZTs. The latent genetic influence in dizygotic twins is .50 because DZ twins on average only share one-half of their segregating genes in common.

These models imply that the genetic influence is all additive-fraternal twins are on average half as similar as identical twins. If this is not so then some other influence is at work. If the fraternal twins are less than half as similar as identical twins then there is some thing that the identical twins share and the fraternal twins do not. There are two possibilities: non-additive genetic effects, which are shared completely by identical twins but only partially shared by fraternal twins (r_{GG}=.25) influence the trait, or the identical twins for some reason share a more similar environment. These ideas can be represented in a path model but we will not describe it here to keep things simple.

From examination of the above diagrams it should be clear that under the right circumstances the TRA design provides a powerful and direct way to detect genetic influences on behavioral traits. The MZA correlation directly estimates the influence of genetic effects. Nevertheless, as we will show below the most powerful implementation of the TRA design is in conjunction with the twin reared together (TRT) design. The TRA method is not a unique method in that it yields findings than cannot be found using other methods. Its great virtue is that it is conceptually simpler and more direct than the TRT design and is encumbered by somewhat different assumptions and biases. As in any other domain of science if two somewhat different methods converge on the same results we can have much greater confidence in the findings. We will illustrate this fact with examples.

BEHAVIOR GENETIC FINDINGS CONFIRMED BY THE TWINS REARED APART METHOD: MINIMUM INFLUENCE OF THE SHARED REARING ENVIRONMENT

One of the most important discoveries in behavior genetics has been that shared rearing environments have much less influence on personality than almost any psychologist would have imagined only a few years ago. The magnitude of the influence of shared rearing environment on a trait can be estimated simply by comparing MZA and MZT or DZA and DZT twins. This same finding has been demonstrated using only twins reared together. Generalizing from the formulas shown in Figure 1 we see that (r_{MZT} - (2 * r_{DZT})) = c^2 or the common environmental influence. The modest effect of c^2 had been hinted at by Galton quite a long time ago and had been quantitatively estimated by behavior genetics and reported on extensively in the 1970's (Loehlin & Nichols, 1976; Nichols, 1978), but it simply was not accepted by the psychological community. The reason it was not accepted is simple. Most psychologists believed the argument (with no real evidence in its favor) that MZT twins experienced far more similar trait relevant environments (environments such as child rearing patterns that supposedly shaped personality) than DZT twins because of their greater physical similarity. The TRT method makes the infamous "equal trait relevant environment" assumption--c^2 is the same for both MZT and DZT twins. A great deal of evidence has accumulated over the years to demonstrate that the assumption is probably correct for many traits (Bouchard, 1984; Loehlin & Nichols, 1976; Rowe, 1994, Chap. 2), but against a hostile belief system statistical arguments that are largely indirect carry very little weight. The TRA method must, of course, make the assumption that the twins are placed in environments that are uncorrelated for features that are determinants of the traits-they are not placed in similar trait relevant environments. We will return to this assumption shortly. In any event it is clear that TRA's are not raised in environments that are anywhere near as similar as TRT's.

Table 1 shows the personality findings from SATSA that address the question of heritability (additive genetic variance--G_a plus non-additive genetic variance--G_d), common family environmental influence (E_s) and non-shared environmental variance plus error (E_{ns}+e).

If one combines the additive genetic variance and the non-additive genetic variance for all the traits the average degree of genetic influence on these personality traits is 28.4%. The average amount of shared environmental variance is 10.2%. These results differ somewhat from those reported for MISTRA. Since different tests were used in the two studies it is not possible to compare the findings directly. We can, however, make some comparisons. Following Loehlin (1992), we can match the Big Five to some of the scales of the Multidimensional Personality Questionnaire used in MISTRA to generate comparable correlations. The findings are shown in Table 2.

Table 1. Intraclass Correlation for MZA, DZA, MZT and DZT Twins and Models Fitting Results for Personality Measures use in the Swedish Adoption Twin Study of Aging

Trait	Twin Type				% of Variance			
	MZA	DZA	MZT	DZT	G_a	G_d	E_s	En_s+e
Big Five[a]								
Extraversion (E)	.30	.04	.54	.06	0	41	7	52
Neuroticism (E)	.25	.28	.41	.24	31	0	10	58
Openness (NEO)	.43	.23	.51	.14	38	2	6	54
Conscientiousness (NEO)	.19	.10	.47	.11	0	29	11	60
Agreeableness (NEO)	.15	-.03	.41	.23	0	12	21	67
EAS[b]								
Emotionality-Distress	.30	.26	.52	.16	36	3	6	55
Emotionality-Fear	.37	.04	.49	.08	0	39	4	57
Emotionality-Anger	.33	.09	.37	.08	13	15	12	60
Activity	.27	.00	.38	.18	0	23	12	65
Sociability	.20	.19	.35	.19	24	0	13	64
Type A[c]								
Framingham	.23	.19	.37	.23	27	0	11	62
Pressure	.20	.19	.42	.23	28	0	12	60
Hard-Driving	.39	.10	.47	.00	0	43	0	57
Ambitious	.40	.08	.30	.11	0	37	0	63
Hostility	.21	.21	.33	.40	20	0	20	59
Inhibition of Aggression	.16	-.08	.32	.20	0	12	19	69
Locus of Control[d]								
Luck	.02	.18	.28	.32	0	0	31	69
Responsibility	.36	.30	.30	.18	34	0	0	66
Life Direction	.32	.23	.30	.15	31	0	0	69
Symptoms of Depression								
CES-D Scale	.22	-.01	.51	.30	16	0	27	55
Miscelaneous Scales[a]								
Impulsivity	.40	.15	.45	.09	2	43	0	55
Monotony Avoidance	.20	.14	.26	.16	23	0	5	72

a. Extraversion, Neuroticism, Impulsivity and Monotony avoidance data from Pedersen, et al. (1988), Openness, Agreeableness and Conscientiousness data from Bergeman, Chipuer, Plomin, Pedersen, McClearn, Nesselroade, et al. (1993), E means abbreviated Eysenck scales, NEO means abbreviated NEO scales.

b. EAS data from Plomin, Pedersen, McClearn, Nesselroade, & Bergeman (1988).

c. Type A data from Pedersen, Lichtenstein, Plomin, DeFaire, McClearn, & Matthews (1989b).

d. Locus of control data from Pedersen, Gatz, Plomin, Nesselroade, & McClearn (1989a).

e. Symptoms of depression from Gatz, Pedersen, Plomin, Nesselroade, & McClearn (1992).

The most parsimonious fit to the Minnesota data is a simple additive genetic model for all five traits with an estimate of genetic influence of 46%. There is not, however, a significant change in the fit of the model when nonadditive genetic and shared environmental parameters are allowed (both are modest in size), and those data are shown in Table 2 for comparison with the Loehlin analysis to be discussed below.

Table 2. Intraclass Correlation for MZA, DZA, MZT and DZT Twins and Models Fitting Results for Personality Measures used in the Minnesota Study of Twins Reared Apart

Trait	Twin Type				% of Variance			
	MZA (59)	DZA (47)	MZT (522)	DZT (408)	G_a	G_d	E_s	En_s+e
Big Five[a]								
Extraversion	.41	-.03	.54	.19	9	29	15	47
Neuroticism	.49	.44	.48	.19	41	9	0	50
Openness	.57	.27	.43	.14	29	15	0	56
Conscientiousness	.54	.07	.54	.29	29	13	13	45
Agreeableness	.24	.09	.39	.11	5	25	9	61
Mean	.45	.17	.48	.18	23	18	7	52

a. Data from Bouchard (in press). The MISTRA measures are the same Multidimensional Personality Questionnaire factors and scales used by Loehlin (1992) Positive Emotionality = Extraversion, Negative Emotionality = Neuroticism, Constraint = conscientiousness, Adsorbtion = Openness, Aggression = Agreeableness. The MZA and DZA data is based on the most recent MISTRA sample. The MZT and DZT data is from the large Minnesota Twin Registry sample (Lykken, et al. 1990).

It is obvious from examination of the raw correlations that the MISTRA data yield higher estimates of genetic influence that the SATSA data. What is the problem? It has been pointed out that the MISTRA sample is a volunteer sample and that perhaps the twins who have chosen to participate are more similar than a random sample of TRA's would be (Kaprio, et al., 1990; Pedersen, Plomin, McClearn, & Friberg, 1988). This is a real possibility, although a number of the MISTRA twins were recruited before the twins found each other. In addition MISTRA recruits both MZ and DZ twins and thereby never rejects twins because of differences between them (an argument commonly made against the previous MZA studies). When the MISTRA sample size is large enough the influence of method of recruitment will be examined systematically. The SATSA sample was recruited systematically from a registry so this selection bias is much less of a problem. Nevertheless, there are some interesting questions that can be asked about the SATSA data. Recall that the difference between reared apart twins and reared together twins estimates common family environmental influence. For the big five in Table 1., the difference for DZ twins is .032 and the difference for MZ twins is .204. The average estimate of 11% for c^2 in SATSA thus reflects the relatively greater MZT than MZA correlations. This pattern would suggest a special assimilation effect among MZT pairs (i.e. a special MZ twin environment as described below). On the other hand for the big five the MISTRA data are largely consistent, the DZ's yield .02 and the MZ's yield .03. For the 11 MPQ scales reported in Tellegen et al. (1988) the figures are also .02 and .03.

One might suspect that the lower heritabilities in SATSA are a function of the shorter scales used by SATSA (and by inference, lower reliabilities). The Extraversion and Neuroticism scales are made up of only nine items whereas all the MISTRA scales are over 20 items and many are much longer. Scale length can, however, only be a minor part of the explanation of the differences between the two studies, if it is an explanation at all, as all heritability estimates for the short version Eysenck scales from 12,898 pairs in the Swedish Twin Registry are remarkably similar to estimates from other registry based samples (Loehlin, 1992).

It is not at all obvious how to resolve this problem. One approach, taken by Loehlin (1992), is to incorporate measures from as many different kinships as possible. His solution, based on data from twins reared apart, twins reared together, adoption studies and twin family studies is shown in Table 3.

Table 3. Summary: Estimates from Fitting Two Simple Models to Data from Big Five Traits

	Special Twin Environment			Non-additive Genetic		
Big Five Factor	h^2	c^2_{MZ}	c^2_S	h^2	i^2	c^2_S
Extraversion	.36	.15	.00	.32	.17	.02
Neuroticism	.31	.17	.05	.27	.14	.07
Openness	.46	.05	.05	.43	.02	.06
Conscientiousness	.28	.17	.04	.22	.16	.07
Agreeableness	.28	.19	.09	.24	.11	.11
Mean	.34	.15	.05	.30	.12	.07

From: Loehlin (1992) Table 3.20

Loehlin (1992) points out that it is not possible to chose between the two different models shown above, a model that incorporates non-additive genetic variance (i^2) or a model that incorporates a special MZ twin environment (c^2_{MZ}). These solutions, of course, have their own problems (Plomin, Chipuer, & Loehlin, 1990). The model fitting is rough and based on a very miscellaneous collection of scales. In addition a great deal of heterogeneity due to age and sex is not taken account of (Eaves, Eysenck, & Martin, 1989).

While the Loehlin analysis contains the SATSA and MISTRA data, it also contains a great deal of other data as well. SATSA yields a broad heritability of 30% while MISTRA yields a broad heritability of 43%. Loehlin finds a broad heritability of 42% if one chooses the non-additive genetic model.

It would be nice if all studies had been able to use the same instruments. In practice this is never the case. All studies make compromises in order to achieve their specific objectives. Nevertheless, in spite of all the problems discussed above, there is considerable agreement that the additive genetic variance for the big five is sizable (22 to 46%) and that the influence of common family environment is quite modest. This latter point is now based on so many different kinships that the argument regarding the equal trait relevant environment assumption is irrelevant. Indeed as shown in Table 3, when a special environmental parameter is fit for MZ twins (making them a special case), h^2 for all traits drops but so does c^2, indeed c^2 is basically trivial for all variables except Agreeableness.

GENETIC INFLUENCE ON ENVIRONMENTAL VARIABLES

Social Class and IQ: A major finding from Behavior Genetics that has become well known in the last ten years is that environmental measures are not always what they seem (Plomin, 1994; Plomin & Bergeman, 1991). This fact was demonstrated long ago with respect to one widely used "environmental" measure, namely socioeconomic status (SES). Over fifty years ago Barbara Burks demonstrated that differences in IQ from one social class to another was largely due to genetic differences and thus parental SES was largely a genetic marker rather than an environmental marker (Burks, 1938). The mechanism underlying this phenomenon is well known, namely social mobility mediated by IQ (Bouchard, 1976; Gottesman, 1968; Herrnstein, 1973; Herrnstein & Murray, 1994). This phenomenon has recently been replicated by Scarr and Weinberg (1978). As they put it "Burks (1938) estimated that genetic differences among the occupational classes account for about 2/3 to 3/4 of the average IQ differences among the children born into those classes. Our studies support that conclusion" (p. 689). This illustration provides a nice opportunity to highlight the advantage of using multiple methods to demonstrate an empirical finding. Ordinary adoption samples are almost always made up of children and parents who are recruited through adoption agencies. This leads to some screening for successful adoptions and the culling of families with low educational and SES levels. TRA studies, however, recruit the participants themselves (largely) as adults without regard to whether the adoption was successful or the SES of the participants' families was high or low. MISTRA includes many twins whose adoption would be labeled "unsuccessful" and who live very modestly indeed (one twin stepped off the airplane in Minneapolis with only $5.00 in her pocket). What some would argue is a selection bias, participation because of the opportunity to visit the United States (about 40% of MISTRA twins come from overseas, mostly the United Kingdom and Australia), actually serves to extend the range of variation on both the "current environment" and the "rearing environment" of the participants. A number of papers dealing with related aspects of SES based on the SATSA sample have appeared (Hershberger, Lichtenstein, Knox, & McClearn, in press; Lichtenstein, Harris, Pedersen, & McClearn, 1993; Lichtenstein, Hersberger, & Pederson, 1995; Lichtenstein, Pedersen, & McClearn, 1992). These papers, based on both TRA and TRT, provide powerful evidence for the significance of genetic variance for individual differences in measures of SES.

Given that many of the TRA's are adopted by non biological relatives the correlation between rearing parent characteristics and characteristics of the twin (personality, IQ) reflect only environmental influences. Table 4 below shows those correlations for MISTRA.

The important correlations are those labeled "Correlations between IQ and placement variable". The correlations between twins IQ and parental SES indicators (Father's education, Mother's education and Father's SES) are quite modest, one even being zero (none are statistically significant). One should note that these correlations must be squared before they are interpreted as measures of amount of variance in the dependent variable accounted for (we will return to this point in a moment). Under the heading physical facilities two of four measures yield significant correlations. One of these measures correlates in the counter intuitive direction. No one would predict that the number of cultural amenities reported to have been available in the rearing home (World atlas, Collection of classical records, Reproductions of famous paintings, Example of original artwork, Encyclopedia set, Books in a foreign language, One or more musical instruments, etc.) would correlate negatively with IQ. The most likely interpretation is that the correlation is a chance deviation from zero, or a modest positive value, as this measure does correlate positively (but modestly) with various measures of verbal ability [see McGue and Bouchard (1989) for a factor analysis of this instrument and the correlation of its scales with special mental abilities]. A nice complementary finding to the inability of SES indicators to explain variance in IQ is the fact that unrelated individuals who grow up in the same family (for whom the SES indicators correlate 1.00) have an IQ correlation of zero in adulthood (McGue, Bouchard, Iacono, & Lykken, 1993). The path diagram for this kinship is shown as Figure 1d.

Table 4. Placement coefficients for environmental variables, correlations between IQ and the environmental variables, and estimates of the contribution of placement to twin similarity in WAIS IQ.

Placement variable	MZA similarity (R_{ff})	Correlation between IQ and placement(r_{ft})	Contribution of placement to the MZA correlation (R_{ff} x r^2_{ft})
SES indicators			
Father's education	.134	.100	.001
Mother's education	.412	-.001	.000
Father' SES	.267	.174	.008
Physical facilities			
Material possessions	.402	.279**	.032
Scientific/technical	.151	-.090	.001
Cultural	-.085	-.270**	-.007
Mechanical	.303	.077	.002
Relevant FES scales			
Achievement	.11	-.103	.001
Intellectual orientation	.27	.106	.003

**r_{ft} significantly different from zero at $P < 0.01$.
From: Bouchard et al. (1990)

The underlying genetic correlation between IQ and social class is scarcely recognized by psychologists. In a classic review of the literature of the correlation between socioeconomic status and achievement (White, 1982) does not even mention the possibility that the correlation is in part genetically mediated. In a recent article on the relationship between socioeconomic status and health, Adler, Boyce, Chensey, Cohen, Folkman, Kahn and Syme (1994) simply brush aside the possibility that the association is mediated by genes (via intelligence and/or social mobility) in spite of evidence to the contrary (Lichtenstein et al., 1993; Lichtenstein & Pedersen, submitted; West, 1991).

The most interesting feature of Table 4 is that it allows us to examine a purported bias that is often called the "fatal flaw" of TRA studies (Lewontin, Rose, & Kamin, 1984, p. 108), namely placement bias. It is simply asserted without quantitative examination of the facts that placement vitiates the conclusions that heredity is an important determinant of the trait when a large and significant MZA correlation is found. This argument can also be put in the form of a path diagram and subjected to empirical evaluation. Figure 2 shows the model that underlies the placement bias argument. We have filled in the numbers for the most influential variable in Table 3, material possessions. Multiplying the components of the paths yields (.278 * .279) * .402 = .032. This is the predicted MZA correlation for MZA IQ's in this sample based on placement for the most influential variable. It should also be kept in mind that since we have picked the variable with the most extreme influence it is also the variable most likely regress back towards the mean on cross-validation. Bouchard (Bouchard, 1983) has shown how serious a problem this can be with small data sets. The solutions to the path diagram for all nine variables in Table 4 are given in the last column of the table. The important point to keep in mind here is that the MZA correlation depends on components that are multiplied, consequently they must be of considerable magnitude before the correlation will reach any appreciable size; a small value of any one of them causes the model to fail to explain any significant amount of variance. Few psychologists intuitively estimate such effects accurately. SATSA has also shown that various measures of separation, including age at separation, degree of separation and number of years separated prior to first contact, can explain at most 3% of trait variance (Pedersen, McClearn, Plomin, & Nesselroade, 1992).

This same model can be used to demonstrate the extreme weakness of the commonly made argument that MZA twins are similar in personality because they are very similar in physical attractiveness (Lewontin et al., 1984, p. 115). The correlations between physical attractiveness and personality traits (the equivalent of r_{ft} in Table 3) are even smaller than the values in Table 3 (Feingold, 1992). The true correlation between twins for physical attractiveness during the developmental period in which their personalities are developing is not known, but it can be estimated. Rowe, Clapp, & Wallis (1987) found a correlation of .54 for pooled ratings (four raters) of physical attractiveness for college age MZ twins. He corrected this correlation by using some crude estimates of rater reliability and concluded the "true" correlation for physical attractiveness is about .94. We will use this figure.

Feingold reports that the mean correlation between attractiveness and Sociability (Extraversion) for 25 studies and 1,710 cases is .04. Consequently the expected correlation between MZA twins on Extraversion would be (.04 * .04) * .94 = .0015. There are some traits reported by Feingold that do show an effect due to attractiveness. Popularity in females yields the highest correlation in his entire meta-analysis--.35. Consequently the expected correlation between female MZA twins on popularity would be (.35 * .35) * .94 = .115. We have chosen the most extreme value in a table of 138 possible values, consequently, even though this is a meta-analysis we are capitalizing on chance to a some degree, and we should therefore expect this figure to regress back towards a lower value in future surveys. It seems safe to conclude that no reliably measured individual differences variable is likely to yield an MZA correlation higher than .10 due to the twins' similarity in physical attractiveness. We note here that all of Feingold's correlation between intelligence and physical attractiveness are zero or negative, consequently physical attractiveness is unable to explain any of the MZA correlation for IQ.

Life Events. One of the recent findings in behavior genetics that challenges a core belief of many psychologists is that life events have a genetic basis. The typical view in psychology is that life events or life changes are important environmental "causal" influences on mental and physical health (Holmes & Rahe, 1967; Paykel, 1971; Rabkin & Struening, 1976), although compare Hudgens (1974). While a number of authors recognize that stress for life events is sometimes initiated by the individual him or herself, it is still assumed that the mediation was via personality traits, etc. that were themselves largely under environmental influence.

A paper from SATSA provided the first demonstration that life events were to some extent under genetic influence and that controllable life events showed the most extensive genetic influence (Plomin, Lichtenstein, Pedersen, McClearn, & Nesselroade, 1990). The same four group design as used to study personality was applied to responses to a 25 item self-report life events questionnaire. The most recent analysis of that data, from Plomin (1994) is in Table 5.

Table 5. Twin Intraclass Correlations and Heritabilities for Four Older Adult Twin Groups for Ratings of Life Events

Life Events Scale	Twin Type				Model-Fitting Heritability
	MZA (45-49)	DZA (107-125)	MZT (90-98)	DZT (112-127)	
Total score	.43	.07	.24	.19	.31*
Events to self	.27	.21	.27	.10	.30*
Events to others	.22	.06	.11	.15	.18*

*p<.05
Data Plomin (1994, p. 91)

As would be expected, events to self (controllable events) yielded a higher heritability than events to others (uncontrollable events). The findings reported in Table 5 have been replicated by MISTRA (Moster, 1991) using only twins reared apart. The results are shown in Table 6 below.

Table 6. Intraclass Correlations for MZA and DZA Twins and Model Fitting Estimates of Genetic and Environmental Sources of Variance for Life Event Scales

Scale	Twin Type		% Variance	
	MZA (37-39)	DZA(23-25)	G	E+e
Uncontrollable events	.24	.00	.23	.77
Controllable event	.50	.25	.53	.47
Holmes & Rahe	.32	.21	.42	.58
Paykel	.46	-.03	.41	.59
Total Life Events	.51	.23	.50	.50

From: Moster (1991)

The findings from the two studies are similar even though MISTRA used much longer scales. Due to much smaller sample sizes MISTRA is not powerful enough to test for nonadditive genetic effects. Both uncontrollable and controllable events show a genetic influence, and the later show twice as strong an effect as the former, exactly as was shown in SATSA. Total life events also show similar heritabilities in both studies. MISTRA also generated scores for scales very similar to the widely used scales devised by Holmes and Rahe (1967) and Paykel (1971). Both scales show sizable heritabilities. Finally somewhat similar results have been reported by Kendler, Neale, Kessler, Heath, & Eaves (in press) using conventionally reared twins.

These findings have profound consequences for the interpretation of findings in the domain of life stress research as they reverse the direction of causation for many of the findings.

Genotype x Environment Interactions

Human behavior genetics has been criticized regularly by "environmentalists" for failing to take account of genotype x environment interactions (Feldman & Lewontin, 1975; Schiff & Lewontin, 1986; Wachs, 1992, chap. 8). There are actually a number of issues that underlie the "interaction" problem. First, it is sometimes implied that the possible existence of interactions makes the analysis of main effects of genes and environment impossible. This assertion, of course, explains nothing and could be brought against research in any domain of science. "Everything in the world can be explained by factors about which we know nothing" (Urbach, 1974, p. 253). Secondly, G x E interactions do not take place in the absence of additive genetic effects. "Since armchair examples of significant interactions in the absence of an additive effect are pathological and have never been demonstrated in real populations, we need not be unduly concerned about interaction effects. The investigator with a different view should publish any worthwhile results he may obtain" (Rao, Morton, & Yee, 1974, p. 357). Thirdly, generalizing from animal to human studies is dangerous. Interactions can easily be demonstrated using inbred organisms as they have been selected precisely for that purpose. It is another matter to demonstrate similar effects in human beings (Crow, 1990; Henderson, 1990). Finally, interactions do not constitute a domain of research in their own right. If findings are simply based on the screening of large amounts of data then their repeatability is suspect and it is necessary to replicate the findings before taking them seriously. While it is fashionable to argue that the search for interactions should be theoretically driven (Plomin & Wachs, 1991), it should be kept in mind that psychological theories are sufficiently flexible to be capable of incorporating almost any set of findings. This warning should be kept in mind as we present the methodology used to illustrate the use of the MZA method to test for interactions.

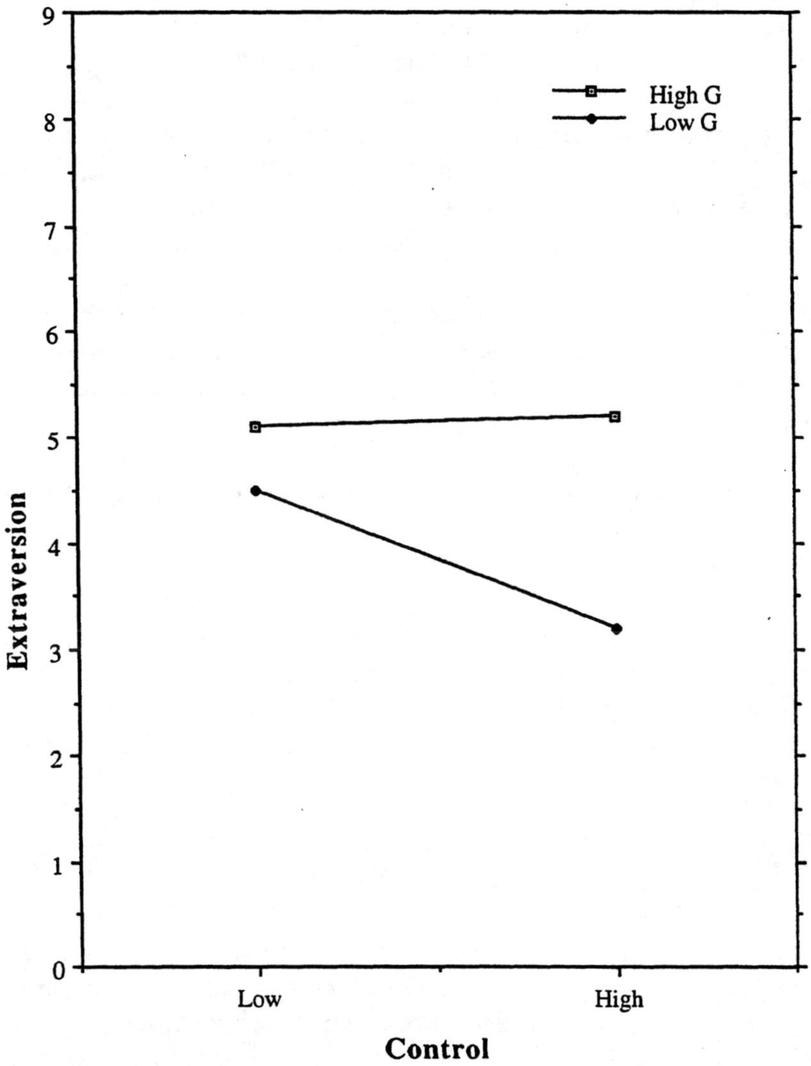

From Bergeman et al., 1988

Figure 2. Genotype-environment interaction of Eysenck Personality Inventory Extraversion with Family Environment Control Dimension

In 1970, Jinks and Fulker (1970) suggested that the presence of genotype by environment interaction could be assessed by comparing within-pair means and differences in monozygotic twins reared apart. While this technique is valuable for indicating the presence of such effects, they do not provide a means of testing for interactions with specific measures of the environment. Bergeman, Plomin, McClearn, Pedersen, & Friberg (1988) have applied a simple hierarchical multiple regression model to personality data in SATSA, in an attempt to evaluate whether there are interactions between personality "genotypes" and childhood environments as measured by the Moos Family Environment (FES) scales. One MZA twin's score on personality traits was predicted by the other twin's score on the trait, a measure of twin A's childhood environment and the interaction of these. The idea is that the co-twin's phenotype provides a measure of the twins' joint genotype. The analysis yielded evidence for 11 significant interactions. Figure 2 depicts the interaction between Control in the family environment and extraversion.

Individuals with a predisposition toward high extraversion were extraverted regardless of the family environment. On the other hand, individuals with a predisposition toward low extraversion were more extraverted in families exerting little control than in families exerting much control. Thus, a controlling environment affected the level of extraversion for some genotypes, that is, those with a disposition to low extraversion, but not for others. The most striking aspects of this study were both the lack of significant results and the relatively small effect sizes. These results are not unique to findings using the MZA design. Using an adoption design to assess possible interactions between temperament and environment, Plomin, DeFries and Fulker (1988) point out that "The results are so meager that they do not suggest hypotheses as to the form or substance of interactions" (p. 249). Arguments calling for the evaluation of G x E interactions should be tempered by these very modest results.

One of the most sever limitations of this model is, as discussed earlier, the evidence that these purported measures of the environment are self-reported perceptions and reflect some genetic variance. This model would be most appropriately applied to environmental indicators for which there is no genetic variance. Likely candidates are noncontrolable, undesirable life events or quality of social support which, unlike controllable life events or perceived adequacy of support show little or no significant genetic variance.

CONCLUSION

While we believe that the TRA method is one of the most powerful methods in human behavior genetics and that the findings generated by the method provide very persuasive evidence for extensive genetic influence on human behavior we would like to re-emphasize our earlier admonition that like any other method it has both strengths and weaknesses. It is best utilized in conjunctions with the TRT design and all findings should be subjected to replication utilizing other designs--particular the URT design. The striking finding of very little c^2 influence for IQ in adulthood using twins has been strongly verified by the lack of c^2 effects for unrelated individuals reared together

(McGue, et al., 1993). Similar studies of adult unrelated individuals reared together, utilizing a wide range of psychological measures, are badly needed. Twin studies and adoption studies are experiments, but they are experiments of nature and therefore they cannot meet the same rigorous methodological requirements of laboratory experiments. They must be complemented by as many other designs as possible. There is no substitute for constructive replication. This admonition in no way detracts from the importance of twin and adoption studies. Laboratory experiments cannot, and with human beings never will, capture a lifetime of experience. Our understanding of the importance of genetic influences on human behavior and the mechanisms that mediate it will depend on these methods for a long time to come.

The Minnesota Study of Twins Reared Apart has been supported by grants from the Pioneer Fund, the David H. Koch Charitable Trust, the Seaver Institute, the Spencer Foundation, the National Science Foundation (BNS-7926654), the University of Minnesota Graduate School and the Harcourt Brace Javnovich Publishing Co. The Swedish Adoption/Twin Study of Aging is supported in part by grants from the National Institute of Aging (AG 04563, AG 10175), the John D. and Catherine T. MacArthur Foundation and the Swedish Council for Social Research.

REFERENCES

Adler, N.E., Boyce, T., Chesney, M.A., Cohen, S., Folkman, S., Kahn, R.L., & Syme, S.L. (1994). Socioeconomic status and health. The challenge of the gradient. *Am Psychol, 49*, 15-24.

Bergeman, C.S., Chipuer, H.M., Plomin, R., Pedersen, N.L., McClearn, G.E., Nesselroade, J.R., Costa, P.T.,Jr., & McCrae, R.R. (1993). Genetic and environmental effects on openness to experience, agreeableness, and conscientiousness: an adoption/twin study. *J Pers, 61*, 159-179.

Bergeman, C.S., Plomin, R., McClearn, G.E., Pedersen, N.L., & Friberg, L.T. (1988). Genotype-environment interaction in personality development: identical twins reared apart. *Psychol Aging, 3*, 399-406.

Bouchard, T.J.,Jr. (1976). Genetic factors in intelligence. In A.R. Kaplan (Ed.). *Human behavior genetics* Springfield, IL: Charles C. Thomas

Bouchard, T.J.,Jr. (1982a). Review of identical twins reared apart: A reanalysis. *Contemporary Psychology, 27*, 190-191.

Bouchard, T.J.,Jr. (1982b). Review of the intelligence controversy. *American Journal of Psychology, 95*, 346-349.

Bouchard, T.J.,Jr. (1984a). Twins reared together and apart: What they tell us about human diversity. In S. Fox (Ed.). *Individuality and determinism: Chemical and biological bases* New York: Plenum

Bouchard, T.J.,Jr. (1984b). Do environmental similarities explain the similarity in intelligence of identical twins reared apart? *Intelligence, 7*, 175-184.

Bouchard, T.J.,Jr. (1993a). The genetic architecture of human intelligence. In P.A. Vernon (Ed.). *Biological approaches to the study of human intelligence* Norwood, NJ: Ablex

Bouchard, T.J.,Jr. (1993b). Twins: Nature's twice told tale. In T.J. Bouchard,Jr. & P. Propping (Eds.). *Twins as a tool of behavior genetics* Chichester, England: Wiley and Sons, Ltd.

Bouchard, T.J.,Jr. (1994). Genes, environment, and personality. *Science, 264*, 1700-1701.

Bouchard, T.J.,Jr., Lykken, D.T., McGue, M., Segal, N.L., & Tellegen, A. (1990). Sources of human psychological differences: the Minnesota Study of Twins Reared Apart. *Science, 250*, 223-228.

Burks, B.S. (1938). On the relative contributions of nature and nurture to average group differences in intelligence. *Proceedings of the National Academy of Sciences, 24*, 276-282.

Burks, B.S. (1942). A study of identical twins reared apart under differing types of family relationships. In Q. McNemar & M.R. Merril (Eds.). *Studies in personality* (pp. 35-69). New York: McGraw-Hill

Burks, B.S. & Roe, A. (1949). Studies of identical twins reared apart. *Psychological Monographs, 63*, 1-61.

Burt, C. (1955). The evidence of the concept of intelligence. *Br J Psychol, 25*, 158-177.

Burt, C. (1958). The inheritance of mental ability. *American Psychologist, 13*, 1-15.

Burt, C. (1966). The genetic determination of differences in intelligence: a study of monozygotic twins reared together and apart. *Br J Psychol, 57*, 137-153.

Burt, C. (1972). Inheritance of general intelligence. *Am Psychol, 27*, 175-190.

Crow, J.F. (1990). How important is detecting interactions? *Behavior and Brain Sciences, 13*, 126-127.

Eaves, L.J., Eysenck, H.J. & Martin, N.G. (1989). *Genes, culture and personality: An empirical approach.* New York: Academic Press.

Fancher, R.E. (1985). *The intelligence men: Makers of the IQ controversy*. New York: W.W. Norton & Co..

Farber, S.L. (1981). *Identical twins reared apart: A reanalysis*. New York: Basic Books.

Feingold, A. (1992). Good-looking people are not what we think. *Psychol Bull, 111(2)*, 304-341.

Feldman, M.W. & Lewontin, R.C. (1975). The heritability hang-up. *Science, 190*, 1163-1168.

Fletcher, R. (1991). *Science, ideology, and the media: The Cyril Burt scandal.* New Brunswick, NJ: Transaction Books.

Galton, F. (1869). *Hereditary genius: An inquiry into its laws and consequences.* London: Macmillan.

Gardner, I.C. & Newman, H.H. (1940). Mental and physical traits of identical twins reared apart: Case XX. The twins Lois and Louise. *Journal of Heredity, 31*, 119-126.

Gatz, M., Pedersen, N.L., Plomin, R., Nesselroade, J.R., & McClearn, G.E. (1992). Importance of shared genes and shared environments for symptoms of depression in older adults. *J Abnorm Psychol, 101*, 701-708.

Gottesman, I.I. (1968). Biogenics of race and class. In M. Deutsch, I. Katz & A. Jensen (Eds.). *Social class, race and psychological development* (pp. 11-51). New York: Holt, Rinehart & Winston

Hayakawa, K. (1987). Smoking and drinking discordance and health condition: Japanese identical twins reared apart and together. *Acta Genet Med Gemellol (Roma), 36*, 493-501.

Hayakawa, K. & Shimizu, T. (1987). Blood pressure discordance and lifestyle: Japanese identical twins reared apart and together. *Acta Genet Med Gemellol (Roma), 36*, 485-491.

Hayes, W.L. (1973). *Statistics for the social sciences.* New York: Holt, Rinehart & Winston.

Henderson, N. (1990). Why do gene-environment interactions appear more often in laboratory animal studies than in human behavioral genetics? *Behavior and Brain Sciences, 13*, 136-137.

Hernshaw, L.S. (1979). *Cyril Burt: Psychologist.* London: Hodder & Stoughton.

Herrnstein, R.J. (1973). *IQ in the meritocracy.* Boston: Little Brown & Company.

Herrnstein, R.J. & Murray, C. (1994). *The bell curve: Intelligence and class structure in American life.* New York: Free Press.

Hershberger, S.L., Lichtenstein, P., & Knox, S.S. (1994). Genetic and environmental influences on perceptions of organizational climate. *J Appl Psychol, 79*, 24-33.

Heston, L.L. (1982). Review of Farber, S. L. Identical twins reared apart: A reanalysis. *The Sciences,*

Holmes, T.H. & Rahe, R.H. (1967). The Social Readjustment Rating Scale. *J Psychosom Res, 11*, 213-218.

Hudgens, R.W. (1974). Personal catastrophe and depression: A consideration of the subject with respect to medically ill adolescents, and a requiem for retrospective life-event studies. In B.S. Dohrenwend & B.P. Dohrenwend (Eds.). *Stressful life events: their nature and effects* (pp. 119-134). New York: Wiley

Jensen, A.R. (1971). Note on why genetic correlations are not squared. *Psychol Bull, 75*, 223-224.

Jensen, A.R. (1974). Kinship correlations reported by Sir Cyril Burt. *Behav Genet, 4*, 1-28.

Jinks, J.L. & Fulker, D.W. (1970). Comparison of the biometrical genetical, MAVA, and classical approaches to the analysis of human behavior. *Psychol Bull, 73*, 311-349.

Joynson, R.B. (1990). *The Burt affair*. New York: Academic Press.

Juel-Nielsen, N. (1965). *Individual and environment: A psychiatric-psychological investigation of MZ twins reared apart*. Copenhagen: Munksgaard.

Kamin, L.J. (1974). *The science and politics of IQ*. Potomac, MD: Erlbaum.

Kaprio, J., Koskenvuo, M., & Rose, R.J. (1990). Population-based twin registries: illustrative applications in genetic epidemiology and behavioral genetics from the Finnish Twin Cohort Study. *Acta Genet Med Gemellol (Roma), 39*, 427-439.

Kendler, K.S., Neale, M., Kessler, R., Heath, A., & Eaves, L. (1993). A twin study of recent life events and difficulties. *Arch Gen Psychiatry, 50*, 789-796.

Langinvainio, H., Kaprio, J., Koskenvuo, M., & Lonnquist, J. (1984). Finnish twins reared apart III: Personality factors. *Acta Geneticae Medicae et Gemellolgiae, 33*, 259-264.

Langinvainio, H., Koskenvuo, M., Kaprio, J., Lonnquist, J. & Tarkkonen, L. (1981). Finnish twins reared apart: Preliminary characterization of rearing environment. In P. Gedda, P. Parisi & W.E. Nance (Eds.). *Twin research 3: Part B, Intelligence, personality and development* New York: Alan R. Liss

Lewontin, R.C., Rose, S. & Kamin, L.J. (1984). *Not in our genes: Biology, ideology, and human nature*. New York: Pantheon.

Lichtenstein, P., Harris, J.R., Pedersen, N.L., & McClearn, G.E. (1993). Socioeconomic status and physical health, how are they related? An empirical study based on twins reared apart and twins reared together. *Soc Sci Med, 36*, 441-450.

Lichtenstein, P., Hersberger, S.L., & Pederson, N.L. (1995). Dimensions of occupations: Genetic and environmental influences. *Journal of Biosocial Science*, 27, 193-206.

Lichtenstein, P. & Pedersen, N.L. (1995). Sibling similarity for educational achievement and occupational status: The importance of genetic influences in common with cognitive abilities. *American Sociological Review*,

Lichtenstein, P., Pedersen, N.L., & McClearn, G.E. (1992). The origins of individual differences in occupational status and educational level: A study of twins reared apart and together. *Acta Sociologica, 35*, 13-31.

Loehlin, J.C. (1981). Review of Farber, S. L. Identical twins reared apart: A reanalysis. *Acta Geneticae Medicae et Gemellolgiae, 30*, 297-298.

Loehlin, J.C. (1992). *Genes and environment in personality development*. Newbury Park, CA: Sage Publications.

Loehlin, J.C. & Nichols, R.C. (1976). *Heredity, environment, & personality: A study of 850 sets of twins*. Austin, TX: University of Texas Press.

Lykken, D.T., Bouchard, T.J.,Jr., McGue, M., & Tellegen, A. (1990). The Minnesota twin family registry: Some initial findings. *Acta Geneticae Medicae et Gemellolgiae, 39*, 35-70.

McGue, M. & Bouchard, T.J.,Jr. (1995). Genetic and environmental determinants of information processing and special mental abilities: A twin analysis. In R.J. Sternberg (Ed.). *Advances in the psychology of human intelligence* (pp. 7-44). Hillsdale, NJ: Erlbaum

McGue, M., Bouchard, T.J.Jr., Iacono, W.G. & Lykken, D.T. (1993). Behavior genetics of cognitive ability: A life-span perspective. In R. Plomin & G.E McClearn, (Eds.). *Nature, nurture and psychology* (pp. 59-76). Washington, D.C.: American Psychological Association

Merriman, C. (1924). The intellectual resemblance of twins. *Psychological Monographs, 33 Whole No.152*, 1-58.

Moster, M. (1991). *Stressful life events: Genetic and environmental components and their relationship to affective symptomatology*. Ph.D. Dissertation, University of Minnesota.

Muller, H. (1925). Mental traits and heredity. *Journal of Heredity, 16*, 433-448.

Neale, M.C. & Cardon, L.R. (1992). *Methodology for genetic studies of twins and families*. Dordrecht: Kluwer Academic Publishers.

Newman, H.H., Freeman, F.N. & Holzinger, K.J. (1937). *Twins: A study of heredity and environment*. Chicago: University of Chicago Press.

Nichols, R.C. (1978). Twin studies of ability, personality and interests. *Homo, 29*, 158-173.

Paykel, E.S., Prusoff, B.A., & Uhlenhuth, E.H. (1971). Scaling of life events. *Arch Gen Psychiatry, 25*, 340-347.

Pearson, K. (1924). *The life, letters, and labours of Francis Galton, Vol.2*. Cambridge: Cambridge University Press.

Pedersen, N.L., Lichtenstein, P., Plomin, R., DeFaire, U., McClearn, G.E., & Matthews, K.A. (1989). Genetic and environmental influences for type A-like measures and related traits: a study of twins reared apart and twins reared together. *Psychosom Med, 51*, 428-440.

Pedersen, N.L., McClearn, G.E., Plomin, R., Nesselroade, J.R., Berg, S., & DeFaire, U. (1991). The Swedish Adoption Twin Study of Aging: an update. *Acta Genet Med Gemellol (Roma), 40*, 7-20.

Pedersen, N.L., Gatz, M., Plomin, R., Nesselroade, J.R., & McClearn, G.E. (1989). Individual differences in locus of control during the second half of the life span for identical and fraternal twins reared apart and reared together. *J Gerontol, 44*, P100-P105.

Pedersen, N.L., McClearn, G.E., Plomin, R., & Nesselroade, J.R. (1992). Effects of early rearing environment on twin similarity in the last half of the life span. *British Journal of Developmental Psychology, 10*, 255-267.

Pedersen, N.L., Plomin, R., McClearn, G.E., & Friberg, L. (1988). Neuroticism, extraversion, and related traits in adult twins reared apart and reared together. *J Pers Soc Psychol, 55*, 950-957.

Pedersen, N.L., Plomin, R., Nesselroade, J.R., & McClearn, G.E. (1992). A quantitative genetic analysis of cognitive abilities during the second half of the life span. *Psychological Science, 3*, 346-353.

Plomin, R. (1994). *Genetics and Experience: The interplay between nature and nurture.* Thousand Oaks, CA: Sage.

Plomin, R. & Bergeman, C.S. (1991). The nature of nuture: Genetic influence on "environmental" measures. *Behavior and Brain Sciences, 14*, 373-427.

Plomin, R., Chipuer, H.M. & Loehlin, J.C. (1990). Behavioral genetics and personality. In *Handbook of personality: Theory and research* New York: Guilford Press

Plomin, R., DeFries, J.C. & Fulker, D.W. (1988). *Nature and nurture during infancy and early childhood.* New York: Cambridge University Press.

Plomin, R., Lichtenstein, P., Pedersen, N.L., McClearn, G.E., & Nesselroade, J.R. (1990). Genetic influence on life events during the last half of the life span. *Psychol Aging, 5*, 25-30.

Plomin, R., Pedersen, N.L., McClearn, G.E., Nesselroade, J.R., & Bergeman, C.S. (1988). EAS temperaments during the last half of the life span: twins reared apart and twins reared together. *Psychol Aging, 3*, 43-50.

Plomin, R. & Wachs, T.D. (1991). *Conceptualization and measurement of organism-environment interaction.* Washington, D.C.: American Psychological Association.

Popenoe, P. (1922). Twins reared apart. *Journal of Heredity, 5*, 142-144.

Rabkin, J.C. & Struening, E.L. (1976). Life events, stress and illness. *Science, 194*, 1013-1020.

Rao, D.C., Morton, N.E., & Yee, S. (1974). Analysis of family resemblance. II. A linear model for familial correlation. *Am J Hum Genet, 26*, 331-359.

Rende, R.D., Plomin, R., & Vandenberg, S.G. (1990). Who discovered the twin method? *Behav Genet, 20*, 277-285.

Rose, R.J. (1981). Review of Farber, S. L. Identical twins reared apart: A reanalysis. *Science, 215*, 959-960.

Rowe, D. (1994). *The limits of family influence: Genes, experience, and behavior.* New York: Guilford Press.

Rowe, D. & Plomin, R. (1978). The Burt controversy: A comparison of Burt's data on IQ with data from other studies. *Behav Genet, 8*, 81-83.

Rowe, D.C., Clapp, M., & Wallis, J. (1987). Physical attractiveness and the personality resemblance of identical twins. *Behav Genet, 17*, 191-201.

Saudek, R. (1934). A British pair of identical twins reared apart. *Character and Personality, 3*, 17-39.

Scarr, S. & Weinberg, R.A. (1978). The influence of family background on intellectual attainment. *American Sociological Review, 43*, 674-692.

Schiff, M. & Lewontin, R. (1986). *Education and class: The irrelevance of IQ genetic studies*. Oxford: Clarendon Press.

Sheilds, J. (1962). *Monozygotic twins: Brought up apart and brought up together*. London: Oxford University Press.

Siemens, H. (1924). *Die Zwillingspathologies*. Berlin: Springer-Verlag.

Stephens, F.E. & Thompson, R.B. (1943). The case of Millan and George, identical twins reared apart. *Journal of Heredity, 34*, 109-114.

Tellegen, A., Lykken, D.T., Bouchard, T.J.,Jr., Wilcox, K.J., Segal, N.L., & Rich, S. (1988). Personality similarity in twins reared apart and together. *J Pers Soc Psychol, 54*, 1031-1039.

Townsend, M. (1874). Letter to Francis Galton. *Galton Archives in the Library of University College London, List #112/3*,

Urbach, P. (1974). Progress and degeneration in th "IQ debate" (II). *British Journal for the Philosophy of Science, 25*, 235-259.

Wachs, T.H. (1992). *The nature of nurture*. Newbury Park, CA: Sage.

West, P. (1991). Rethinking the health selection explanation for health inequalities. *Soc Sci Med, 32*, 373-384.

White, R.K. (1982). The relation between socioeconomic status and academic achievement. *Psychological Bulletin, 91*, 461-481.

Yates, N. & Brash, H. (1941). An investigation of the physical and mental characteristics of a pair of like twins reared apart from infancy. *Annals of Eugenics, 11*, 89-101.

CHAPTER 5

THE IOWA ADOPTION STUDIES: METHODS AND RESULTS*

William R. Yates, M.D.
Remi J. Cadoret, M.D.
Edward P. Troughton, B.S.

INTRODUCTION

One of the most important and pressing methodological issues in behavior genetics today is finding ways to correlate the extensive findings of molecular genetics with the epidemiology of psychiatric conditions. The problem grows larger with the passage of time and the exponential increase in information from molecular genetic studies such as the human genome project. The purpose of this chapter is to provide evidence that adoption or separation studies can play a significant role in correlating the clinical epidemiology of psychiatric conditions with the findings from human genome research. Adoption studies provide the opportunity to study several etiologic mechanisms including analysis for genetic factors (G), environmental factors (E), genetic and environmental interactions (G x E). Additionally, adoption studies provide the opportunity to investigate the heterogeneity of psychiatric disorders like alcoholism and drug abuse.

The scenario of the adoption study involves an adoptee who has been separated at some time in his/her life (ideally at birth) from the biologic home and family and raised in an adoptive home by adoptive, biologically-unrelated parents. Information about the biologic parents and the adoptive environment are correlated with adoptee outcome to determine which factor, the absentee biologic parent representing "heredity," or the adoptive home representing "environment" is more important in determining outcome.

---

* This work was supported by grant R01 DA05821 from the National Institute on Drug Abuse, Rockville, MD.

This approach was first suggested by Richardson (1912-3) to weigh the relative importance of heredity versus environment (nature versus nurture) as the cause of intelligence.

Genetic (G) Effects in Adoption Studies: The primary purpose of adoption studies has often been to examine the genetic effects for medical, psychiatric and behavioral disorders. Interest in the use of adoption studies was stimulated by the power of such an approach to evaluate for genetic effects. Genetic effects from a biologic parent will likely express a phenotype whether the child is raised by the biologic parent or an adoptive parent (assuming there is no G x E effect in expression of the phenotype). Phenotype is defined as "the external appearance or discernible characteristics of an organism resulting from the interaction of a genotype and it's environment" (Barnhart, 1986). A series of photographs of a person over a lifetime would represent the physical phenotype of an individual.

Suppose for illustrative purposes, a researcher was interested in whether blue eyes were determined by a genetic effect. A adoption sample paradigm could be used to identify a sample of adoptees with biologic parents who had blue eyes and a control adoptee sample with biologic parents without blue eyes. Adoptees would be assigned to adoptive families randomly (some with blue-eyed parents and some without blue-eyed parents). The researcher would examine the adoptees at some age in which the adoptee phenotype could be accurately assessed. Since eye color changes during the neonatal period this assessment should occur after infancy. Using the results, the researcher would compare the rates of blue eyes in the adoptees with a blue-eyed biologic parent to the rates of blue eyes in adoptees without blue-eyed biologic parents.

Adoption studies have been used to investigate the role of genetics in the genesis of psychiatric conditions such as alcoholism (Goodwin et. al., 1973, Goodwin et. al, 1974, Cadoret and Gath, 1978, Cadoret et. al., 1980, Cadoret et. al., 1985b, Cloninger, et. al. 1981, Bohman et. al., 1981, Cutrona et. al, 1994), drug abuse (Cadoret, et. al. 1986, Cadoret,R.J. et. al, 1995a), schizophrenia (Heston, 1966, Rosenthal and Kety, 1968, Rosenthal, et. al., 1971, Kety, et.al, 1971, Wender, et. al., 1974,), bipolar affection disorder (Mendlewicz, J. and Rainer, J., 1977), unipolar depression (Cadoret, R.J., 1978, von Knorring, A.L, et. al, 1983, Cadoret, R. J., et. al., 1985a, Ingraham, L.J. and Wender, P.H., 1992), the "schizophrenia spectrum" of schizotypal, paranoid conditions (Kety, et. al., 1978), antisocial personality (Crowe, R. R., 1974, Cadoret R.J., et.al., 1987, Cadoret, R.J., et. al. 1985b, Cadoret, R.J., et. al., 1990), criminality (Mednick, S.A., et. al., 1984, Bohman, M., et. al. 1982, Cloninger, C.R., et. al., 1982, Sigvardsson, S., et. al., 1982) and somatoform disorders (Sigvardson, S., et.al., 1984, Cloninger, C.R., et. al., 1984).

Environmental Effects in Adoption Studies: Although the initial impetus in psychiatric adoption studies was to demonstrate a genetic factor, the adoption design proved to also have a purpose for examining environmental effects. Suppose for an illustrative example that a researcher believed that smoking was due primarily to being raised in an environment where parental smoking occurred. An adoption paradigm could be utilized to examine this hypothesis. A group of adoptees with biologic parents who were smokers could be compared to a group of adoptees without smoking in their biologic parents. Adoptees from both groups would likely be placed in some homes with and without adoptive parents who smoked. The researcher's hypothesis in this example would be supported if adoptees who smoked were more likely to be from the homes with adoptive parent smoking. Genetic effects in smoking risk could concurrently be examined in this example.

The adoption paradigm and the Iowa Adoption Studies have been successful in identifying environmental effects for psychiatric and substance use disorders. Environmental factors have been found in such conditions as alcoholism (Cadoret, et. al., 1985b, Cloninger, et. al., 1981), drug abuse (Cadoret, et al, 1995a) criminality (Bohman, et. al, 1982), schizophrenia (Wender, et. al, 1974), and major depression (Cadoret, et. al, 1985a). This ability to detect environmental effects means that phenocopies (phenotype look-alikes which may be created by environmental influences) can be identified by their correlation with environmental factor(s) while controlling for genetic factors in the analysis. This ability to identify phenocopies can be put to use in genetic analyses of linkage and association as will be developed later in this chapter.

Genetic-Environmental Interaction Effects in Adoption Studies: Once the purpose of adoption studies expanded to environmental as well as genetic effects a third purpose for adoption studies emerged-- the elucidation of (G x E) interaction effects. Suppose for illustrative purposes, a researcher wanted to prove the hypothesis that alcoholism was due to having both a genetic risk to alcoholism *and* being raised in a home with a father who had alcoholism. This hypothesis could be studied using the adoption paradigm. Adoptees with and without a biologic parent with alcoholism would be placed in adoptive homes with and without an adoptive father who had an alcohol problem. The hypothesis would be supported if adoptee alcoholism rates were highest in adoptees with both a biologic parent with alcoholism and an adoptive father with alcohol problems. Three comparison groups would be available: adoptees with a biologic parent with alcoholism and without an adoptive father with alcohol problems, adoptees without a biologic parent with alcoholism and without a father with alcohol problems, and adoptees without a biologic parent with alcoholism and with a father with alcohol problems. Another example of G x E interaction in a medical illness is phenylketonuria. The clinical manifestation of the recessive gene would be masked by the absence in the diet of phenylalanine, but clinical phenylketonuria would appear when the environment (the diet) contained phenylalanine.

The adoption paradigm can detect G x E interactions provided that the environment effect can be specified. This ability would also be important in detecting correlations between genotype and behavioral outcomes in that the G x E effect can in some cases magnify the effect of a gene and so better detect its presence. G x E interaction occurs when the expression of a genetic effect is influenced by environmental factors.

G x E interaction effects in psychiatric and substance abuse disorders using the adoption paradigm have been elicited. Adoption studies demonstrating GxE interaction effects have been published for criminality (Cloninger, et. al. 1982), alcoholism (Cloninger, et. al, 1981, Cutrona, et. al., 1994), and for conduct disorder and aggressivity (Cadoret, et. al., in press). These studies demonstrate the importance of considering not only a genetic or an environmental model for psychiatric and substance use disorders. Both effects as well as the interaction effect are important issues to measure and examine statistically.

Heterogeneity and Adoption Studies: Adoption studies can clarify genetic and environmental factors in complex human behaviors such as substance abuse by examining issues of genetic heterogeneity. Heterogeneity is a term that is defined as being " made up of dissimilar elements or parts; unlike or varied in kind" (Barnhart, 1986). For the purpose of adoption and genetic studies, heterogeneity is limited to the issue of multiple genetic and environmental pathways to a particular phenotype. For example, if one examined all people with blond hair, this group could be consider heterogeneous. One part of the group would be those with natural blond hair (due to a genetic factor) and another would be those with artificial blond hair (an environmental factor). One might even be confused by a group with blond wigs. These last two groups could be considered phenocopies for the blond gene. If a researcher were trying to identify the gene for blond hair, including the last two groups would reduce the power to find the genetic site producing the phenotype.

There is a great deal of evidence showing that common conditions such as depression or substance abuse are clinically heterogeneous (Anonymous, 1993, Merikangas, et.al., 1994, Keller and Hanks, 1994). In some cases different clinical conditions occur in the same individual, thus contributing to clinical non-uniformity. An example of the latter is the commonly reported finding of a number of different psychiatric conditions in individuals diagnosed as alcoholic (Helzer, et al, 1991). The most common condition associated with alcoholism is antisocial personality (ASP) in both men and in women. The odds ratio for comorbidity of antisocial personality disorder given alcoholism is 12.0 for men and 29.6 for women. Other psychiatric conditions that have links to alcoholism include from the ECA data (Helzer, et al, 1991) (first number=Odds ratio for men, second number = odds ratio for women): mania (Odds ratio=6.5,9.3), drug abuse, dependence (OR =4.8, 8.80), schizophrenia (OR= 6, 5.6), panic disorder (OR=4.2, 4.4), obsessive compulsive disorder (OR=3.0, 2.1), dysthymia (OR=2.5, 2.2), major depression (OR= 2.4, 2.7), phobic disorders (OR=1.8, 2.1), any psychiatric diagnosis (OR=2.4,2.2). This frequent concurrence of psychiatric disorder with alcoholism demonstrates the difficulty in studying the specific genetic and environmental contributors to alcoholism. We shall show how adoption studies can dissect the relationship between ASP and substance abuse in terms of the genetic and environmental factors involved. The question of heterogeneity becomes important when the search is for causal factors (such as genetic or environmental). In the case of substance abuse the presence of clinical non-uniformity as defined above suggests the possibility that there could be more then one causal "pathway" to substance abuse, that is, heterogeneity of etiologic genetic factors as well. Such a possibility is very likely because many of these psychiatric conditions comorbid with substance abuse have been demonstrated to have genetic etiologies of their own.

Both family and adoption studies show separate and independent inheritance of factors etiologic both to ASP and to alcoholism (Cadoret, et. al, 1985a, Cadoret, et. al., 1987). The temporal development of ASP starting in childhood and leading through adolescent conduct disorder to adult ASP suggests that these behavioral disturbances lead to early use of addicting substances and the subsequent development of abuse/dependency. The predisposing role of ASP in alcoholism has led to the development of models of inheritance by two independent groups studying adoptees. Two genetic types of alcoholism have been proposed based upon analyses of the Swedish adoption sample(Cloninger, et. al, 1981). Type I alcoholism is characterized by a genetic diathesis for alcoholism which is influenced by environmental factors; in contrast, in Type II alcoholism, the genetic diathesis was predominant and did not appear to be influenced by environment. Evidence from the Swedish studies suggest that ASP alcoholics are most likely included in the Type II alcoholic genetic typology.

Using the Iowa adoption study a model of alcohol abuse has been developed based upon studies of adult adoptees from the state of Iowa which demonstrate two independent genetic "pathways" to alcoholism (Cadoret, et. al, 1978, 1980, 1985b, 1987): (1) a direct correlation between biologic parent alcoholism and adoptee alcoholism, and (2) a pathway from biologic parent ASP to ASP and thence to alcoholism. The Iowa studies have also described a third pathway which is independent of both genetic paths and involves a variety of environmental factors such as alcoholism or depression, and other psychiatric conditions in adoptive parents and siblings. Thus adoption studies have shown an ability to disentangle the tangled skein of causality in complex human behaviors by identifying the heterogeneous genetic (and environmental factors) which enter into such behaviors. This point will be developed more when we describe in more detail the results of the Iowa adoption studies of substance abuse.

Behavioral Disorders and Adoption Studies: Adoption studies can be applied to behavioral disorders as well as psychiatric disorders. A recent adoption study using the Iowa Adoption Study sample examined the genetic contribution to human obesity (Price, et. al., 1987). In this study, an index of body fat, the body mass index, was calculated for biologic parents, adoptive parents and adoptees. Other environmental effects that could affect eating behavior such as a disturbed adoptive home were also examined for an effect on adoptee obesity. This study found a strong relationship between biologic mother BMI and adoptee BMI. A similar genetic effect on human body fat has been found in family and twin studies supporting the validity of the Iowa Adoption Study sample. There was no evidence of a selective placement based on body weight from this study. The next section will detail the rationale of the adoption paradigm as well as practical considerations necessary to carry out a separation study.

RATIONALE AND PRACTICAL APPLICATION OF THE ADOPTION SEPARATION PARADIGM

Adoption Methodological Issues: Practical access to adoptees in the USA is accomplished through the social agencies which are involved in the adoption process as part of their mission. These are private agencies (often under the aegis of a religious organization) or public agencies which are part of state governments. In the USA adoption is regulated by state laws which generally require that confidentiality of biologic parents identity be maintained. Confidentiality is accomplished by restricting access to adoption documents. The problem of obtaining access to adoptees under these circumstances can be facilitated by collaboration with an adoption agency. Agencies usually have considerable documentation in their own files about adoptees' biologic background, and especially about adoptive parents. This information can be useful in delineating genetic and environmental factors. There are additional advantages in accessing adoptees through an agency. The adoption agency has the ability to contact adoptive parents to determine their willingness to cooperate in a study and can facilitate getting adoptee and adoptive parent permission to release their identity to outside researchers. Agencies can also cooperate with an institutional record search in which names of biologic parents are submitted to the state correction system, juvenile justice system, or state mental hospital system, etc. Such a search is easier because from their records an agency can prepare an alphabetized list of biologic parents for the research process. The record search can be done in such a way that individual biologic parent identity is kept from investigators. Institutional records of biological parents are released to research personnel only after removal of names or social security numbers from the files. In this approach, individual biological parent records are linked to an individual adoptee without discovery of the biological parent's identity.

Adoptions occur "independently" as well as through agencies. In Iowa in the first 6 months of 1993 there were 496 adoptions to non-related families out of a total of 1,239 cases (40%), and of these 496 cases, 138 were adopted through the public agency, 197 through licensed private agencies, and 158 were adopted to non-related individuals by the independent route (personal communication). The independent adoption is accomplished through an attorney who makes the legal arrangements. Adoption agencies are not involved in the placement process even though a social worker from an agency may perform the required home study (which can be done by a licensed social worker, not necessarily employed by a social agency). Independent adoptions are difficult to study because there are no centralized records as with an agency, and legal records of the adoption are sealed by law in the county where the adoption was finalized with access possible only with a court order. However, as can be seen from the figures given above, a substantial number of adoptees placed with non-relatives are available from agencies. Of these, the majority are placed at less than one month of age.

The adoption separation paradigm disrupts the usual link between an individual's genetic background from their environmental exposure. This is usually accomplished by separation of the biologic parents offspring from the biologic parents and their environment as close to birth as possible, thus limiting biologic environment effects to the prenatal period as much as possible. Figure 1 shows the various factors influencing the adoptee. We shall discuss each of these factors focusing upon their measurement. The separation between biologic background and adoptive environment is indicated by the wavy vertical line. If placement of an infant in an adoptive home is random, then the separation has been successful in keeping the biologic background independent from the adoptive environment. However, the adoption process as actually carried out can lead to confounding because children may not be assigned at random to adoptive homes.

Selective Placement Issues: Deliberate placement confounding can occur when characteristics of the biologic parents or their environment are used to determine where a child is placed. This process is usually termed "selective placement." An example of selective placement is the effort of some agencies to consider the child's developmental potential as a factor in placement, such that a child from poorly educated parents and an impoverished social background would be less likely to be placed in a very high socioeconomic home. Selective placement can be a problem in interpreting findings from a separation study when it confounds again what the separation was designed to avoid, as might occur if a child from biologic alcoholic parents were deliberately placed in an adoptive family where alcoholism was present in a parent. Selective placement influenced by knowledge of biologic factors can be controlled to a certain extent by statistical methods which will be discussed later.

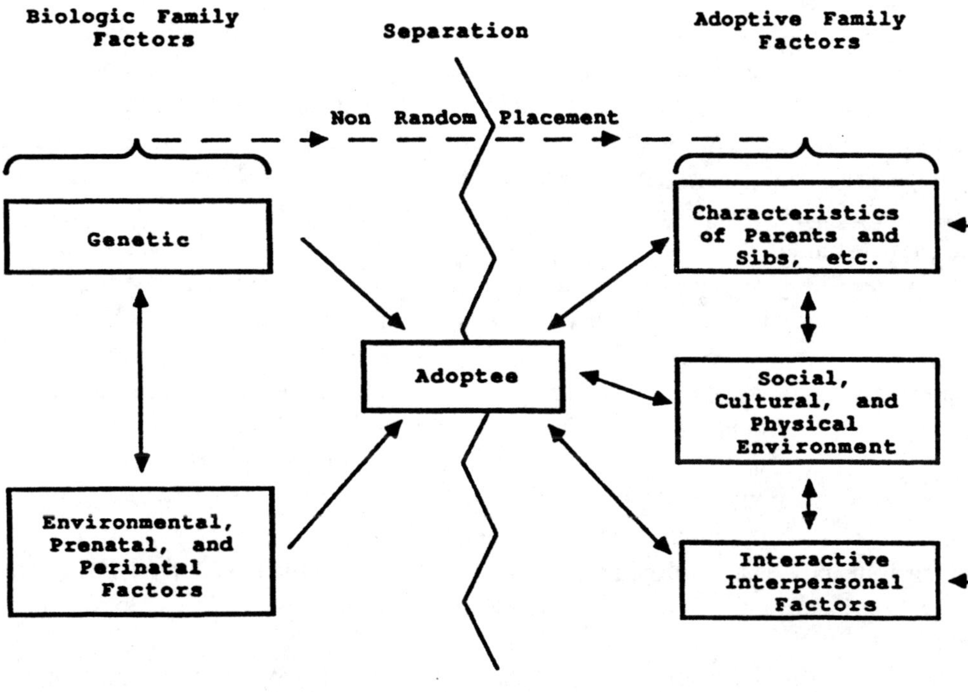

Figure 1. Adoption paradigm showing important genetic and environmental factors impinging on adoptee.

Genetic and Biological Factors: In Figure 1, the first box on the biologic family side represents genetic factors contributed to the adoptee by the mother and father. Genetic contributions can be defined by a variety of measures and sources; documents evidencing parental conditions which are hypothesized to be genetic such as alcoholism, depression, intelligence, etc. The presence of these conditions in biologic parents can be determined from records such as hospital, prison, juvenile justice, or adoption agency records. In the Iowa adoption studies, agency records were first used to determine biologic parent conditions, but the most recent study (Cadoret, et. al, 1995a) has used institution records (prison, hospital) to supplement agency records in identifying parental psychiatric disorders. This was accomplished by developing a method to maintain confidentiality. Adoption agencies submitted names of biologic parents directly to prison and mental hospital medical record personnel. Copies of medical records were then made and reviewed for names or social security numbers that could identify the biologic parents. Any identifying information was deleted or covered to protect confidentiality. Each parental record was then linked to an specific adoptee.

In some cases biologic parents can be interviewed and even tested at the time of giving up a child for adoption, thus allowing for a better and more specialized assessment of biologic parent conditions. This approach has been used by the Colorado Adoption Project (Plomin & Defries, 1985) which is designed to determine the inheritance of temperament and cognitive abilities. If biologic parents are available blood could even be drawn for molecular genetic analyses, however, as will be developed later, adoptee blood is also available for molecular genetic analyses, allowing association studies of candidate genes without the necessity of obtaining biologic parent DNA.

Separation does not remove the adoptee from many factors in the biologic environment which are of import to the social and physical development of the adoptee. Prenatal variables such as maternal drinking, smoking, poor prenatal care, diet, medications, illegal drugs, can influence the health of the future adoptee. In the most recent Iowa adoption study evidence was found that mothers who drank during their pregnancy gave birth to infants who were significantly lower in birth weight and were more likely as adolescents to manifest symptoms of conduct disorder (Cadoret, et. al, 1995b). Birth and delivery add their own risks, though these are generally documented by adoption agencies that are concerned by evidence of possible birth complications and generally are cautious about placing children until it is evident that a birth problem has been shown to be without significant sequelae. Infants who are injured at birth or demonstrate obvious developmental defects often may not be adopted but placed in foster care indefinitely. Such individuals could still be followed as part of a separation study.

In recent years, adoption agencies place newborns in permanent adoptive homes at the time of discharge from the hospital newborn nursery, thus minimizing contact with biologic parents. However, many infants may be given up by biologic parents later in life, as in situations where a biologic mother is unable to cope with the responsibility of caring for a child. Infants who remain with a biologic parent or relative for weeks, months or years, are exposed necessarily to that environment, making it difficult to draw conclusions as to the importance of genetic versus environmental effects in the case of later life separation and adoption. Our Iowa studies have demonstrated that the social milieu during the first year of life is associated with significant behavior deviations later in life (Cadoret, et. al, 1990).

Environmental Factors: In the not too distant past with American adoptions it was popular to place newborns into temporary foster homes at birth until an adoptive family had been selected, paper work done, and an opportunity made to observe the infant to make sure that normal development was occurring. As noted above, this is no longer the policy of most adoption agencies and infants are generally placed without delay in the adoptive home. If delays have occurred, such as placement in foster care pending legal work, etc., this can be factored into the analysis of results. For example, in the early 1960's an Iowa adoption agency boarded a number of newborns with a university home economics department pending adoption. During their 3-6 months stay in foster care at the university facility, the infants were used in a home management program to teach young women how to care for an infant. This involved daily contact with each infant by a number of different student caregivers, although nighttime care was the responsibility of one individual. These infants were placed in permanent adoptive homes when they were about 6 months old. Studies were done comparing these infants with control adoptees who had been placed at birth, and no differences in socialization or adjustment reported through the 8th year of life. However, when examined as adults, these infants exposed to the discontinuous mothering at the home management situation were found to have increased antisocial behaviors (Cadoret,et. al, 1983) which manifested during adolescence, and as adults reported more psychiatric symptoms of an affective nature (Cadoret, et. al, 1990). As a general rule, adoption agency records detail birth, delivery, and neonatal conditions in some detail. Prenatal conditions are more likely to be less well documented, in large part because of a tendency for many mothers of these infants to seek medical care late in their pregnancy or only at time of delivery.

Infants are placed by agencies into carefully researched adoptive homes. Homes and adoptive family are required to meet certain standards which can result in a truncation of the full range of variability that would occur in a population. For example, in the Iowa studies, adoptive homes in the highest and lowest socioeconomic are less frequently represented in adoption studies than one would expect from their frequency in the general population. Adoptive family selection also can remove parents with certain psychopathology. For example, in the Iowa studies antisocial personality disorder in an adoptive parent has never been diagnosed. Conditions such as depression and anxiety occur at rates which are similar to those found in the general population (Myers,et. al., 1984). Although the selection results in a more stable home for the adoptee (for example, divorce rates are generally much lower than the general population) the truncation of environmental variability can result in a systematic underestimation of the importance of the environment in determining adoptee outcome.

On the adoptive family side of Figure 1, personal characteristics of parents and siblings in the adoptive home can be determined in a variety of ways. The fact that adoptive families are usually available in person for study facilitates the amount and quality of information available about the environment. Sources of information vary from adoption agency records made at the time of adoption to contemporary personal interviews of family members which can include structured psychiatric instruments to measure individual characteristics such as personality, temperament, and psychopathology. Other instruments can be given adoptive family members (including the adoptee) to determine the social, cultural, and physical environment. Interaction of the adoptee with the environment can be assessed by a wide variety of instruments, including direct observation of modes of interacting with the adoptive family. Adoptees outcome can be measured by interviews or questionnaires given to family members, teachers, spouse, and other informants, and by personal interview of the adoptee using structured instruments to measure the condition of interest. School and hospital records can add to the information about adoptee adjustment. As in any study of humans, consideration must be taken of the subject burden associated with being administered a large number of instruments and the effect of the resulting fatigue on subject cooperation and validity of information. One approach to the problem of subject fatigue is to split the tasks up in time, as by sending questionnaires to subjects for filling out prior to a personal interview.

Shared versus Non-shared Environmental Effects: Recent discussion of environmental factors (Plomin & Daniels, 1987, Gatz, et.al., 1992, Bergeman, et. al., 1993,Plomin, et. al, 1994,) has emphasized the importance of distinguishing between environmental factors which are shared by all the individuals in a family (such as the size of the family home) and non-shared or unique factors (such as differential treatment of siblings by parents). Techniques for assessing the importance of shared versus non-shared environmental factors have been developed in analyses of twin data (Pickens, et. al. 1991). In intact families shared and non-shared factors can be measured by establishing differential exposure of two or more sibs to environmental factors. In most adoption studies, including the Iowa studies, a single adoptee per adoptive family is the focus of the analysis. Thus, environmental factors which are described from this type of analysis could be either shared or non-shared. The methodology when one child only is studied does not allow a distinction between shared and non-shared but does identify a specific environmental factor. However, methodology could be introduced which could detect differential treatment of children in adoptive families where sibs are present. Adoptees and sibs in the family and adoptive parents could be administered instruments to measure specific and nonspecific environmental factors. Interactional observations of the family could also be used to establish differential reactions between parents and sibs.

Summary: The availability of the adoptive family for study allows the use of a wide variety of methods and instruments to obtain information relevant to the effect of environment. The principal limitation imposed on the environmental assessment is the retrospective nature of data about conditions prior to the study period. There are even more limitations imposed by the retrospective nature of information obtained about biologic parents and biologic family environmental conditions during pregnancy. The availability of prison, hospital, and juvenile justice records improves the behavioral diagnosis of biologic parents; however, the fact that a parent is not found in an institutional record search does not mean that the parent is "clean" behaviorally since very many individuals with significant psychiatric illness (including substance abuse) never present for treatment. There is also the additional fact that a biologic parent could be treated only at a private facility and a search of public institutions would not reveal the fact of this treatment. This caveat applies mainly to the behavioral "cleanness" of any comparison or control group which might be selected from the ranks of adoptees whose biologic parents failed to appear at a public institution.

TECHNIQUES OF ANALYSIS OF ADOPTION-SEPARATION DATA

The most important consideration in the analysis of adoption-separation data is the acknowledgment of the multifactorial causes of human behavior. The analysis and its interpretation must consider not only many genetic and environmental variables but also the many factors which are confounded with the variables of interest. Fortunately, in recent years a number of appropriate multivariate techniques are available which can be used to disentangle factors statistically and control for identified confounds.

One of the most frequently used multivariate techniques to disentangle genetic from environmental effects is log-linear model building (Bishop, et. al., 1975, Fienberg, 1980). This method allows the use of discrete, usually dichotomous, variables such as psychiatric diagnoses. Both genetic and environmental factors can be put into these models. One of the advantages of log-linear analysis is the ability to control for selective placement which has been based upon knowledge about biologic parent and potential adoptee home which is available to the agency at the time of placement. For example, information about a biologic parent in prison for a felony conviction could lead an agency to select a lower socioeconomic (SES) adoptive home for his offspring. The relationship between adoptive home SES and the biologic factor of criminality can be forced into the log-linear model to control for such a selective placement factor. The analysis can determine the significance of such a factor as well as the effect of this factor on remaining relationships in the model. Although the log-linear model can detect genetic and environmental effects, gene-environment effects are more difficult to assess because of the logarithmic scaling used in the analysis. The concept of gene-environment interaction holds that interaction represents a non-linear result when adding together arithmetically the genetic and environmental effects. What is non-linear on an arithmetic scale may be linear on a logistic scale. Thus the log-linear analysis may not detect all interactions on an arithmetic scale but only those which deviate markedly from arithmetic scale linearity.

Another disadvantage of the log-linear analysis is the loss of information inherent in dichotomizing variables which are measured on an ordinal, interval, or ratio scale as opposed to the nominal measurement of diagnosis. For example, diagnosis of substance abuse such as alcoholism can be "quantitated" by constructing a symptom count of behaviors consistent with a diagnosis of alcoholism such as an excessive amount of alcohol consumed/day or other time period and could include other symptoms such as psychosocial complications due to alcohol, or medical problems related to alcohol. The number of such symptoms could provide a measure of severity as well. In constructing log-linear models a rule of thumb suggests that for every factor in the model there should be fifteen observations. This limits model making considerably. The use of techniques which are more powerful in utilizing more of the information in the system would get around this limitation. Models could be constructed using some variant of path analysis such as LISREL (Cardon, et.al., 1991) and using latent variables (Melton, et. al. , 1994, Farone & Tsuang, 1994). Additional types of multivariate approaches have been used to elucidate genetic and environmental effects such as the use of discriminate function analysis and canonical correlations in the Swedish adoption studies of alcoholism and criminality (Bohman et al., 1982; Cloninger et al., 1982; Sigvardsson et al., 1982).

SPECIFIC TYPES OF SEPARATION-ADOPTION PARADIGMS

It is possible to classify separation designs on the basis of two characteristics: (1) the completeness of information on both the genetic background and the adoptive environment, and (2) whether the adoptee or the biologic parent is the starting point for the study.

Rosenthal (1970) classified adoption designs according to the second point (above). If the starting point were the biologic parent he termed this the "adoptees' study method." From the biologic parent starting point, the investigation followed up the adoptee and the adoptive environment. When the starting point was the adoptee, Rosenthal named this design the "adoptees' family method," and the follow-through investigated the biologic and adoptive parents. Both these designs are "complete" in that information is obtained about both biologic and adoptive families (and, of course, the adoptee outcome). Incomplete designs can be used to advantage in certain situations. Because of the remoteness of biologic parents, the most common incomplete design is one with information about adoptive families and the adoptee. Studies of I.Q. have used this approach by determining correlations between I.Q. of adoptees and that of their adoptive parents, and comparing these correlations to those between parents' I.Q. and the I.Q. of their biologic offspring (Scarr and Weinberg, 1981; Speer, 1940).

Table 1 shows some of the main characteristics of the separation paradigms: their starting points and information available. In doing studies there is usually a hypothesis to be tested and a comparison group is used to detect differences. In the adoptees study method typically a high-risk strategy is employed to test the hypothesized genetic factor by comparing adoptees with a contrasting biologic background (usually one without psychopathology). In the adoptees family method a high contrast strategy is usually used by comparing the adoptee with pathology to a control without pathology. Both designs have potential biases introduced by the selection of adoptees, or of the biologic parent. Possible biases associated with the adoptee or biologic parent selection appear in Table 1. The "incomplete" designs shown in Table 1 are occasionally found in the adoption literature and are usually characterized by missing information about biologic background.

Table 1

ADOPTION-SEPARATION PARADIGMS

	Complete Designs		Incomplete Designs	
	Adoptees' Study Method	Adoptees' Family Method	Adoptees' Family Method	Separated Twin Study
Starting Point	biologic parent	adoptee	adoptee	adopted away twins (MZ and DZ)
Information Available on Biologic Parent	+	+	0	0
Adoptive Home	+	+	+	+
Comparison Group	(high risk strategy) Adoptees from contrasting biologic background ("normal" or different Pathology unrelated)	(high contrast strategy) normal or contrasting adoptees	biologic offspring of same or matched parents	contrasts between MZ and DZ concordance
Possible Biases	"normal" biologic background may have undetected pathology	environmental factors could bring adoptee into study (phenocopy)	environmental factors could bring adoptee into study (phenocopy)	recruiting techniques could lead to selection based on outstanding concordances
	biologic parent with pathology could be extreme and unrepresentative case	comorbid conditions could lead to increased chance of inclusion in study	comorbid conditions could lead to increased chance of inclusion in study	twins selected for pathology could introduce biases similar to those under 1st and 3rd columns of this table.
	biologic parent could have comorbid condition			

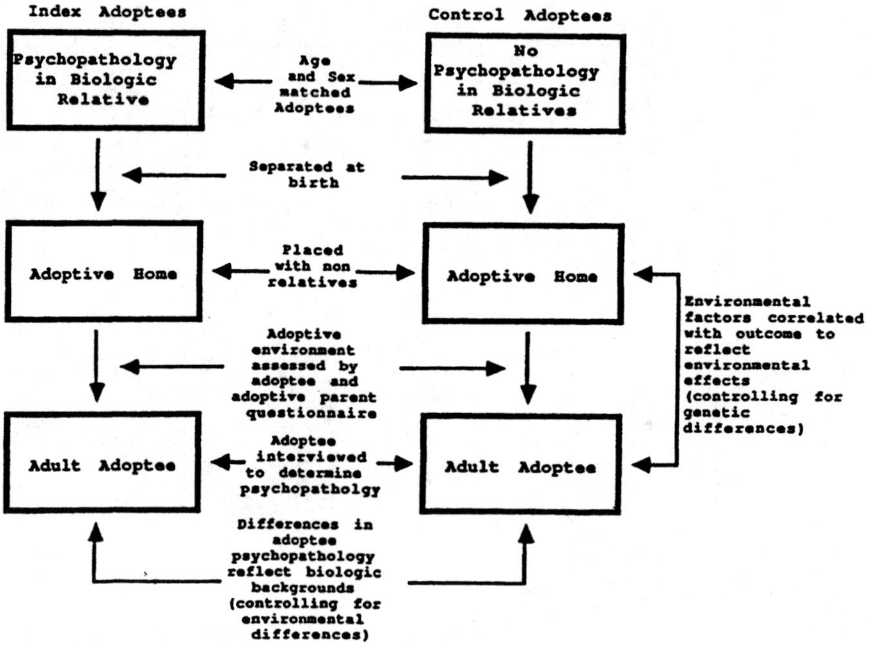

Figure 2. Adoption study paradigm used in Iowa Studies

In the remainder of this section the adoptees' study method used in the Iowa Adoption Studies will be detailed to provide a practical example of the application of this paradigm. Figure 2 shows the paradigm. The starting point for the high risk group of adoptees is the biologic parent with psychopathology (with a psychiatric condition hypothesized to be etiologic in adoptee outcome). The offspring is separated at birth and placed in an adoptive environment - ideally as close to a random placement as possible. Comparison adoptees are selected from the same agency and matched to the high risk adoptees by age, sex, and age of biologic mother. To maximize the high-risk paradigm the comparison group should not have a genetic diathesis that would lead to the pathologic outcome hypothesized for the high risk group. In practical terms this usually is accomplished by picking as "clean" a biologic parentage as possible. As indicated above, biologic parents with an apparently "clean" retrospective record might still have significant psychopathology both treated and untreated. In cultures where there is centralized record-keeping (as in Scandinavian countries) and record linkages are practical, a clean record may mean more. However, in the USA confidentiality rules and impractical or nonexistent record linkage may preclude a lifetime search of an individual's psychopathology.

There are ways to "clean" control groups that are of practical use. For example, alcoholics in a control group of biologic parents might be objectionable in an experimental design. To find and eliminate alcoholics the names of control parents (if available) could be compared to death certificates indicating probable alcoholism as a factor in death (such as Laennec's cirrhosis).

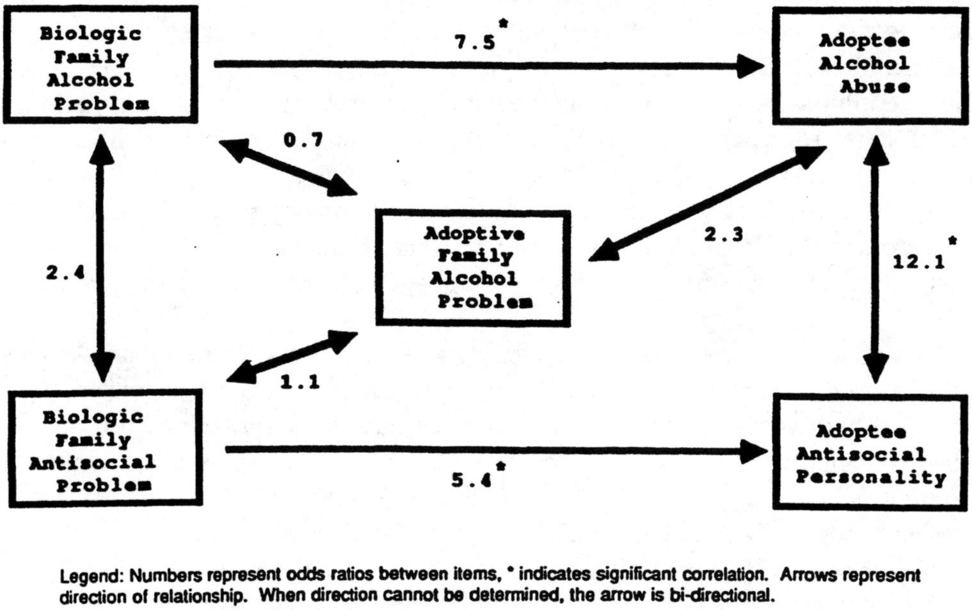
Biologic Family Alcohol Problem
7.5*
Adoptee Alcohol Abuse
0.7
2.4
Adoptive Family Alcohol Problem
2.3
12.1*
1.1
Biologic Family Antisocial Problem
5.4*
Adoptee Antisocial Personality
Legend: Numbers represent odds ratios between items, * indicates significant correlation. Arrows represent direction of relationship. When direction cannot be determined, the arrow is bi-directional.

Figure 3. Interaction diagram for ICFS males.

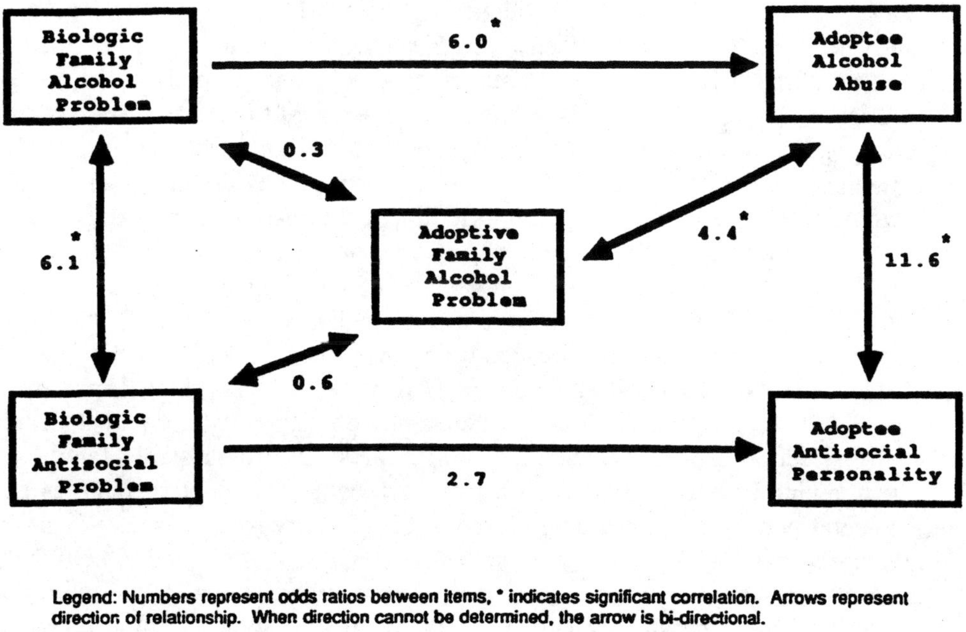
Biologic Family Alcohol Problem
6.0*
Adoptee Alcohol Abuse
0.3
6.1*
Adoptive Family Alcohol Problem
4.4*
11.6*
0.6
Biologic Family Antisocial Problem
2.7
Adoptee Antisocial Personality
Legend: Numbers represent odds ratios between items, * indicates significant correlation. Arrows represent direction of relationship. When direction cannot be determined, the arrow is bi-directional.

Figure 4. Interaction diagram for LSS males.

In the Iowa studies adoption agency records were used initially to determine biologic parent psychopathology. However, the most recent studies (Cadoret, et. al., 1995a) have used institutional records to make biologic parent diagnoses. The use of medical and prison records allowed us to check the accuracy of control and high risk designations in a previous study which used only adoption records to determine membership in a high risk group: a biologic parent with alcohol problem. In the high risk group of 50 adoptees as determined from adoption agency records, 10 parents turned up in institutional records with treatment for alcohol abuse/dependency. None of the control male adoptees had a parent with a record of treatment. This finding suggests that adoption agency records can be a source of valid information.

The adoptive environment is assessed by a variety of instruments which are administered to adoptive parents and adoptees. Adoptive parent characteristics are determined by a structured psychiatric interview and by information from the other spouse and adoptee. The adoptee outcome is determined by a personally administered structured psychiatric instrument (the DIS-Robins, et. al, 1981)) and supplemented by information about development, socialization, etc., from adoptive parents.

As can be seen in Figure 1, differences in outcome are a measure of the effect of the hypothesized genetic factor. Correlations between the adoptive environment and outcome for each risk group measure the effect of factors in the environment on outcome. For each environmental factor there are two correlations with a particular output: one for the high risk group, the second for the low risk group. If the two correlations are similar there is no evidence of interaction; if the correlations are significantly different, then this is evidence of gene environment interaction (defined as a different effect of the environment depending on the genetic background).

As mentioned above, environments in adoptive homes tend to be less extreme than in the general population. However, some psychosocial conditions with the potential for significant impact upon adoptees can be found in adoptive environments. In the most recent Iowa Adoption study, rates of psychiatric illness and marital problems in adoptive parents were: alcoholism-17.8%, depression, 38.0%, anxiety-36.0%, marital problem-20.3%, and divorce-8.1%. The lifetime prevalence of common psychiatric conditions such as alcoholism and depression is as high or higher than ECA reported data (Myers, et.al, 1984). These parental psychiatric conditions have been associated with adoptee outcomes of conduct disorder and aggressivity (Cadoret, et., al., in press). However, the number of divorces is only 8%, well under the population figure and consistent with the fact that the selection process for adoptive homes picks out more stable families. In addition to showing increased stability, other characteristics set adoptive families apart from the general population. Since demonstration of infertility (often a pre-requisite for adopting a newborn) takes a number of years, adoptive parents often tend to be older. In part increased age may account for their increased stability. In the Iowa studies the adoptive families are generally over-represented by middle socioeconomic class and underrepresented by lowest socioeconomic class. That there is some selective placement based upon social class can be demonstrated by the most recent study (Cadoret, et. al, 1995a) where a small but significant regression was found between biologic family SES and adoptive family SES. In biologic families, low SES is more correlated with high genetic risk. In the latest study biologic parents with antisocial personality, alcohol abuse/dependency, or other psychiatric problems were significantly more likely to represent lower SES backgrounds. This fact provides an excellent rationale for the forced inclusion in the log-linear analyses of biologic parent diagnosis-adoptive environmental factor relationships to control for selective placement.

THE IOWA ADOPTION STUDIES: ADOPTEE SUBSTANCE ABUSE

This section will review the sample characteristics and outcomes for the main finding for adoptee substance abuse using the Iowa Adoption Study sample.

Sample Characteristics: The Iowa Adoption Studies have taken place over 20 years and include samples from three primary groups (Table 2). In total, 643 adoptees have been studied with focus on the genetic, environmental and interaction effects in alcoholism, drug abuse, antisocial personality and major depression. Table 2 highlights the sample characteristics for the Iowa Adoption Studies samples. Over the study period samples have been recruited from a public agency--the Iowa Children and Family Services and two agencies with a religious link--Lutheran Social Services and Catholic Charities. The Lutheran Social Service sample was selected for both an antisocial target phenotype and an alcoholism phenotype. The Catholic Charities sample originated from three separate sites in Iowa. Adoptees were usually identified and recruited for participation when they were in young adulthood. Case and control adoptees were identified based on biological parent information. Case adoptees had evidence from adoption agency or other records of an alcoholism or antisocial behavior problem. Controls without evidence of biological parent disorder were identified using similar record information.

Sample recruitment has targeted similar numbers of men and women to examine the effect of adoptee gender on the development of target phenotypes. Rates for psychopathology in adoptees have been high consistent with the use a high-risk strategy.

Methodological Strategies: Table 3 outlines the key methods used in the Iowa Adoption Studies over the various adoptee sample groups. The studies span nearly two decades. Diagnostic classification systems have changed dramatically over this time. Despite this, the Iowa Adoption Study has consistently utilized structured adoptee psychiatric diagnostic interviews with diagnosis based on standard criteria such as DSM-III (American Psychiatric Association, 1980) DSM-IIIR (American Psychiatric Association, 1987) and the Feighner criteria (Feighner et al, 1972).

Adoptees and their adoptive parents have been interviewed in person by research assistants blind to the adoptee biologic parent status. Structured interviews using the Diagnostic Interview Schedule (DIS) have been most frequently used (Robins, et. al, 1981 and 1989) although a for a minority of adoptees, the SADS-L interview was utilized (Spitzer & Endicott, 1979). Agency adoption records have provided the majority of information about biologic parent background, although as previously discussed the most recent adoptee biologic parents were checked against statewide mental health and correctional records.

Table 2. Iowa Adoption Study: Adoptee and Biologic Parent Characteristics

	Iowa Children and Family Services (ICFS)	Lutheran Social Services ASP Study (LSSAS)	Lutheran Social Services ALC Study (LSSALC)	Catholic Charities (CC)
Adoptee Sample				
Total N	152	83	108	300
Years of Data Collection	1975-1977	1978-1979	1980-1981	1986-1989
Males (%)	133 (53%)	55 (66%)	53 (49%)	140 (47%)
Age Range (in years)	18-40	18-39	18-40	17-48
Age Mean (S.D.)	23.5 (5.6)	25.0 (6.5)	25.9 (6.2)	27.1 (7.3)
Biologic Parent Sample				
Alcoholism	13 (5%)	9 (17%)	18 (17%)	42 (14%)
Antisocial Personality	16 (6%)	36 (43%)	12 (11%)	90 (30%)
Other Psychopathology	39 (15%)	3 (4%)	13 (12%)	20 (7%)
Any Psychopathology	76 (30%)	37 (45%)	36 (33%)	155 (52%)

Table 3. Iowa Adoption Studies Methodology by Adoptee Agency

	Iowa Children and Family Services (ICFS)	Lutheran Social Services ASP Study (LSSAS)	Lutheran Social Services ALC Study (LSSALC)	Catholic Charities (CC)	NIDA-DHS Hillcrest (NIDA)
Target Phenotype	Any psychopathology Alcoholism, ASP, Depressionand MR	Antisocial Personality	Alcoholism	Alcoholism	Drug Abuse Alcoholism ASP
Adoptee Interview	DIS	SADS-L	DIS	DIS	DIS
Diagnostic Criteria	Feighner	DSM-III	DSM-III	DSM-IIIR	DSM-IIIR
Adoptive Parent Interview	Feighner	SADS-L	DIS	DIS	DIS
Biologic Parent Data	Adoption Agency	Adoption Agency	Adoption Agency	Adoption Agency	Adoption Agency Correctional and Mental Health Facility Records
Statistical Analysis Model	Multiple Regression	Log-linear Modeling	Log-linear Modeling	Log-linear Modeling	Log-linear Modeling

The primary method of analysis for examination of genetic, environmental and interaction effects has been log-linear modeling (Bishop et. al., 1975, Fienberg, 1980) although regression analysis has been employed in early studies. Commonly available statistical packages such as BMDP4F (Dixon, 1981) provide log-linear models that can be applied to adoption data. The objective in the log linear analysis is to define an equation that predicts the log of the frequency of occurrence in each cell of a multidimensional contingency table. The general procedure for BMDP4F is : (1) configurations representing "independent" variables (genetic or environmental factors) were included to condition the analysis on these variables, (2) dependent variables (such as adoptee alcoholism, drug abuse or antisocial personality disorder) higher-order effects are added by using the ADD=MULTIPLE options. This provides a method to determine potential significant interactions, (3) the selection of terms in the model was stopped when no other term had P values less than .05, as calculated from the differences in the likelihood chi-square statistic, (4) the DELETE=SIMPLE option was used on the model obtained in step 3 to delete unnecessary higher-order or interaction terms, (5) estimates of cell frequency and terms to estimate odds rations were obtained from the final model. Results are then presented as odds ratios used to describe relationships between two nominal variables such as an environmental condition and an adult diagnosis.

Overview of the Iowa Adoption Studies Results: The Iowa Adoption Study results have been summarized in Table (4). This table identifies significant relationships between genetic, environmental and GXE interaction effects for various adoptee diagnoses or disorders. Significant results are identified by adoption sample studied. Some analyses have combined samples to search for specific effects. Some analyses have looked at only men, some at only women, and some have combined the genders for analysis. Combining samples can increase the power to detect significant differences.

Genetic Factors: The table demonstrates the several effects of a biologic parent with alcoholism. The LSS sample and ICFS sample both found adoptee alcoholism associated with biologic parent alcoholism--the LSS sample found this association for both genders, the ICFS for males only. Additionally, a biologic parent with alcoholism has been linked to adolescent and adult antisocial behavior as well as drug abuse. The combined sample has demonstrated two pathways to alcoholism: one pathway to adoptee drug abuse travels through a biologic parent with alcoholism (Cadoret, et. al, 1995a).

Genetic effects via a biologic parent with antisocial personality disorder have also been demonstrated. An antisocial biologic parent predicts adoptee adult antisocial behavior and adolescent antisocial behavior for women. The combined sample second pathway to drug abuse start with a biologic parent with antisocial behavior (Table 4).

Environmental Factors: Various environmental factors have been demonstrated to influence risk of adoptee diagnosis for various substance abuse and depressive conditions. Environmental factors studied have included adoptive parent psychopathology, adoptive sib psychopathology, socioeconomic status, death of an adoptive parent, a drinking or behavior problem in the adoptive home, adoptive parent divorce, and a combined adverse environmental effect that combines several adverse effects for a summary "bad home" variable.

Table 4. Summary of Major Findings from Iowa Adoption Studies by Adoption Sample

	Adoptee Diagnosis/Disorder				
	Alcoholism	Adult Antisocial Behavior	Adolescent Antisocial Behavior	Drug Abuse	Depression
Genetic Factors (G)					
Biologic Parent Alcoholism	LSS (B) ICFS (M)	ICFS+LSS(M) ICFS (M)	ICFS(M)	ICFS+LSS(B) Total (M)	
Biologic Parent Antisocial		LSS (M) ICFS (B) ICFS+LSS(M) Combined (B)	ICFS(F)	ICFS+LSS(B) Total (M)	
Environmental Factors (E)					
Adoptive Home Environment					
Drinking Problem	ICFS (M)				ICFS+LSS (M)
Behavior Problem					ICFS+LSS(F)
Death of Adoptive Parent				ICFS+LSS(B)	ICFS+LSS(F)
Psychiatric Disturbance				ICFS+LSS(B)	ICFS(B)
Divorce					
Combined Adverse Environment			ICFS(M) Combined(B)		
Gene x Environment Interaction Factors (GxE)					
Biologic parent alcoholism x adoptive home conflict or psycopathology		CC (F)			
Biologic parent affective disorder and later age of adoptive placement					ICFS+LSS(M)
Biologic parent alcoholism or antisocial x adverse environment			ICFS(B)		
Biologic parent alcoholism x disturbed adoptive parent					Combined(F)

ICFS = Iowa Children and Family Services, LSS = Lutheran Social Services, CC = Catholic Charities. M = male adoptees, F = Female adoptees, B = Both.

Table 4 demonstrates that environmental effects have been demonstrated more for depression than for alcoholism. The ICFS sample demonstrated a male adoptee risk for alcoholism when an adoptive home drinking problem was present. Despite this finding, there is limited support in the Iowa Adoption samples to support a large environmental influence on risk to the development of alcoholism. Adoptee drug abuse did appear to be associated with death of an adoptive parent and adoptive parental divorce. More individual environmental effects have been demonstrated for depression in both men and women.

Gene X Environmental Factors: Several gene x environment interaction effects have been noted in the Iowa Adoption studies. Adoptee adolescent antisocial behavior increases with a combination of biologic parent alcoholism or antisocial behavior and an adverse adoptive environmental experience. Biologic parent affective symptoms and later age of adoptions predicts male adoptee depression in the ICFS and LSS sample. Using the combined Iowa adoption sample, female adoptee alcoholism is linked an interaction between biologic parent alcoholism and adoptive home conflict or adoptive home psychopathology. The combined sample also demonstrated a link between female adoptee depression and an interaction between biologic parent alcoholism and a disturbed adoptive parent.

Multifactorial Etiologic Models: As indicated in this chapter's introduction, the adoption paradigm can be used to sort out the tangled causality leading to complex human behaviors such as substance abuse. Early in the 1980's Bohman and his group (1981) suggested on the basis of Swedish adoption data that not only was alcohol genetically inherited but that there were two kinds of genetic inheritance.

This model of alcoholic inheritance received further support from the LSS and ICFS sample. The Iowa Adoption Studies model of alcoholism is explored in further detail in figures 3 and 4. The interaction diagrams of results from both agencies represent best fitting models and are remarkably similar to each other. Each indicates at least 3 independent "paths" to alcohol abuse/dependence. One is a direct path from alcohol abusing biologic parents. The second is indirect and begins with an antisocial parent which predisposes to adult antisocial personality (ASP), which itself predisposes to alcohol abuse/dependency. The adoptee ASP alcohol abuse relationship is very strong and the causal direction is most likely from ASP to alcohol abuse. This interpretation is supported by conduct disorder chronically preceding adult ASP.

The third influence on alcohol abuse shown in these analyses is an environmental factor: an alcohol problem in an adoptive family member increased the likelihood of adoptee alcohol abuse. This environmental factor was independent of the two genetic factors and free to operate on adoptees from either genetic background.

This model of alcohol abuse suggested a hypothesis of genetic and environmental etiology of drug abuse/dependency, a condition which is commonly associated with the antisocial personality diagnosis. The LSS and ICFS data sets were combined and it was found that adoptee drug abuse/dependency correlated very strongly with ASP in the adoptee. However, a subgroup of non-antisocial adoptees were found who were nevertheless drug abusers and whose biologic parentage was that of alcoholism (Cadoret,et. al, 1986). This analysis suggested the hypothesis that drug abuse was caused by two genetically transmitted factors: one starting with ASP parents and leading through antisocial personality to drug abuse; and the second genetic factor starting with alcoholic biologic parents and leading to drug abuse (not through antisocial personality).

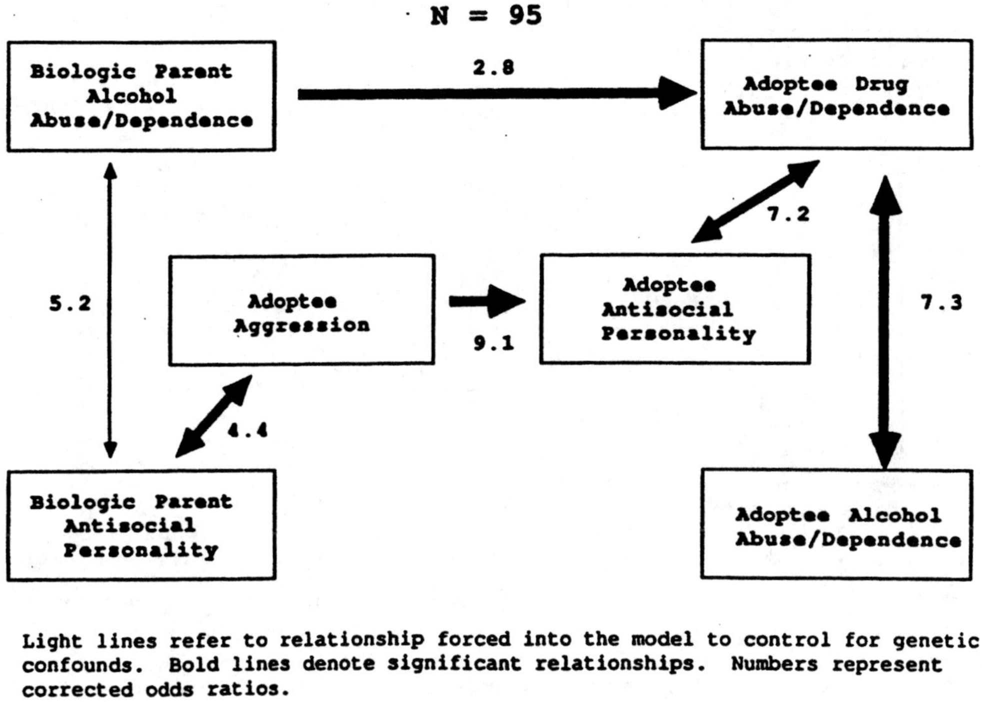

Figure 5. Interaction diagram for male model with adoptee aggrssion as intervening variable

The most recent Iowa adoption study tested this hypothesis of drug abuse etiology using the improved methodology of obtaining prison and hospital records of biologic parents so that more valid diagnoses of biologic parents were possible (Cadoret, et. al, 1995a). The results of this study are shown as interaction diagrams in Figure 5 for males and Figure 6 for females. The log-linear analyses confirm the two-factor genetic hypothesis in male adoptees (Figure 5) with a direct effect of alcoholic biologic parent upon drug abuse/dependency, and an indirect pathway to drug abuse starting with ASP in a biologic parent with intervening (developmental) conditions of aggressivity and conduct disorder. The model for females (Figure 6) is very similar with an indirect path to drug abuse starting with biologic ASP parents leading through aggressivity and conduct disorder to drug abuse/dependency. Although the log-linear model in Figure 6 does not show a direct effect upon adoptee drug abuse of an alcoholic biologic parent, one is present in the form of a gene-environment interaction. Previous work with female adoptees had shown that alcohol abuse was more likely to develop in an adoptee with a biologic background of alcoholism if the environment was seen by the subject as providing poor social support (Cutrona, et. al., 1994). In these new data a similar interaction effect was found for females but related to drug abuse. Given a biologic background of alcoholism a female adoptee was significantly more likely to be a drug abuser if poor social support was part of the environmental picture. Thus the female model for drug abuse consists of two genetic effect as well.

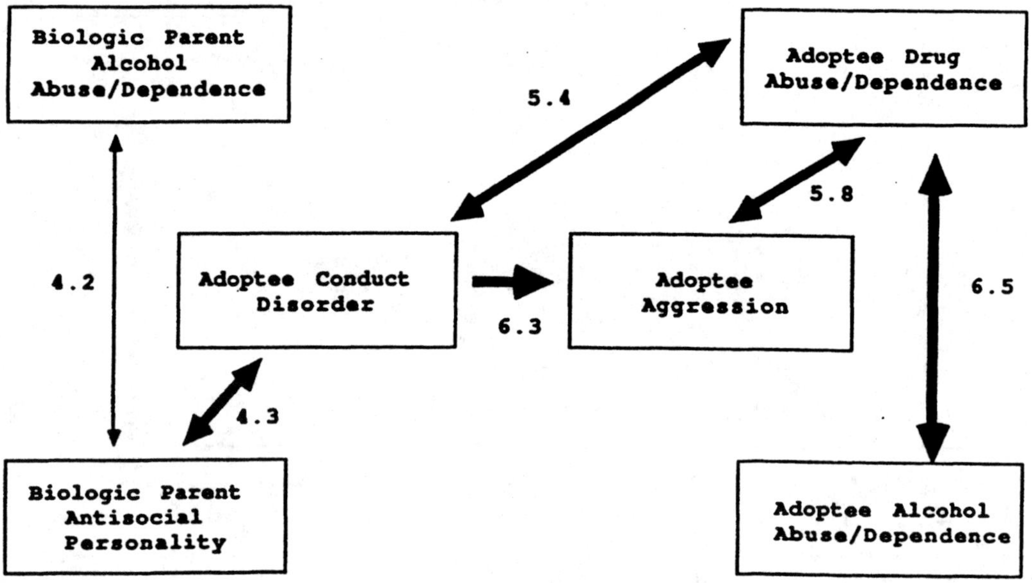

Light lines refer to relationship forced into the model to control for genetic confounds. Bold lines denote significant relationships. Numbers represent corrected odds ratios.

Figure 6. Interaction diagram for female model with adoptee conduct disorder and aggressivity as intervening variables (N=102)

Environmental factors in the male and female models shown in Figures 5 and 6 can impinge directly on drug abuse, or on the intervening variables of aggressivity and conduct disorder. In a study of environmental factors which affect aggressivity and conduct disorder it was found that exposure to an "impaired" adoptive parent increased both aggressivity and conduct disorders, especially in female adoptees. "Impaired" adoptive parent was defined as a parent with any psychiatric problem (depression, panic disorder, generalized anxiety, social phobia, alcoholism) or marital problems including divorce (Cadoret, et. al., 1995a). The effect of the parental environment was further influenced by the presence of a particular biologic background in the adoptee (whether ASP or alcoholism). Females showed a very significant gene-environment interaction effect. In the presence of "impaired" parent and ASP biologic background, aggressivity and conduct disorder in females increased more than one would expect from both factors acting independently (Cadoret, et. al, in press).

Thus the way to drug abuse is influenced not only by specific genetic factors (ASP and alcoholism in biologic parents) but by environmental effects which not only act independently of the genetic factors but in some cases interact with the genetic factors to produce their effects. The advantage of the adoption-separation paradigm is its ability to separate out these different effects, quantitate them, and assess their significance.

REPRODUCIBILITY AND GENERALIZABILITY OF ADOPTION FINDINGS

Most reported adoption studies have involved samples collected from one population, as the Danish and Swedish studies which utilized central registries for crime, psychiatric treatment, etc. Heterogeneity of findings is something which has not been systematically studied. An opportunity to examine heterogeneity presented itself in the Iowa Adoption studies because of the wide variety of adoption agencies which participated, each with its own population of client biologic and adoptive families. The early findings from two agencies reported above showed similar results for transmission of alcoholism for males. The two agencies provided a contrast in clients with one agency (ICFS) dealing with lower SES biologic parents, and the second agency (LSS) with higher SES families. In addition, there were obvious differences in religion of clients served.

In 1986-1989 a new adoption study was done with Catholic adoption agencies in Iowa to confirm the findings of the ICFS and LSS studies in alcoholism inheritance. The results of the new study showed no evidence for a genetic factor in male alcoholism with a sample size of 118 adoptees (Cutrona, et. al., 1994). A number of hypotheses were entertained to account for this unexpected finding, among them the possibility that in sampling a new agency a very different population containing different genes or environmental factors was responsible (and not just a sampling error). In order to test this heterogeneity hypothesis, a further study was done during 1989-1992 with four agencies which contributed adoptees. One of these was the LSS agency whose adoptees had shown a significant genetic effect of alcoholism (odds ratio = 7.6 with 95% confidence limits 2.2 - 26.3); the other contrasting agency was the Dubuque Diocese which had contributed the most male adoptees to the Catholic agency study (N=91) and whose findings for a genetic effect on adoptee alcoholism were negative (odds ratio = 0.7 with 95% confidence limits 0.3 - 1.6). The new study involved the use of hospital and prison records to diagnose biologic parent ASP and alcoholism, thus eliminating the possibility that faulty adoption agency records led to an erroneous biologic parent diagnosis.

The results of the new study showed that the Dubuque Catholic agency again gave negative results for a genetic effect on adoptee alcohol abuse (sample N=22, odds ratio = 0.5, 95% confidence limit 0.1 - 3.1); while the LSS agency, which had shown significant alcoholism genetic effect in the past, again demonstrated a significant genetic effect with an independent sample (N=15, odds ratio = 8.3, 95% confidence limit 1.2 - 5.9).

Examination of odds ratios for the alcoholic biologic parent-alcoholic adoptee relationship for all of the nine separate agencies used in the Iowa studies subsequent to the initial ICFS and LSS studies shows significant heterogeneity, with the principal source of heterogeneity difference detailed above between LSS and the Catholic agency. These results suggest strongly that the manifestation of a genetic factor can vary widely between agencies which represent different ethnic, religious, socioeconomic backgrounds for both biologic and adoptive parents. Which factor (or factors) is responsible is not clear, but the results suggest that environmental and gene-environment effects could also show significant heterogeneity as well. Such heterogeneity could explain other "negative" findings in adoption studies such as Ann Roe (1944). Heterogeneity raises the question of generalizability of results and suggests that large samples from varied sources would enhance generalizability.

FUTURE DIRECTION OF ADOPTION-SEPARATION PARADIGMS

Adoption designs can be administratively cumbersome and expensive in time and money. When centralized records are available (as in Scandinavian countries) the complication of getting data is less than when multiple agencies are involved as in the USA Since the adoption-separation paradigm as discussed here leads to measurement of specific genetic and environmental factors, it is the method, par excellence to determine if a psychiatric condition has a genetic transmission, while at the same time environmental factors and their interaction are measured. The major use of the adoption-separation paradigm in psychiatry has been to establish the importance of genetic factors . However, the ability of the same paradigm to detect environmental factors and gene-environmental effects is equally important and leads to the possibility of using the adoption paradigm in future molecular genetic studies.

One of the current techniques used to demonstrate a genetic trait is the linkage study (Ott, 1991). Linkage studies require ascertainment of extensive pedigrees and are subject to the influence of factors such as variable penetrance, gene-environment interaction, and phenocopies (genotype look-alikes which are due to non-genetic influences such as environment). This type of linkage analysis is usually not possible in adoption studies because of the difficulty in ascertaining and obtaining retrospective information on biologic parents and their pedigrees. Despite this limitation, adoption studies could be useful in linkage analyses by measuring environmental effects and gene-environment interaction and incorporating this information into the analyses.

There is a further usefulness for the adoption separation paradigm in today's information explosion from human genome research. Through the use of DNA determination from blood samples it is possible to be specific about selected candidate genes that are present in an adoptee. If this genetic information were combined with information about the adoptive environment an efficient association study could be done taking into account phenocopies (at least those determined by the environment) and gene-environment interaction effects which could enhance or reduce the effects of a particular gene. DNA information could take the place of the "genetic" factor hypothesized to be present in the biologic parent and transmitted to the adoptee. The advantage of such a study would include: (1) exact information about genetic makeup of adoptee determined from DNA analysis; (2) no need to have information about conditions in biologic parents, especially in cases of dubious paternity; (3) blood from adoptees could be stored indefinitely for future DNA extraction as tests for more candidate genes become available; (4) adoptees are frequently found in the general population (estimated 1-2% of population "adopted") so that large representative samples are more readily obtained; (5) fewer confidentiality problems - the identity of biologic parents does not enter into such a design; (6) in sampling from general population all kinds of designs are possible to test hypotheses (e.g. clinical samples of depressed adoptees matched on a case control basis with non-depressed adoptees; or random samples of adoptees to do an ECA type study looking at environmental factors and genotype environment interactions); (7) if records of biologic parents are not requested and adoption agencies not involved, the cost of a study determining psychiatric morbidity and environmental factors could be no more than a standard survey of the type used for the ECA data collection.

As more and more genes are discovered by molecular genetics, more candidate genes will be available for testing in association studies, and a large adoptee sample could be followed for years to estimate psychiatric illness in the sample and assess the influence of environment. One blood sample could be obtained and used for future DNA extraction. Obviously, the adoption-separation paradigm will be useful for some time to come with its potential value in elucidating the connections between molecular biology and behavior.

REFERENCES

American Psychiatric Association, (1980). *Diagnostic and Statistical Manual of Mental Disorders, Third Edition.* Washington, DC: American Psychiatric Association.

American Psychiatric Association, (1987). *Diagnostic and Statistical Manual of Mental Disorders. Third Edition-Revised.* Washington, DC: American Psychiatric Association.

Anonymous, (1993). Project MATCH (Matching Alcoholism Treatment to Client Heterogeneity): Rationale and methods for a multisite clinical trial matching patients to alcoholism treatment. *Alcoholism, Clinical & Experimental Research, 17,* 1130-1145.

Barnhart, R.K. (1986). *The American Heritage Dictionary of Science.* Boston: Houghton Mifflin Company.

Bergeman, C.S., Chipuer, H.M., Plomin, R., Pedersen, N.L., McClearn, G.E., Nesselroade, J.R., Costa, P.T.,Jr., & McCrae, R.R. (1993). Genetic and environmental effects on openness to experience, agreeableness, and conscientiousness: an adoption/twin study. *J Pers, 61,* 159-179.

Bishop, Y.M., Fienberg, S.E. & Holland, P.W. (1975). *Discrete multivariate analysis theory and practice.* Cambridge, MA: MIT Press.

Bohman, M., Cloninger, C.R., Sigvardsson, S., & von Knorring, A.L. (1982). Predisposition to petty criminality in Swedish adoptees. I. Genetic and environmental heterogeneity. *Arch Gen Psychiatry, 39*, 1233-1241.

Bohman, M., Sigvardsson, S., & Cloninger, C.R. (1981). Maternal inheritance of alcohol abuse. Cross-fostering analysis of adopted women. *Arch Gen Psychiatry, 38*, 965-969.

Cadoret, R.J. (1978). Evidence for genetic inheritance of primary affective disorder in adoptees. *Am J Psychiatry, 135*, 463-466.

Cadoret, R.J. & Cain, C. (1980). Sex differences in predictors of antisocial behavior in adoptees. *Arch Gen Psychiatry, 37*, 1171-1175.

Cadoret, R.J. & Cain, C. (1981). Environmental and genetic factors in predicting adolescent antisocial behavior in adoptees. *Psychiatric Journal of the University of Ottawa, 6(4)*, 220-225.

Cadoret, R.J., Cain, C.A., & Grove, W.M. (1980). Development of alcoholism in adoptees raised apart from alcoholic biologic relatives. *Arch Gen Psychiatry, 37*, 561-563.

Cadoret, R.J. & Gath, A. (1978). Inheritance of alcoholism in adoptees. *Br J Psychiatry, 132*, 252-255.

Cadoret, R.J., O'Gorman, T.W., Heywood, E., & Troughton, E. (1985). Genetic and environmental factors in major depression. *Journal of Affective Disorders, 9*, 155-164.

Cadoret, R.J., O'Gorman, T.W., Heywood, E., & Troughton, E. (1986). An adoption study of genetic and environmental factors in drug abuse. *Arch Gen Psychiatry, 43*, 1131-1136.

Cadoret, R.J., Troughton, E., Bagford, J., & Woodworth, G. (1990). Genetic and environmental factors in adoptee antisocial personality. *European Archives of Psychiatry and Neurological Sciences, 239*, 231-240.

Cadoret, R.J., Troughton, E., Merchant, L.M. & Whitters, A. (1990). Early psychosocial events and adult affective symptoms. In L.N. Robins & M. Rutter (Eds.). *Straight and devious pathways from childhood to adulthood* Cambridge, England: Cambridge University Press

Cadoret, R.J., Troughton, E., & O'Gorman, T.W. (1987). Genetic and environmental factors in alcohol abuse and antisocial personality. *Journal of Studies on Alcohol, 48*, 1-8.

Cadoret, R.J., Troughton, E., O'Gorman, T.W., & Heywood, E. (1985). Alcoholism and antisocial personality: Inter-relationships, genetic and environmental factors. *Arch Gen Psychiatry, 42*, 161-167.

Cadoret, R.J., Yates, W.R., Troughton, E., Woodworth, G., & Stewart, M.A. (1995a). Adoption study demonstrating two genetic pathways to drug abuse. *Arch Gen Psychiatry, 52*, 420-452.

Cadoret, R.J., Yates, W.R., Troughton, E., Woodworth, G., & Stewart, M.A. (1995b). Gene-environment interaction in the genesis of aggressivity and conduct disorder. *Arch Gen Psychiatry*, *52*, 916-924.

Cardon, L.R., Carmelli, D., Fabsitz, R.R., & Reed, T. (1994). Genetic and environmental correlations between obesity and body fat distribution in adult male twins. *Hum Biol*, *66*, 465-479.

Cardon, L.R., Fulker, D.W., & Joreskog, K.G. (1991). A LISREL 8 model with constrained parameters for twin and adoptive families. *Behav Genet*, *21*, 327-350.

Cloninger, C.R., Bohman, M., & Sigvardsson, S. (1981). Inheritance of alcohol abuse. Cross-fostering analysis of adopted men. *Arch Gen Psychiatry*, *38*, 861-868.

Cloninger, C.R., Sigvardsson, S., Bohman, M., & von Knorring, A.L. (1982). Predisposition to petty criminality in Swedish adoptees. II. Cross-fostering analysis of gene-environment interaction. *Arch Gen Psychiatry*, *39*, 1242-1247.

Cloninger, C.R., Sigvardsson, S., von Knorring, A.L., & Bohman, M. (1984). An adoption study of somatoform disorders. II. Identification of two discrete somatoform disorders. *Arch Gen Psychiatry*, *41*, 863-871.

Crowe, R.R. (1974). An adoption study of antisocial personality. *Arch Gen Psychiatry*, *31*, 785-791.

Cutrona, C.E., Cadoret, R.J., Suhr, J.A., Richards, C.C., Troughton, E., Schutte, K., & Woodworth, G. (1994). Interpersonal variables in the prediction of alcoholism among adoptees: evidence for gene-environment interactions. *Compr Psychiatry*, *35*, 171-179.

Dixon, W.J. (1981). *BMDP statistical software*. Berkley, CA: University of California Press.

Faraone, S.V. & Tsuang, M.T. (1994). Measuring diagnostic accuracy in the absence of a "gold standard". *Am J Psychiatry*, *151*, 650-657.

Feighner, J.P., Robins, E., Guze, S.B., Woodruff, R.A.,Jr., Winokur, G., & Munoz, R. (1972). Diagnostic criteria for use in psychiatric research. *Arch Gen Psychiatry*, *26*, 57-63.

Frenberg, S.E. (1980). *The analysis of cross-classified categorical data*. Cambridge, MA: MIT Press.

Gatz, M., Pedersen, N.L., Plomin, R., Nesselroade, J.R., & McClearn, G.E. (1992). Importance of shared genes and shared environments for symptoms of depression in older adults. *J Abnorm Psychol*, *101*, 701-708.

Goodwin, D.W., Schulsinger, F., Moller, N., Hermansen, L., Winokur, G., & Guze, S.B. (1974). Drinking problems in adopted and nonadopted sons of alcoholics. *Arch Gen Psychiatry*, *31*, 164-169.

Goodwin, D.W., Schulsinger, F., Hermansen, L., Guze, S.B., & Winokur, G. (1973). Alcohol problems in adoptees raised apart from alcoholic biological parents. *Arch Gen Psychiatry, 28,* 238-243.

Helzer, J.E., Burnam, A. & McEvoy, L.T. (1991). Alcohol abuse and dependence. In L.N. Robins & D.A. Regier (Eds.). *Psychiatric disorders in America* New York: The Free Press

Heston, L.L. (1966). Psychiatric disorders in foster home reared children of schizophrenic mothers. *Br J Psychiatry, 112,* 819-825.

Ingraham, L.J. & Wender, P.H. (1992). Risk for affective disorder and alcohol and other drug abuse in the relatives of affectively ill adoptees. *J Aff Disorders, 26,* 45-51.

Keller, M.B. & Hanks, D.L. (1994). The natural history and heterogeneity of depressive disorders: implications for rational antidepressant therapy. *J Clin Psychiatry, 55 Suppl A,* 25-31; dis.32-3,98-100.

Kety, S.S., Rosenthal, D., Wender, P., Schulsinger, F. & Jacobsen, B. (1978). The biologic and adoptive families of adopted individuals who became schizophrenic: Prevalence of mental illness and other characteristics. In L.C. Wynne, R.L. Cromveel & Mathysses (Eds.). *The Nature of Schizophrenia* New York: Wiley

Kety, S.S., Rosenthal, D., Wender, P.H., & Schulsinger, F. (1971). Mental illness in the biological and adoptive families of adpoted schizophrenics. *Am J Psychiatry, 128,* 302-306.

Mednick, S.A., Gabrielli, W.F., & Hutchings, B. (1984). Genetic influences in criminal convictions: Evidence from an adoption cohort. *Science, 224,* 891-894.

Melton, B., Liang, K.Y., & Pulver, A.E. (1994). Extended latent class approach to the the study of familial/sporadic forms of a disease: Its application to the study of the heterogeneity of schizophrenia. *Genetic Epidemiology, 11(4),* 311-327.

Mendlewicz, J. & Rainer, J.D. (1977). Adoption study supporting genetic transmission in manic--depressive illness. *Nature, 268,* 327-329.

Merikangas, K.R., Wicki, W., & Angst, J. (1994). Heterogeneity of depression. Classification of depressive subtypes by longitudinal course. *Br J Psychiatry, 164,* 342-348.

Myers, J.K., Weissman, M.M., Tischler, G.I., Holzer, C.E., Leaf, P.J., Orvaschel, H., Anthony, J.C., Boyd, J.H., Burke, J.D., Kramer, M., & Stolzman, R. (1984). Six-month prevalence of psychiatric disorders in three communities. *Arch Gen Psychiatry, 41,* 959-967.

Ott, J. (1991). *Analysis of human genetics linkage.* Baltimore: Johns Hopkins Univeristy Press.

Pickens, R.W., Svikis, D.S., McGue, M., Lykken, D.T., Heston, L.L., & Clayton, P.J. (1991). Heterogeneity in the inheritance of alcoholism. A study of male and female twins. *Arch Gen Psychiatry, 48,* 19-28.

Plomin, R., Coon, H., Carey, G., DeFries, J.C., & Fulker, D.W. (1991). Parent-offspring and sibling adoption analyses of parental ratings of temperament in infancy and childhood. *J Pers, 59*, 705-732.

Plomin, R. & Daniels, D. (1987). Why are children in the same family so different from one another? *Behavioral and Brain Sciences, 10*, 1-60.

Plomin, R. & DeFries, J.C. (1985). Origins of individual differences in infancy. In *The Colorado Adoption Project* New York: Academic Press

Plomin, R., Owen, M.J., & McGuffin, P. (1994). The genetic basis of complex human behaviors. *Science, 264*, 1733-1739.

Price, R.A., Cadoret, R.J., Stunkard, A.J., & Troughton, E. (1987). Genetic contributions to human fatness: an adoption study. *Am J Psychiatry, 144*, 1003-1008.

Richardson, L.F. (1912). The measurement of mental "nature" and the study of adopted children. *Eugenics Review, 4*, 391-394.

Robins, L., Helzer, J., Cottler, L. & Goldring, E. (1989). *NIMH Diagnostic Interview Schedule, version III, revised (DIS-IIIR).*

Robins, L.N., Helzer, J.E., Croughan, J.L, Williams, J.B. & Spitzer, R.L. (1981). *NIMH Diagnostic Interview Schedule, version III.* St. Louis, MO: Washington University.

Roe, A. (1944). Adult adjustment of children of alcoholic parents raised in foster homes. *Quarterly Journal of Studies on Alcohol, 5*, 378-393.

Rosenthal, D. & Kety, S. (1968). *Transmission of Schizophrenia.* Oxford: Pergamon Press.

Rosenthal, D., Wender, P.H., Kety, S.S., Welner, J., & Schulsinger, F. (1971). The adopted-away offspring of schizophrenics. *Am J Psychiatry, 128*, 307-311.

Sigvardsson, S., Cloninger, C.R., Bohman, M., & von Knorring, A.L. (1982). Predisposition to petty criminality in Swedish adoptees. III. Sex differences and validation of the male typology. *Arch Gen Psychiatry, 39*, 1248-1253.

Sigvardsson, S., von Knorring, A.L., Bohman, M., & Cloninger, C.R. (1984). An adoption study of somatoform disorders. I. The relationship of somatization to psychiatric disability. *Arch Gen Psychiatry, 41*, 853-859.

Spitzer, R.L. & Endicott, B. (1979). *Scedule for affective disorders and schizophrenia-lifetime version (SADS-L).* New York: New York State Psychiatric Institute.

von Knorring, A.L., Cloninger, C.R., Bohman, M., & Sigvardsson, S. (1983). An adoption study of depressive disorders and substance abuse. *Arch Gen Psychiatry, 40*, 943-950.

Wender, P.H., Rosenthal, D., Kety, S.S., Schulsinger, F., & Welner, J. (1974). Crossfostering. A research strategy for clarifying the role of genetic and experiential factors in the etiology of schizophrenia. *Arch Gen Psychiatry, 30*, 121-128.

Section II
Family, Twin and Adoption Studies: Diversity of Approaches

CHAPTER 6

THE USEFULNESS OF FAMILY STUDIES

Some Results on Spouse Resemblance and Familial Transmission of Anxiety and Depression in a Large Norwegian Sample

Kristian Tambs
National Institute of Public Health
Department of Epidemiology
Oslo, Norway

INTRODUCTION

Biometric genetic analyses of data from families with parents and offspring (nuclear families) have long traditions in medicine, especially in the study of cardiovascular diseases and their risk factors like blood pressure (see e.g. Tambs et al., 1992a). Traits that pertains to psychology have often been examined in adoptive families, or, even more commonly, in twins, whereas family studies are much less common (Eaves, Eysenck, and Martin, 1989). There are several possible explanations for this low representation of family data. The most probable explanation is that data from parents and offspring alone are not sufficient to disentangle genetic effects and effects from the family environment. This disadvantage is probably not critical for some medical variables, where sometimes the effect of the family environment can be assumed to be modest, whereas in psychology most theoretical frameworks - from psychoanalysis to social learning theory - assumes that some of the major causes of individual variation are associated with the family. Straight-forward assessment of family aggregation effects in parents and offspring is also complicated because behavior changes during life, and valid measuring of the same trait in persons from two generations, whose ages vary considerably, can be problematic.

Some properties of nuclear family data that may be disadvantageous for some purposes may be helpful for other purposes, however. During the last decade there has been a growing appreciation of the need for types of information that can be provided from nuclear families but not from twin studies. Data sets from extended twin studies, usually including parents of twins and sometimes the twins' spouses and offspring together with the MZ (monozygotic = identical) and DZ (dizygotic = fraternal) twin pairs are increasingly more common. For instance, researchers in Richmond, Virginia, USA have obtained a variety of data from a large number of twin pairs and the twins' relatives, the "Virginia 30,000" (Truett et al., 1993). But studies including nuclear family data are still less common than "classical" twin studies (where MZ and DZ pairs reared together are compared). From 33 biometric genetic analyses of human data (28 real data sets, 5 analyses of simulated data) presented in Behavior Genetics 1990-1992 only 5 used extended twin samples (twins and relatives of twins), 5 used adoption data, and the remaining 23 used pure twin data.

This chapter presents some questions that may be appropriately addressed using data from large family samples. These questions are widely recognized and have been frequently discussed but are still basically unanswered, at least for most traits that have been studied in behavior genetics. These questions pertains to degree and type of *non-random mating*, what *types of family environment* that may affect behavior, and whether there are *age-specific* genetic or environmental effects. These phenomenons deserve to be studied simply because of their general interest in the behavioral, social, and medical sciences. Furthermore, if present, they affect resemblance between relatives, and, if ignored, may bias results from genetic analyses. Self-reported data from a large family sample will be used to illustrate how data from types of relatives other than twins can be helpful in addressing these questions. The data base includes most of the adult population of the Norwegian county of Nord-Trøndelag.

How Data From Family Studies Can Contribute Knowledge in Behavior Genetics

Non-random Mating

For most behavioral traits that can be measured, there is a correlation between spouses. Spouse resemblance in western societies is usually very high (correlations from 0.5 to almost unity) for demographic variables like age, race (McLemore, 1980), geographic rearing area (Clark, 1952), education (Heath et al, 1985), and attitudes (Martin and Jardine, 1986), including political and religious affiliation (Cavalli-Sforza et al., 1982; Basavarajappa et al, 1988). Spouse correlations are also high (typically around 0.5) for most socioeconomic variables (Price and Vandenberg, 1980; Tambs et al., 1989; Watkins and Meredith, 1981) and life style variables (Price and Vandenberg, 1980; Tambs and Vaglum, 1990) and somewhat lower (mostly in the range 0.2 to 0.4) for cognitive abilities (Johnson et al., 1980) and psychiatric disorders (Merikangas, 1982; Tambs, 1991). The values for resemblance on measures of personality are usually below 0.2 (Eaves, et al., 1989, Farley and Davis, 1977; Farley and Mueller 1978, Lesnick-Oberstein et al. 1984; Price and Vandenberg, 1980; Tambs et al., 1991a; 1992b). There is little or no evidence of negative spouse correlations.

This spouse resemblance is usually assumed to reflect *assortative mating*; persons who resemble each other for a certain trait tend to mate. Another explanation is the notion of *social homogamy*, according to which spouse selection tends to happen within social groupings or strata. This causes spouse similarity for characteristics that vary between social groups. A third possibility is *convergence* before or during marriage which results from direct reciprocal influence or from common household and family environment. Of course, none of these explanations are mutually exclusive, and they may be more or less true for different traits.

Models utilizing data for a certain trait in relatives and spouses of relatives permit a resolution of the effect of social homogamy and assortative mating (Heath and Eaves, 1985). The basic idea is that relatives, for instance siblings or twins, will have more similar spouses if they marry within the same social groups (social homogamy) than if they choose spouses that resembles themselves (assortative mating). In fact under perfect social homogamy the expected correlation between the spouses of two siblings and the in-law correlations are as high as the expected spouse correlation because the siblings and their spouses all belong to the same social groups, and that is the only cause of the resemblance. In the case of assortative mating the expectation for the in-law correlation is the product of the spouse correlation and the sibling correlation, and the correlation between each of the siblings' spouses is the product of the squared spouse correlation and the sibling correlation. Models can be specified for several relatives, like MZ and DZ twins, siblings, and half-siblings together with their spouses. Where such data are available, it is also possible to analyze data from *chains of relatives*, for instance a pair of siblings with spouses, each of the spouses' siblings with their spouses etc. (Such analyses of the Nord-Trøndelag data are under preparation.)

The explanation of convergence is intuitively appealing but generally, the available evidence does not support it (Ditto and France, 1990; Eysenck and Wakefield, 1981; Guttman and Zohar, 1987; Lesnik-Oberstein and Cohen, 1984; Lynum and Skau Dahl, 1987; Martin, 1978; Mascie-Taylor, 1989; Schumm et al., 1980, Watkins and Meredith, 1981). Price and Vandenberg (1980) have shown some convergence for alcohol consumption and nutritional habits, however. To detect convergence that mimics non-random mating, longitudinal data are well suited. Nonetheless, almost all the extant evidence is from cross-sectional studies with data on duration of marriage. Cross-sectional studies using information of marital duration require large samples, especially if short-time effects of cohabitation are to be detected.

The choice of assumptions about assortative mating/homogamy has important consequences for the understanding and estimation of genetic and environmental transmission. One consequence is increased phenotypic, and, in case of assortative mating, genetic variance in the population. More important for biometric genetic analyses is the increase of the expected resemblance between first degree relatives. Under *social homogamy* (and environmental transmission from parental phenotype to offspring's environment, which will be described later in this chapter) the part of the correlations in parent-offspring, siblings, and MZ and DZ twins caused by environmental transmission increases by the proportion $(1+r_M e^2)$, where r_M is the observed spouse correlation and e^2 is the fraction of variance caused by environmental factors. Unless there are genetic differences between the groups within which homogamy takes place, social homogamy leaves the correlation due to shared genes unchanged. Under *assortative mating* the parent-offspring correlation due to shared genes is increased by the proportion $(1+hr_M)$, where h is the square root of the heritability, h^2, and the parent-offspring correlation due to environmental transmission is increased by the proportion $(1+r_M)$. In siblings and DZ twins the genetic correlation increases by the proportion $(1+h^2r_M)$ and the correlation due to environmental transmission from parents increases by the proportion $(1+r_M)$. In MZ twins the genetic correlation is already unity, and cannot increase, but the correlation due to environmental transmission increases as for DZ twins and siblings. In biological parents, offspring and siblings separated by adoption the genetic correlations are affected as in non-separated relatives and the environmental correlations remain zero under social homogamy as well as assortative mating. In the corresponding adoptive relations the correlations due to environmental transmission are changed as for biological relatives and the genetic correlations remain zero. The genetic correlation in halfsiblings is not affected by non-random mating. The environmental correlations in halfsiblings reared together is increased as in ordinary siblings.

Assortative mating has more serious consequences than social homogamy for biometric genetic analyses. Firstly because the increases of the expected correlations due to assortative mating are larger unless the heritability is very low. Secondly because the increases due to social homogamy are similar for most types of relatives, whereas the magnitude of the increases due to assortative mating varies strongly with type of relative. It is, for instance, easy to realize that a much larger increase of the correlation in DZ than in MZ twins can affect the results from classical twin studies, where the estimates are largely based on the difference between the correlations in MZs and DZs. Unlike assortative mating, social homogamy may be ignored unless environmental transmission is very strong, and it seldom appears to be strong for psychological variables (Eaves et al., 1989). Effects of convergence for resemblance in relatives are similar to those of social homogamy.

Effects of Family Environment

Effects of environmental factors which are shared by two or more members of a family can be classified in two or more types. Environmental effects may be *transmitted from parents to offspring*, usually called *cultural transmission.* Such transmission can lead from paternal phenotype (via the offspring's environment) to the offspring's phenotype due to parental education, identification or model learning. It is easy to imagine that such phenotype-to-phenotype transmission takes place for attitudes, personality, and most types of behavior. Another kind of cultural transmission is from the paternal environment to the offspring's environment. It is likely that for instance obesity in offspring is caused partly by environmental factors which are "inherited" from the parents, such as nutritional habits and other life style factors, but less likely that obesity in offspring is caused directly by obesity in parents. It is not possible to specify both kinds of cultural transmission in one single model for genetic and environmental effects in families. Usually the phenotype-to-phenotype transmission type is assumed for behavior because it tends to give more precise results. In most cases the choice of type does not affect the estimates of genetic and environmental effects seriously (Eaves et al. 1989).

Siblings share some experience due to cultural transmission, but may also share environmental effects which are unrelated to parental phenotype and environment. Such *environmental sibling effects* may still be caused by the parents, who sometimes encourage behavior which they don't exhibit themselves. Furthermore, siblings share environmental factors independent of parental influence, such as common friends, and, perhaps most important of all, siblings directly influence each others behavior. Mutual influence that increase sibling resemblance is termed "cooperation", influence that increases differences is termed "competition". Twins may share more of the environment than siblings, and sometimes a specific *environmental twin effect* is specified in the causal molds.

Twins reared together share all kinds of family environment. Therefore classical twin studies are not suited to separate cultural transmission from sibling (twin) effects. Here, the effect of common (family) environment is to increase the correlations equally in MZ and DZ twins, and, since the DZ correlation is usually lower than the MZ correlation, increase the proportion of the DZ correlation to the MZ correlation. In contrast, genetic non-additive effect, due mainly to dominant/recessive inheritance, decreases the proportion of the MZ correlation to the DZ correlation because MZ twins share all genetic variance, whereas DZ twins share 1/2 of the genetic additive variance and only 1/4 of the variance due to genetic dominance. Consequently the effects of genetic dominance and of family environment will counteract and disguise each other when both present in classical twin studies. Models for classical twin studies can be specified either with genetic dominance or with common environment, but not with both parameters simultaneously.

Unlike classical twin studies, adoption studies provide, in principle, specific information about cultural transmission and environmental sibling effects. Cultural transmission increases correlations between adoptive parents and adoptive children and between adoptive siblings. Unless genetic effects are very low and cultural transmission extremely strong, this increase is much larger in adoptive parents-children than in adoptive siblings. Environmental sibling effects only increase the correlation in adoptive siblings, whereas the correlations between separated biological relatives are not affected by any type of family environment. Because the separation of the two types of environmental family effects is only based on the relative correlation values in adoptive siblings compared to those in adoptive parents-offspring, however, larger samples of adoption families than what is usually available is required for a precise separation.

In general, nuclear family data alone do not provide sufficient information to estimate various family environment effects specifically. As with adoption data, cultural transmission contributes to the correlation between parents and offspring and, to a less extent, to the sibling correlation, and the environmental sibling effect contributes to the sibling correlation. In order to separate the two environmental effects from genetic effects additional data from other types of relatives are needed. Nonetheless, nuclear family data are useful in the validation of results from other methods, and in some special cases results from nuclear families give direct evidence of specific environmental family effects. Low parent-offspring correlations compared to sibling correlations can reflect either environmental sibling effects, genetic non-additivity or, as discussed later, certain types of age effects. High parent-offspring correlations compared to sibling correlations are likely to reflect cultural transmission. Some patterns of sex-specific parent-offspring correlations almost certainly imply cultural transmission or environmental sibling effects. For example higher correlations in mother-son and mother-daughter pairs than in father-son and father-daughter pairs can only point to an environmental maternal effect because no types of sex-specific genetic effects can explain such a pattern.

Nuclear family data are useful for assessing the effects of family environment *in combination* with other data like MZ and DZ twin data. In principle, combined nuclear family and twin data permit the separation of genetic additive and dominance effects, cultural transmission and environmental sibling effects. The separation of sibling effects and genetic dominance has proved difficult in practice, however (Tambs et al. 1991b; 1992a), because large twin samples are required. Supplementary data from half-siblings living together, who share the sibling effect but not the effect of genetic dominance, are helpful for this purpose.

AGE-SPECIFIC GENETIC OR ENVIRONMENTAL EFFECTS

Often large age differences between family members are considered a disadvantage of family data. Sometimes one cannot be certain that the same phenotype is measured in persons from different generations, because it may be expressed qualitatively differently at different ages. Trait instability across the life-span can reflect effects of genes or environmental factors which vary systematically with age. Several traits, like male baldness, are affected by genes that start acting at a certain age. It is, perhaps, harder to find examples of age-specific environmental effects, but cigarette smoking as an environmental factor for cancer, usually starting in the late teens or early twenties, may serve as one. A developmental model in which the effects of specific genes and environmental factors switch on at different ages and are transmitted forward in time through the phenotype is shown in Figure 1. When different genes affect a trait at different periods of life, relatives distant in age will, on average, be less similar for this trait than relatives close in age. It can be shown (Eaves, Long, and Heath, 1986) that under the first order time-series model illustrated here, the correlations between relatives have the form $Te^{-A\delta_h} + \varepsilon e^{-A\delta_e}$. A denotes age difference in years, and t and e are the correlations in relatives between additive genotypes and environmental values, respectively, when all age-specific effects are shared (when A=0). e is the base of the natural logarithm. δ_h and δ_e are parameters expressing the decay in correlation with age differences. The model assumes there is no genotype-environment correlation or genetic dominance, and that residual variance is caused by individual environmental factors and measurement error.

Age-specific genetic or environmental effects wash out some of the observed resemblance between relatives distant in age. If ignored, such age-specific effects may lower the parent-offspring correlations substantially, lower the sibling correlations moderately, and leave the twin correlations unchanged (compared to a situation in which age-specific factors are substituted by age-stable factors). A larger reduction of parent-offspring resemblance than of sibling and twin resemblance causes deflated estimates of cultural transmission or inflated estimates of genetic dominance or environmental sibling effects. Reduced sibling resemblance together with unchanged twin resemblance may upwards bias environmental effects shared only by twins or - due to the increased difference between the MZ twin resemblance and resemblance in other relatives - may upwards bias genetic effect, especially genetic dominance.

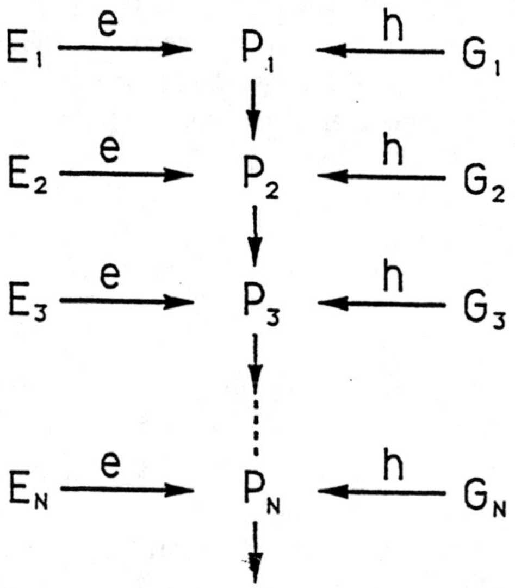

E_1 to E_N are age-specific environmental factors acting at time 1 to time N.
G_1 to G_N are age-specific genotypic values acting at time 1 to time N.
P_1 to P_N are age-specific phenotypic values observed at time 1 to time N. The relative strength of the environmental effect, e, and the genetic effect, h, is assumed to be constant from time 1 to time N.

Figure 1. A Model for Age-specific Genetic and Environmental Effects

A family data set with noticeably low parent-offspring correlations compared to sibling correlations and, if included, high twin correlations may suggest that age-specific effects should be specified. Large samples of families are needed to detect such age-specific effects of genes or environmental factors, however.

Applications from a Family Sample

From 1984 to 1986, the National Health Screening Service requested adult persons over 19 years living in the county of Nord-Trøndelag, Norway, to participate in a health screening. The governmental census agency, Central Bureau of Statistics (CBS), was able to provide information on first degree family relationships and spouses/cohabiting partners for all but the oldest part of the screening population. The data files were matched with the family identification data at the CBS, and stripped of personal identification numbers before being released to our research group.

Up till now, family data from the Nord-Trøndelag sample on obesity (body mass = weight/height^2) (Tambs et al., 1991b), height (Tambs et al., 1992c), systolic and diastolic blood pressure (Tambs et al., 1992a, 1993), and mental health (Tambs and Moum, 1993) have been analyzed. Here we will briefly describe a test of whether spouse resemblance increases with marital duration (convergence) for a set of variables in pairs of spouses (Tambs and Moum, 1992). The Nord-Trøndelag study is well suited to test models of assortative mating vs. models of social homogamy, but these analyses are still to be finished. Results based on family data on mental health (Tambs and Moum, 1993) will be used to illustrate the usefulness of family data for the detection of effects of family environment and age-specific genetic or environmental factors. Data from a recent twin study are analyzed together with the family data in order to examine the consistency of twin data with family data.

METHOD

SAMPLE

74,994 of a total population of 85,125 persons over 19 years participated. Of the 11.9% (10.1% females, 13.7% males) who did not attend the screening, more than one third returned a questionnaire with information of reason for not meeting (working at sea, military service, staying abroad, sickness etc.). At the screening, before the medical testing, the participants completed a one-page questionnaire (Questionnaire 1). They were asked to return a second questionnaire (Questionnaire 2) by mail within a few weeks after the medical examination. 62,784 subjects, or 84% of those originally screened, returned Questionnaire 2.

It was possible to identify the following number of pairs of relatives with valid medical screening data: 83 pairs of MZ and 92 pairs of DZ twins, 19,168 pairs of siblings, 43,613 pairs of parents and offspring, 24,281 pairs of spouses, 1,318 pairs of grandparents-grandchildren, 1,218 pairs of biological uncle/aunt-nephew/niece, 849 pairs of non-biological uncle/aunt-nephew/niece, and lower numbers (<200) of pairs of ordinary cousins, pairs of cousins related through MZ twins, and half-siblings. The proportion of missing data varied somewhat across items. Numbers of pairs for the various types of relatives with valid Questionnaire 1 data are close to 90%, and numbers of pairs with Questionnaire 2 data are close to 70% of those with data from the medical examination. The population and ascertainment have been described more thoroughly elsewhere (Holmen et al, 1990; Tambs and Moum, 1993).

MEASURES

Data included in the tests of *marital convergence* were obtained partly from the physical examination, partly from the questionnaires. The variables from the examination were *systolic and diastolic blood pressure*, resting *heart rate*, and *body mass index*, all based on physical examination. The measuring procedures and age corrections of these variables are described elsewhere (Tambs et al., 1991a; 1992a).

Questionnaires 1 and 2 both include questions about health and sickness, life style, quality of life, and mental health. With the exception of subjectively perceived health all questionnaire information utilized in the convergence testing is either from Questionnaire 2 or from Questionnaires 1 and 2 in combination, reducing the effective sample size to 15,925 pairs of spouses for these variables. All items (except those where the appropriate responses are exact values or "yes/no") are of the Likert type. *Alcohol consumption* was measured as the weighted sum (first unrotated factor from principal components analysis) of three items, tapping frequency of drinking, frequency of heavy drinking, and feeling of having been drinking too much in periods. *Smoking* was measured as number of reported cigarettes per day. *Exercise* was measured as the weighted sum of three items on frequency, intensity, and duration of each exercise period. *Stress at work* was measured as the weighted sum of 3 items, including stress and tension, physical demands, and requirements of high concentration and attention at work. *Nervousness* was computed as a weighted sum of 4 items (Numbers 1, 2, 6, and 12 in Table 1). *Type A* was measured as the weighted sum of 3 items: Taking one's duties more seriously than other people, "pushing or steadily urging oneself forward", and constant shortness of time. *Life satisfaction* was measured as a composite of 5 items: overall satisfaction, included in both questionnaires, strength and energy or tiredness, feeling "calm and good about oneself", and being mostly cheerful or dejected. Subjectively perceived overall *health* was measured by a single item: "How is your health at the moment?"

Information on *income*, originally obtained from the governmental bureau of taxation, and information on *marital duration* was provided by the Central Bureau of Statistics. Precise information on marital duration was not available for spouses married more than 20 years. For this group marital duration was estimated on the basis of age. 252 unmarried, cohabiting couples with at least one of the spouses younger than 25 years were grouped together with couples married 1 year or less.

Test-retest correlations for the variables included in the convergence study ranged from .55 to .82 over 1-3 year time lags in a subsample of 4,621 persons. A more detailed description of the test-retest results and of the data used in this study is given elsewhere (Tambs and Moum, 1992).

The *family study* utilized 12 questions from Questionnaire 2 pertaining to life satisfaction, personality, and mental health. 6,380 persons from the Nord-Trøndelag population participated in a follow-up study in 1989. The follow-up questionnaire included the said 12 questions together with the SCL-25 questionnaire (Hesbacher et al., 1980; Winicur et al., 1984), designed to measure symptoms of anxiety and depression. The intercorrelation between the separate SCL-25 anxiety and depression scores was 0.75, suggesting that in a normal population sample not much is gained by classifying anxiety and depression separately. A global score for anxiety and depression is computed as a sum of all the 25 items.

The logarithmic transforms of the SCL-25 global score (the transformation was due to a skewed distribution) were regressed on the 12 life satisfaction/mental health questions in the follow up sample. Based on the results from the regression analysis, the 12 items were combined in a new indicator of mental health. The correlation between the new 12-item indicator and the SCL-25 sum-score was 0.83. A principal component factor analysis of the 12 items revealed a correlation of 0.977 between the un-rotated first factor and the 12-item indicator based on multiple regression analysis, demonstrating that the correlation structure inherent in the 12 items corresponds very well with the SCL-25 structure.

The theta reliability (see e.g. Zeller and Carmines) for the indicator was 0.83. Quite reassuring test-retest results are presented elsewhere (Tambs and Moum, 1993). The alpha reliability for the SCL-25 sum-score was 0.91. The intercorrelation between the SCL-25 and the 12-item indicator is almost as high as expected in a situation in which the two indicators tap exactly the same symptoms and the intercorrelation only deviates from unity because of random measurement errors.

Table 1 shows the 12 questions and their relative contribution to the indicator (standardized beta-values). Each question has three to seven response alternatives with verbal labels along a graded scale (like "all the time", "often", "now and again" or "never"). Moderate mean and variance differences between sexes and age groups were removed before the biometric analyses of the family data.

Table 1. 12 items and their relative contributions to (β), and correlations (r) with the indicator of anxiety/depression

Question	β	r
1. Over the last month, have you suffered from nervousness (irritability, anxiety, tension or restlessness)?	0.30	0.85
2. Do you mostly feel strong and fit, or tired and worn out?	0.16	0.69
3. Do you often feel lonely?	0.16	0.65
4. Do you suffer from any long time illness or complaint of a physical nature which impair your functions in your day-to-day life [From a list of different types of impairment] Impairment due to psychological complaints?	0.12	0.50
5. Do you by and large feel calm and good about yourself?	0.12	0.74
6. Have you had any problems falling asleep or sleep disorders during the course of the last month?	0.12	0.62
7. Would you say that you over the last year have pushed yourself or steadily urged yourself forward?	0.07	0.46
8. Are you constantly short of time even when it comes to day-to-day tasks?	0.06	0.35
9. Do you have a tendency to take your duties more seriously than other people?	0.05	0.33
10. When you think about your life at the moment, would you say that you by and large are satisfied with life, or are you mostly dissatisfied?	0.04	0.60
11. Would you say you are usually cheerful or dejected?	0.03	0.60
12. How often have you taken tranquilizers/sedatives or sleepiness medication over the course of the last month?	0.02	0.50

Table 2. Spouse correlations in groups with various marital duration.

	< 1 year[a]	1-3 years	3-5 years	5-10 years	10-15 years	15-20 years	>20yrs, age <60	>20yrs, age 60-69	>20yrs, age >69
N (pairs)	363-616	547-934	658-1062	1992-3032	2468-3616	2346-3452	2691-3789	3267-4458	1593-2239
Body Body mass	.144	.126	.050	.119	.143	.079	.108	.101	.098
Systolic BP	.036	.111	.085	.109	.104	.070	.068	.072	.070
Diastolic BP	.087	.089	.075	.096	.095	.135	.081	.112	.064
Heart rate	.033	-.021	.065	.057	.042	.035	.076	.044	.045
Alcohol consump.[b]	.384	.452	.450	.486	.480	.496	.472	.508	.499
Smoking[b]	.344	.364	.289	.319	.317	.331	.311	.316	.294
Exercise	.285	.229	.213	.255	.285	.293	.303	.313	.379
Working stress[c]	.034	.150	.151	.217	.209	.130	.145	.208	.247
Nervousness [b,d]	.336	.205	.235	.231	.238	.194	.187	.240	.202
Type A	.166	.209	.173	.194	.142	.131	.127	.168	.171
Quality of life	.377	.457	.333	.382	.376	.346	.345	.367	.372
Subjective health[b]	.120	.313	.316	.254	.267	.224	.258	.251	.260
Income[b]	.236	.149	.085	-.032	.002	.045	.090	.070	.100

* $p < .05$; ** $p < .001$; [a]Includes some pairs married after the information was recorded and a few not married, cohabiting pairs with females aged 22 or younger; [b] Polychoric correlations; [c] Responses are missing for some persons defining themselves as not working, mainly retirees.

RESULTS

SPOUSE RESEMBLANCE: CONVERGENCE OR NON-RANDOM MATING?

Spouse correlations for the various variables were computed in groups of pairs classified by marital duration. The correlations by marital duration are shown in Table 2. Spouse correlations for *alcohol consumption* increase significantly with marital duration (X^2_8=18.08, p=.021, testing for heterogeneity among groups with different marital duration), mainly during the first five years of marriage. For *exercise* there is a trend of decreasing correlations during the first five years, and then increasing correlations throughout life (X^2_8=27.62, p<.001). For *work stress* the correlation increases from almost zero to higher than 0.2 during the first eight years, after that it decreases slightly but tends to increase again after 20 years of marriage (X^2_8=30.96, p<.001). A trend for *nervousness* (X^2_8=23.47, p=.003) seems to reflect a negative convergence (divergence) during the first two years of marriage and little or no change after that. For *subjective health* the correlation increases from 0.12 to 0.31 during the first one or two years and slightly decreases after that (X^2_8=21.53, p=.006). This initial convergence is hardly due to any sudden change of "true" health, rather the increased spouse resemblance reflects a convergence for standards of judgement. The correlation for *income* decreases from 0.24 to below zero during the first seven years, and increases somewhat with duration in couples married longer than seven years (X^2_8=90.24, p<.001). It is worth noticing that the correlation is at the lowest during the years when most couples have young children, when one of the spouses may need to stay home to take care of the children. There were no significant effects of convergence for the remaining seven variables.

Figures 2-4 show the spouse correlations as a function of marital duration for alcohol consumption, exercise, and nervousness. The smooth curves show the expected values, specifying a linear or logarithmic function for convergence or, in cases of decreasing correlations with marital duration, divergence.

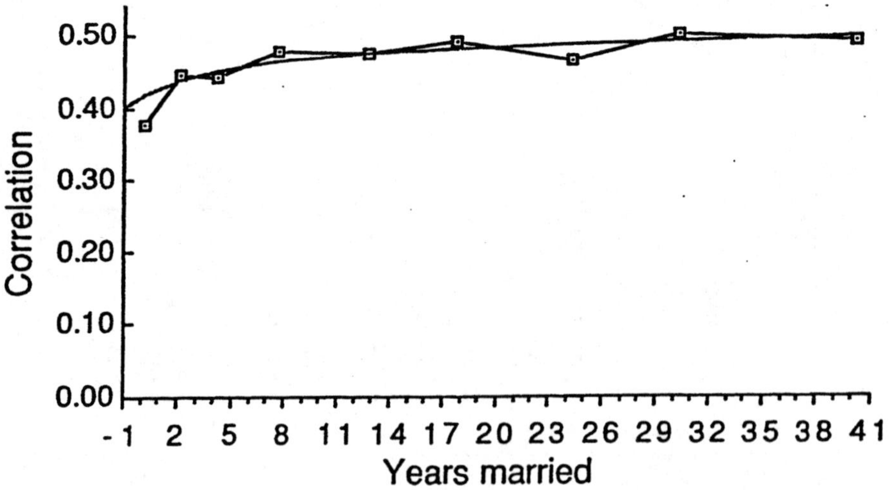

Figure 2. Observed and expected spouse correlations for alcohol consumption as function of marital duration

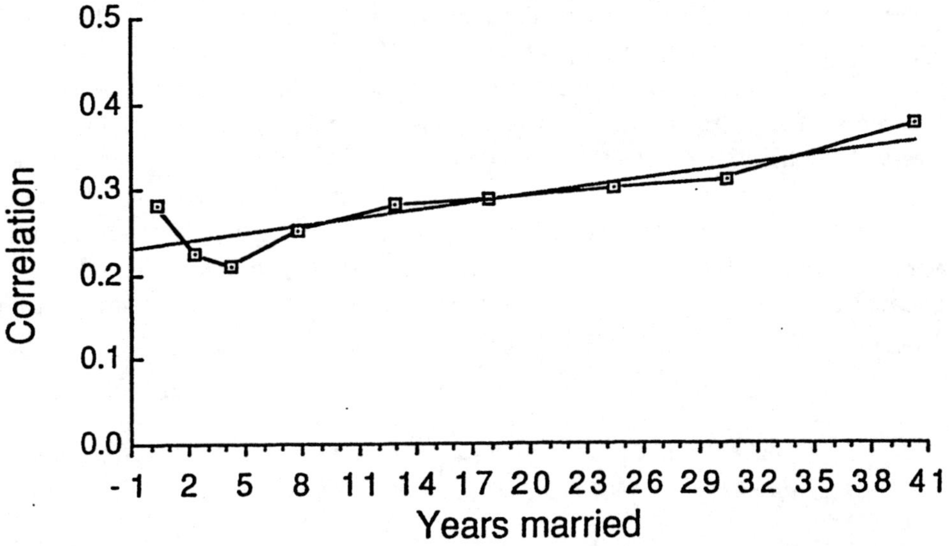

Figure 3. Observed and expected spouse correlations for exercise as function of marital duration

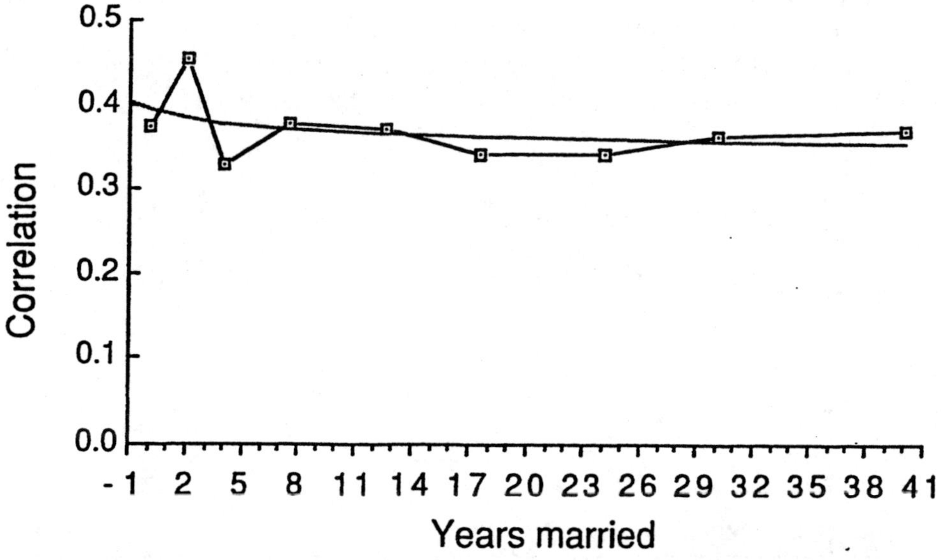

Figure 4. Observed and expected spouse correlations for nervousness as function of marital duration

In General, the results show slight effects of convergence for some life style variables (alcohol consumption, exercise, and work stress), and significant (nervousness) or non-significant (type A, life satisfaction) trends of divergence for variables associated with personality. The tendency of decreasing spouse resemblance with marital duration for some facets of personality supports theories of mate selection that include role complementarity between the mates (Kerckhoff and Davis, 1962; Murstein, 1976). The trends of convergence and divergence are not very strong, however, and may be confounded with secular trends if degree of homogamy have changed during the last decades. Confounding with secular trends is more likely for lifestyle than for personality, and more likely for changes during several decades (as with exercise) than for changes that take place mainly at the beginning of marriage.

By and large the results do not suggest that convergence is a major source of any of the observed spouse correlations. For nervousness the apparent divergence implies that the observed spouse correlation is an underestimate rather than an overestimate of homogamy (social homogamy or assortative mating). The low convergence/divergence during marriage does not fully insure the absence of convergence *prior to* marriage, however. Such early convergence can be tested by matching the Nord-Trøndelag data with updated information on marital duration, since a large number of persons in Nord-Trøndelag have married after the screening in 1984-86.

BIOMETRIC ANALYSIS OF SYMPTOMS OF ANXIETY/DEPRESSION: FAMILY ENVIRONMENT AND AGE- AND SEX-SPECIFIC EFFECTS

The nervousness measure used in the convergence study did not include all the 12 variables utilized in the family study, but the intercorrelation between the two indicators, 0.85, indicates that they measure approximately the same phenotype. The weak divergence for nervousness does not preclude that the observed spouse correlation is used as a proxy of homogamy in the genetic analyses of the family data.

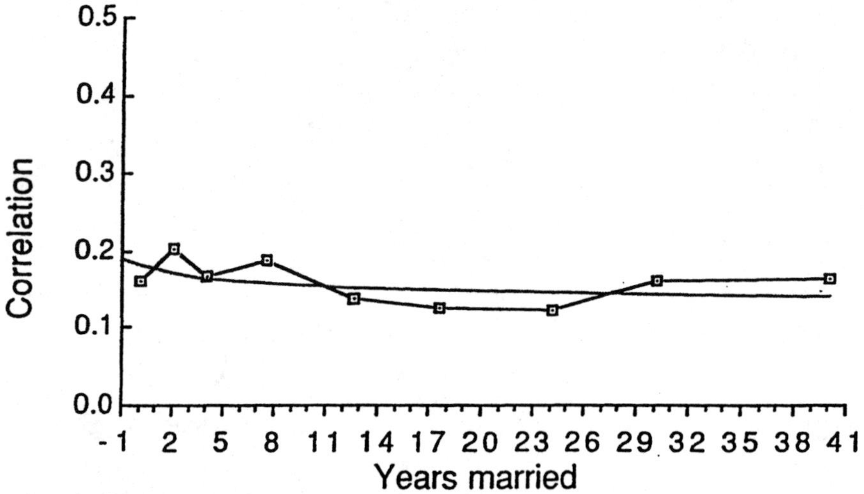

Figure 5. Model for transmission of genes, cultural factors from parental to offspring's environment, common sibling environment, and social homogamy in nuclear families

Correlations were computed separately in groups according to (1) type of relatives, (2) sex combination, and (3) age differences between relatives (for instance, siblings were classified in groups with age difference one year, two years, three years etc.). Full and reduced structural equation models based on the rules of path analysis were fit to these correlation data. A path diagram of a simplified model with only effects common to both sexes is shown in Figure 5. Additive genotypic effect, *h*, is transmitted from the paternal and maternal genotypes, or "genotypic values", G_F and G_M, to the genotypes of children, G_1 and G_2. The correlation between the genotypic value in parents and offspring is 0.5. Environmental (cultural) transmission, c, leads from parental environments, E_F and E_M, to the children's environments, E_1 and E_2. Non-random mating is assumed to stem from social homogamy, an assumption often judged to be less realistic than the alternative assumption of assortative mating. The specification of social homogamy rather than assortative mating cannot seriously affect the results unless the spouse correlation is high and the total environmental effect, e, is low, however. Combined with the specification of "environment to environment" cultural transmission rather than "phenotype to environment" transmission, the assumption of social homogamy simplifies the model by avoiding the implication of a correlation between the genotype and the phenotype. Together with age-specific genetic or environmental effects, such a correlation would have implied an extremely complex model. Some environmental factors, S, are partly shared by siblings, but not by parents, and affect each of the siblings' total environments, E_1 and E_2, through the sibling effect, s. Like-sexed DZ twins are not assumed to depart from ordinary like-sexed siblings, whereas MZ twins depart from siblings by perfectly correlated genotypes. Half-siblings grown up together are assumed to share half the genes transmitted from their common parent (corresponding to a genetic correlation of 0.25), the environmental factors transmitted from their common parent and the environmental sibling effect. Half-siblings grown up apart are only assumed to share half the genes from their common parent. The assumptions about the extent to which half-siblings share environmental family factors may not be fully correct in all cases, but the approximation can only cause negligible errors, especially if the true effects of the family environment are moderate. Not shown in the path diagram are sex-specific environmental effects from fathers and mothers and sex-specific environmental sibling effects, that is, separately for brothers, sisters, and opposite sex siblings.

The diagram also does not show the effects of *age-specific* genes or environmental factors. The parameters for these effects, δ_H and δ_S, reduce the expected correlation between relatives by multiplying the effect of h by $e^{-A\delta_H}$ or the effect of s by $e^{-A\delta_S}$, where e is the base of the natural logarithm and A is the age difference. Only one of these parameters could be specified in the same model. The structural equation models were fitted with a program based on the SAS NLIN procedure (SAS, 1988), described elsewhere (Tambs, 1991; Tambs and Moum, 1993). The observed correlations are not independent because each individual may be included in more than one pair of a certain type of relatives (for instance, first sibling paired with second sibling, first sibling with third sibling, and second sibling with third sibling) and each individual is included in more than one type of relations (mother, wife and, sister). The consequences of assuming independent data when in fact they are not fully independent have been shown to have small effect (McGue et al., 1984; Eaves et al., 1989), and are discussed elsewhere (Tambs and Moum, 1993).

The correlations for symptoms of anxiety and depression between various types of relatives are shown in Table 3. Standard errors vary from 0.007 to 0.019 in pairs of spouses, parent-offspring and siblings. There are obvious variability based upon sex of relative. For example, there is a higher resemblance between mothers and offspring than between fathers and offspring, and a higher resemblance between sisters than between other types of siblings. The remaining groups include too few cases to reveal sex differences, and are collapsed across sex in the table (but not in the further analyses).

Table 3. Sex-specific correlations for symptoms of anxiety/depression in first degree relatives

Relationship	r	N of pairs
Spouses	0.266	18,768
Father-son	0.135	7,712
Father-daughter	0.119	6,281
Mother-son	0.168	9,384
Mother-daughter	0.171	7,511
Brothers	0.124	4,103
Sisters	0.181	2,712
Brother-sister	0.111	6,319
MZ twins	0.298	57
DZ twins	0.033	60
Half-siblings together	0.296	85
Half-siblings apart	0.160	40

Table 4. Correlation between anxiety/depression scores in fathers and offspring by age differences

Age difference (years)	r	Number of pairs
Less than 22	0.113	392
22-23	0.189	716
24-25	0.088	1191
26-27	0.139	1501
28-29	0.147	1761
30-31	0.152	1717
32-33	0.120	1602
34-35	0.130	1396
36-37	0.169	1124
38-39	0.077	887
40-41	0.063	656
42-44	0.092	575
Over 45	0.119	463

Table 5. Correlations between anxiety/depression scores in mothers and offspring by age differences

Age difference (years)	r	Number of pairs
14-19	0.203	692
20-21	0.182	1336
22-23	0.144	1849
24-25	0.177	1993
26-27	0.151	2113
28-29	0.161	1944
30-31	0.173	1707
32-33	0.199	1502
34-35	0.181	1192
36-37	0.206	974
38-39	0.138	693
40-41	0.144	502
Over 41	0.063	383

Table 6. Correlations between anxiety/depression scores in siblings by age differences

Age difference (years)	r	Number of pairs
1	0.210	952
2	0.159	1941
3	0.128	1702
4	0.126	1609
5	0.121	1378
6	0.116	1102
7	0.126	949
8	0.095	804
9	0.106	607
10-11	0.128	899
12-13	0.107	555
14 and more	0.064	629

Tables 4 to 7 show correlations between pairs of parents and offspring, siblings, and half-siblings with various age differences. None of the correlation sets are significantly heterogeneous. Nevertheless, in order to test for a monotonous trend, correlation values were regressed on age differences with each data point weighted by relative proportion of number of pairs. The linear effect reached significance only in siblings (t=3.77, p=.004). The uniform tendency of decreasing correlations with age differences indicates that age-specific effects should be specified in the models.

Table 7. Correlations between anxiety/depression scores in half-siblingsby age differences

Age difference (years)	r	Number of pairs
< 4	0.394	35
4-7	0.240	49
> 7	0.150	41

Several models were fitted to the correlations in 109 groups of pairings classified by type of relationship, sex combination, and age difference. There were 4 twin groups (male and female MZ's and DZ's), 12 groups of brothers, 12 groups of sisters and 12 groups of siblings with opposite sex (each sibling group with age difference from 1 to "14 or more" years), 6 groups of maternal half-siblings grown up together (grouped by sex combination and dichotomized age difference), 5 groups of paternal half-siblings (only 5 pairs of halfsisters did not permit grouping by age difference), 5 groups of half-siblings grown up apart, and 13 age difference groups for each of the four sex combinations of parents and offspring. Spouses were not grouped by age.

The pattern of sex differences between correlations in sibling, parents, and offspring does not suggest X-linked or sex-limited genetic effect. Accordingly, models with sex-specific environmental effects seem more realistic than models with sex-specific genetic effects. Mothers and offspring are more similar than fathers and offspring, but there are no large differences between relations including sons and daughters. For the sake of simplicity, sex-specific parameters were restricted to maternal and paternal environmental effects and sex-specific environmental sibling effects.

The sample sizes and structure of the data do not permit a reliable test of models specified with more than one type of age-specific effect (either genetic or environmental) in the same model. *A priori*, the pattern of a clear age-specific trend for siblings and rather ambiguous tendencies of such age trends in other types of relationships, suggests either age-specific genetic effects or age-specific environmental factors shared only by siblings. Parameter estimates and chi-square values indicating the goodness of fit for various models are shown in Table 8. The total effect of the environment, e, is not a free parameter, but can be expressed as the square root of $(1-h^2)$. The degrees of freedom are determined by number of observations (109) minus the number of free parameters.

Model 1 is a "full" model with age-specific genetic effects. Each parameter was fixed at zero one at a time, starting with those with the lowest estimates in the full model. The results from the worst-fitting models are not shown. The estimate of environmental sibling effect in siblings with opposite sex, s_{BS}, was virtually zero, and this parameter could be excluded without an increased chi-square value (model 2). Fixing the sibling effect in brothers at zero increased the chi square value somewhat, but did not result in a significantly worse model fit (model 3). Fixing the environmental transmission from father to offspring, c_F, at zero caused a significantly worse model fit (difference of chi-square values, model 4 against model 2:X^2_1=6.63, p=0.01). No other parameters could be removed from models 1 to 3 without an even clearer worsening of the model fit.

Table 8. Parameter estimates and goodness of fit to data on anxiety/depression in families

	h	c_F	c_M	s_B	s_S	s_{BS}	m	e^*	δ_H	δ_S	X^2	df	p	AIC
1	.43	.09	.16	.17	.32	.00	.33	.90	.066	-	109.74	100	.238	-90.26
2	.43	.09	.16	.17	.32	-	.33	.90	.067	-	109.74	101	.260	-92.26
3	.43	.09	.16	-	.32	-	.33	.90	.059	-	111.95	102	.235	-92.05
4	.44	-	.10	.18	.31	-	.33	.90	.002	-	116.37	102	.157	-87.63
5	.30	.06	.12	.33	.45	.25	.29	.95	-	.093	109.61	100	.239	-90.39
6	.45	-	.09	.26	.43	.00	.33	.89	-	.159	111.69	101	.220	-90.31
7	.45	-	.09	.27	.43	-	.33	.89	-	.163	111.70	102	.240	-92.30
8	-	.09	.14	.36	.45	.30	.27	1	-	.064	110.10	101	.275	-91.90
9	.43	.02	.09	.17	.32	.00	.33	.90	-	-	114.51	102	.187	-89.49

* This parameter is derived from the other parameters in the model

h = genetic (additive) effect
c_F = environmental ("cultural") transmission from father to offspring
c_M = environmental transmission from mother to offspring
s_B = effect of environmental factors shared by brothers
s_S = effect of environmental factors shared by sisters
s_{BS} = effect of environmental factors shared by brother and sister
m = correlation between paternal and maternal environments
e = total environmental effect
δ_H = the extent to which the genetic effect, h, is age-specific
δ_S = the extent to which environmental sibling effects, s_B, s_S, and s_{BS}, are age-specific

Model 5 is a full model, only different from model 1 in the sense that the parameter for age-specific genetic effects are substituted by a parameter for age-specific environmental sibling effects. The goodness of fit is almost exactly the same as for model 1. Paternal cultural transmission, c_F, could be fixed at zero without a significant worsening of the fit (model 6), and with that parameter removed, s_{BS} was estimated at zero, and could be removed almost without an increase of the chi square value (model 7). The genetic effect could also be removed from model 4 without a significant increase of chi-square value (model 8). No other parameters could be removed from models 5-8 without a highly significant increase of chi square.

Finally, both parameters for the age-specific effects were removed simultaneously (model 9). This restriction resulted in a significant worsening of the fit compared to both model 1 (X^2_1=4.77, p=0.029), and model 5 (X^2_1=4.89, p=0.027). The results demonstrate the existence of age-specific family effects - genetic or associated with sibling environment - but do not give any clue for deciding whether the first, the second, or both types are really in action.

All models which included paternal and maternal cultural transmission were tested with the constraint that these two parameter are equal, or c_F=c_M. The constraint caused a significantly worse model fit for all models, with increases of chi-square values of 10.04 or more (p<0.002). These results show that maternal cultural transmission is stronger than paternal cultural transmission.

Table 8 includes values for Akaike's Information Criterion (Akaike, 1970). The criterion, defined as the chi-square value minus the double value of the degrees of freedom ($AIC=X^2\text{-}2df$) indicates the comparative model fit across models with varying degrees of freedom. According to the criterion a more complex model should be preferred to a more parsimonious model not only when the difference of the chi-square values is significant, but whenever the difference reach a certain value: 2.0 if the more complex model includes one extra parameter, 4.0 if it includes two extra parameters etc. Goodness of fit is expressed as low values. The four best-fitting models, models 2, 3, 7, and 8 all have very similar AIC-values.

The fraction of the phenotypic variance that can be explained by genetic effects can be expressed as the squared of the h parameter, and the proportion attributable to environmental factors as the squared of the e parameter. As shown in Figure 5, environmental factors shared by siblings affect the total environment (E_1 and E_2) directly and the phenotype indirectly through the total environment. Accordingly, the fraction of phenotypic variance explained by the environmental sibling effect can be expressed as s^2e^2. It follows from the tracing rules in path analysis and covariance algebra that the fraction of variance explained by cultural transmission from father and mother is $2c^2e^2(1+m)$, or, when cultural transmission is specific for father and mother, $c_F^2e^2+c_M^2e^2+2mc_Fc_Me^2$.

The heritability, or proportion of genetic variance, h^2, varies from zero to 0.20 in the four best-fitting models in terms of AIC-values. The corresponding proportion of total environmental variance, e^2, varies from 0.80 to 1.00. The evidence of paternal cultural transmission, c_F, is not clear. The estimates of the proportion of variance attributable to the direct effect of paternal environment, determined by the expression $c_F^2e^2$ (ignoring the covariance term, $2mc_Fc_Me^2$, between maternal and paternal environment), varies from zero to 0.8% in the four best-fitting models. There is clear evidence of a maternal cultural transmission, c_M. The estimated proportion of variance due to maternal cultural transmission, $c_M^2e^2$, varies from 0.6% (model 7) to 2.0% (models 2 and 3). The fraction of variance due to both paternal and maternal environments, including the effect of the covariance between them, varies from 0.6% (model 7) to 3.4% (model 8). The sibling effects, s_B and s_S, vary from zero to 0.36 for brothers and from 0.32 to 0.45 for sisters. Explained variances, $s_B^2e^2$ and $s_S^2e^2$, are zero (model 3) to 13.0% (model 8) for brothers and 8.8% (models 2 and 3) to 20.3% (model 8) for sisters. The corresponding effect for opposite sex siblings, s_{BS}, only departs from zero in the purely environmental model (model 8).

The strength of the age-specific genetic or environmental effects may not be readily gleaned from the values of the estimates in Table 8. However, under the assumption of age-specific *genetic* effects (model 2) the estimated value of δ_h corresponds to a correlation of 0.51 between the genotypic value at a particular age and the genotypic value of the same individual 10 years later. After 20 years this genotypic correlation has decreased to 0.26. Under the assumption of age-specific effects of *environmental* factors shared by siblings (model 7) the contribution to the sibling correlation from such factors decreases from unity in twins with no age difference to 52% in siblings 4 years apart and to 20% in siblings 10 years apart.

Table 9. Correlations between siblings born 1955 - 1965, living apart or together in 1980

	Apart			Together				
	r	N	Mean age difference	r	N	Mean age difference	Z	p
Brothers	-0.011	283	3.8	0.206	411	3.2	2.85	0.005
Sisters	0.224	478	3.4	0.300	134	2.8	0.84	0.40
Brother-sister	0.120	745	3.5	0.187	482	3.1	1.16	0.25

Siblings could be classified as living together or not in the census-year 1980. Table 9 shows correlations between siblings living together or apart, aged 14-25 in 1980. The sibling resemblance is significantly higher in brothers living together than in brothers living apart and there are tendencies in the same direction for sisters, for unlike-sexed siblings, and, not tabulated, for parents and offspring. Similar mean age differences in siblings living together and apart do not suggest that the results in Table 9 primarily reflect a confounding between effects of age difference and the effect of cohabitation during late adolescence and early adulthood. Rather, the results contribute supplementary evidence of environmental family effects during this age period. An alternative explanation, that sibling resemblance of mental disorders increases the likelihood of staying home together, cannot be completely ruled out, however.

THE INCLUSION OF DATA FROM A NEW TWIN PANEL

Recently all Norwegian twins born 1967-1972 were requested to return a questionnaire (Harris, Tambs, and Magnus, 1993). Included in the questionnaire were 7 of the items on which the 12-item anxiety/depression indicator used in the family study was based. The correlation between the 12-item indicator and a weighted sum of the 7 items used in the twin study was 0.97, so the two measures can for most purposes be assumed identical. The cotwin correlation for anxiety/depression was 0.421 (*s.e.*=0.040) in 413 male MZ pairs, 0.432 (*s.e.*=0.035) in 526 pairs of female MZ pairs, 0.318 (*s.e.*=0.046) in 386 male DZ pairs, 0.179 (*s.e.*=0.046) in 441 female DZ pairs, and 0.168 (*s.e.*=0.035) in 793 unlike sex pairs. The low unlike sex correlation is in line with the results from the family study, but the significantly higher DZ correlation in brothers than sisters (z=2.14, p=0.032) is in contrast to the higher correlation in ordinary sisters than in brothers. The "double difference" heritability in males (twice the difference between the male MZ and DZ correlations) is 0.21 (*s.e.*=0.12), which corresponds well with the results from the family study, but the heritability based on the female twin correlations, 0.51 (*s.e.*=0.12) is clearly out of range compared to the family data results. Table 10 shows the parameter estimates and goodness-of-fit statistics for the two best-fitting models, analyzing the *combined* data from the family and the twin study. Only one of the models was significantly rejected, but the worsening of the fit compared to the analyses without the data from the twin study (models 1 and 6 in Table 8) was highly significant for both models (X^2_5=19.49, p=0.002 and X^2_5=23.00, p=0.0003). The estimates of the age-specific effects, δ_h and δ_s, are unrealistically high and the genetic effects increase somewhat. The twin data have been analyzed in more detail elsewhere (Tambs, Harris, and Magnus, 1995). The preliminary analyses illustrate the point, however, that family studies and twin studies tend to yield different results for some phenotypes.

Table 10. Parameter estimates and goodness of fit to data on anxiety/depression in families, including data from a separate twin study

h	c_F	c_M	s_B	s_S	s_{BS}	m	e^*	δ_H	δ_S	X^2	df	p	AIC
.49	.11	.18	.34	.44	.24	.34	.87	.850	-	129.23	105	.238	-80.27
.43	-	.10	.47	.41	.18	.32	.90	-	.327	134.69	106	.220	-77.31

This parameter is derived from the other parameters in the model

h = genetic (additive) effect
c_F = environmental ("cultural") transmission from father to offspring
c_M = environmental transmission from mother to offspring
s_B = effect of environmental factors shared by brothers
s_S = effect of environmental factors shared by sisters
s_{BS} = effect of environmental factors shared by brother and sister
m = correlation between paternal and maternal environments
e = total environmental effect
δ_H = the extent to which the genetic effect, h, is age-specific
δ_S = the extent to which environmental sibling effects, s_B, s_S, and s_{BS}, are age-specific

CONCLUSIONS

Some evidence suggest that heritability estimates from classical twin studies, which have contributed most of the results in biometric genetical analyses of behavior, may be biased upwards because MZ twins have more equal environments than DZ twins (Rose et al., 1988), although a large majority of studies testing such possible bias show negative results (Kendler et al., 1993; Loehlin and Nichols, 1976; Matheny, Wilson, and Brown, 1976; Scarr and Carter-Saltzman, 1979; Smith, 1965; Tambs, Sundet, and Berg, 1985; Vandenberg, 1984). Further, results from twin studies may be biased in any direction due to selective recruitment (Martin and Wilson, 1982; Lykken, McGue, and Tellegen, 1987) or other sources of systematic errors. Adoption studies may suffer from other sources of errors, including random fluctuation due to moderate sample sizes. There are probably also pitfalls specific to family studies and various types of "extended" family studies, although these are not extensively discussed in the literature. The likelihood of method specific bias necessitates evidence from all types of family relations, and the larger variety of relatives that can be included in the same analysis, the better. Family data on behavior are scarce, and thus, highly needed.

Like there probably are method-specific types of biases, each method can provide specific types of information. The results presented in this chapter represent only a few examples of the utilization of family data. *Spouse resemblance* can be assessed. Models for non-random mating - either specifying assortative mating or social homogamy - can be tested if spouses of siblings are available. The effect of mutual influence and shared environment (convergence) in spouses can be estimated if data on marital duration is available. Observations of spouse correlations are important because ignoring marked deviations from random mating will bias the results from biometric genetic analyses of relatives. Likewise mis-specifications of assortative mating as social homogamy, social homogamy as assortative mating, or convergence as any of the two may cause biased results. Only the observation of spouse correlations for a set of variables and the estimation of convergence for these variables was described in the present chapter. The results suggest that there may be some convergence for life-style variables like alcohol consumption and exercise, and a slight negative convergence (divergence) for some personality traits, but in general, spouse resemblance does not appear to be much changed after mating.

The results from the biometric analyses of the extended family data showed how large family samples can be useful in separating *cultural transmission* and *environmental effects shared by siblings* and in separating such male and female effects. The results for symptoms of anxiety/depression show cultural transmission from parents to offspring and besides of the effect of cultural transmission, there are other effects of environmental factors shared by siblings. Maternal cultural transmission is stronger than paternal cultural transmission, the latter being weak or absent. An effect of environmental factors shared by sisters was assessed to be substantial (explaining 8-20% of the total variance), and there was a corresponding effect in brothers (2% and 6% of the total variance in the two best-fitting models). However, there is no evidence of an environmental sibling effect in unlike-sexed siblings, implying that males and females siblings share different environmental factors that affect anxiety/depression.

The correlations from the briefly described twin study are not fully convergent with the correlations from the extended family study. Biometric analyses of these twin data showed lower effects of shared family environment than the family study even though twins probably share more of the environment than other relatives (Tambs et al., 1994).

Age-specific genetic or environmental factors can represent a problem in (extended) family studies if left undetected. Age-specific genetic effects cause decreased correlations in siblings, and, even more so, in parents and offspring compared to twins. This pattern can cause bias as described in the introduction. Assuming age-specific genetic effects, the genotypic value at a certain age is only correlated 0.51 with the genetic value 10 years later, and the correlation is reduced to 0.26 after 20 years. Assuming age-specific environmental sibling effects, only a small fraction of environmental factors shared by like-sexed siblings close in age are shared by like-sexed siblings with age difference 10 years or more.

Even with what may be considered quite strong age-specific effects, the heterogeneity for correlations in relatives by age difference can only be detected in large samples. The observed effect of age differences on the correlations in siblings would only have been just-significant (and the corresponding tendency in parents and offspring would have been far from significant) with a sample size reduced to 27% of the actual sample size. Thus, with a sample size one quarter of ours, which is still a large sample, the age effect might not have been detected. Ignoring a possible age-specific *genetic* effect, like in model 9, Table 8, almost did not change the parameter estimates, however, except for lower effects of cultural transmission (compared to the corresponding model with age-specific genetic effects, model 1). Assuming that the age-specific pattern is due to *sibling* effects, neglecting it would cause inflated genetic effect and deflated environmental sibling effect (comparing the estimates from model 9 with those from model 5). Judged from the present data, the detection of age-specific effects requires very large samples, but the failure of detecting such effects does not necessarily affect the parameter estimates dramatically.

Whereas relatively powerful twin and adoption studies may represent a better cost-benefit ratio than nuclear family studies under "traditional" ascertainment, nuclear family data from nation-wide public *registries* may represent inexpensive data bases. Such data are almost free from recruitment bias, and very large sample sizes permit more complex models and more precise estimates than what have been demonstrated in this chapter. In many countries, like the Scandinavian, census and birth registry data may be utilized to group all or almost all members of a population in families. Individual data on for example educational attainment, income, mental illness or cancer from nation-wide registries may be subject to biometric analyses. In order to protect personal integrity, personal identification data can be removed before the files are released for research purposes. Of course, such data are greatly enriched if matched with twin or adoption registries, or if information from broken homes (biological vs. social parents-offspring and half-siblings reared together or apart) is available. The data base described here is very large compared to all existing family samples, but nonetheless stem from a single county comprising only 2.5% of the population of a small country like Norway.

REFERENCES

Akaike, H. (1970). Statistical predictor identification. *Ann Inst Stat Math*, *21*, 243-247.

Basavarajappa, K.G., Norris, M.J., & Hall, S.S. (1988). Spouse selection in Canada, 1921-78: An examination by age, sex and religion. *Journal of Biosocial Science*, *20*, 211-233.

Cavalli-Sforza, L.L., Feldman, M.W., Chen, K.H., & Dornbusch, S.M. (1982). Theory and observation in cultural transmission. *Science*, *218*, 19-27.

Clark, A.C. (1952). An examination of the operation of the residential propinquity as a factor in mate selection. *American Sociological Review*, *17*, 17-22.

Ditto, B. & France, C. (1990). Similarities within young and middle-aged spouse pairs in behavioral and cardiovascular response to two experimental stressors. *Psychosom Med*, *52*, 425-434.

Eaves, L.J., Eysenck, H.J. & Martin, N.G. (1989). *Genes, Culture, and Personality: An Empirical Approach*. London: Academic Press.

Eaves, L.J., Long, J., & Heath, A.C. (1986). A theory of developmental change in quantitative phenotypes applied to cognitive development. *Behavior Genetics*, *16*, 143-162.

Eysenck, H.J. & Wakefield, J.A. (1981). Psychological factors as predictors of marital satisfaction. *Advanced Behavior Research Therapy*, *3*, 151-192.

Farley, F.H. & Davis, S.A. (1977). Arousal, personality, and assortative mating in marriage. *J Sex Marital Ther*, *3*, 122-127.

Farley, F.H. & Mueller, C.B. (1978). Arousal, personality, and assortative mating in marriage: generalizability and cross-cultural factors. *J Sex Marital Ther*, *4*, 50-53.

Guttman, R. & Zohar, A. (1987). Spouse similarity in personality items: Change over years of marriage and implications for mate selection. *Behavior Genetics*, *17*, 179-189.

Harris, J., Tambs, K. & Magnus, P. (1995). *A new Norwegian twin study. Data description and analyses of body mass*. (un pub)

Heath, A.C., Berge, K., Eaves, L.J., Solaas, M.J., Sundet, J.M., Nance, W.E., Corey, L.A., & Magnus, P. (1985). No decline in assortative mating for educational level. *Behavior Genetics*, *15*, 349-369.

Heath, A.C. & Eaves, L.J. (1985). Resolving the effects of phenotype and social background on mate selection. *Behavior Genetics*, *15*, 15-30.

Hesbacker, P.T., Rickels, R., Morris, R.J., Newman, H., & Rosenfeld, M.D. (1980). Psychiatric illness in family practice. *Journal of Clinical Psychiatry*, *41*, 6-10.

Holmen, J., Midthjell, K., Bjartveit, K., Hjort, P.F., Lund-Larsen, P.G., Moum, T., Naess, S. & Waller, H.T. (1990). *The Nord-Trondelag Health Survey 1984-86*. Verdal, Norway: National Institute of Public Health, Community Medicine Research Center.

Johnson, R.C., Ahern, F.M., & Cole, R.E. (1980). Secular change in degree of assortative mating for ability? *Behavior Genetics*, *10*, 1-8.

Kendler, K.S., Neale, M.C., Kessler, R.C., Heath, A.C., & Eaves, L.J. (1993). A test of the equal-environment assumption in twin studies of psychiatric illness. *Behavior Genetics*, *23*, 21-27.

Kerckhoff, A.C. & Davis, E.K. (1962). Value consensus and need complementarity in mate selection. *American Sociological Review*, *27*, 295-303.

Lesnik-Oberstein, M. & Cohen, L. (1984). Cognitive style, sensation seeking, and assortative mating. *Jòurnal of Personality and Social Psychology*, *46*, 112-117.

Loehlin, J.C. & Nichols, R.L. (1976). *Heredity, Environment, and Personality. A Study of 850 Sets of Twins*. Austin: University of Texas Press.

Lynum, A.F. & Skau Dahl, E. (1987). *Ektefellevalg. En undersokelse av selekivt partnervalg pa personlighetsfaktorer*. Thesis at Institute of Psychology, University of Oslo.

Martin, N.G. (1978). Genetics of sexual and social attitudes. In W.E. Nance (Ed.). *Twin Research: Psychology and Methodology* New York: Alan R. Liss

Martin, N.G. & Jardine, R. (1986). Eysenck's contribution to behavior genetics. In S. Modgil & C. Modgil (Eds.). *Hans Eysenck: Consensus and Controversy* (pp. 13-47). Lewis, Sussex: Palmer Press

Mascie-Taylor, C.G. (1989). Spouse similarity for IQ and personality and convergence. *Behavior Genetics*, *19*, 223-227.

Matheny, A.P., Wilson, R.S., & Brown, D.A. (1976). Relations between twins' similarity of appearance and behavioral similarity: Testing an assumption. *Behavior Genetics*, *6*, 343-350.

McGue, M., Wette, R., & Rao, D.C. (1989). Path analysis under generalized marital resemblance: Evaluation of the assumptions underlying the mixed homogamy model by the Monte Carlo Method. *Genetic Epidemiology*, *6*, 373-388.

McLemore, S.D. (1980). *Racial and ethnic relations in America*. New Haven, CT: Allyn & Bacon, Inc..

Merikangas, K.M. (1982). Assortative mating for psychiatric disorders and psychological traits. *Archives of General Psychiatry*, *39*, 1173-1180.

Murstein, B.I. (1976). *Who will marry whom? Theories and research in marital choice*. New York: Springer.

Price, R.A. & Vandenberg, S.G. (1980). Spouse similarity in American and Swedish couples. *Behavior Genetics*, *10*, 59-71.

Rose, R.J., Koskenvuo, M., Kaprio, J., Sarna, S., & Langinvainio, H. (5495). Shared genes, shared experiences, and similarity of personality: Data from 14,288 adult Finnish co-twins. *Journal of Personality and Social Psychology*, *54161*, 161-171.

SAS/STAT, (1988). *User's Guide*. Carey, N.C.: SAS Institute, Inc..

Scarr, S. & Carter-Saltzman, L. (1979). Twin method: Defense of a critical assumption. *Behavior Genetics, 9*, 527-542.

Schumm, W.R., Figley, C.R., & Fuhs, N.N. (1980). Similarity in self-esteem as a function of duration of marriage among student couples. *Psychological Reports, 47*, 365-366.

Smith, R.T. (1965). A comparison of socioenvironmental factors in monozygotic and dizygotic twins testing an assumption. In S.G. Vandenberg (Ed.). *Methods and Goals in Human Behavior Genetics* (pp. 45-61). New York: Academic Press

Tambs, K. (1991). Transmission of symptoms of anxiety and depression in nuclear families. *J Affect Disord, 21*, 117-126.

Tambs, K., Moum, T., Eaves, L.J., Neale, M.C., Midthjell, K., Lund-Larsen, G., & Naess, S. (1991). Genetic and environmental contributions to the variance of body height in a sample of first and second degree relatives. *American Journal of Human Biology, 3*, 257-267.

Tambs, K., Moum, T., Eaves, L.J., Neale, M.C., Midthjell, K., Lund-Larsen, P.G., & Naess, S. (1992a). Genetic and environmental contributions to the variance of body height in a sample of first and second degree relatives. *Am J Phys Anthropol, 88*, 285-294.

Tambs, K., Moum, T., Holmen, J., Eaves, L.J., Neale, M.C., Lund-Larsen, G., & Naess, S. (1992b). Genetic and environmental effects on blood pressure in a Norwegian sample. *Genetic Epidemiology, 9*, 11-26.

Tambs, K., Eaves, L.J., Moum, T., Holmen, J., Neale, M.C., Naess, S., & Lund-Larsen, P.G. (1993). Age-specific genetic effects for blood pressure. *Hypertension, 22*, 789-795.

Tambs, K., Harris, J. & Magnus, P. (1995). Sex specific causal factors and effects of common environment for anxiety and depression in twins. (*Behavior Genetics*, 25, 33-44)

Tambs, K. & Moum, T. (1992). No large convergence during marriage for life-style and personality. *Journal of Marriage and the Family, 54*, 957-971.

Tambs, K. & Moum, T. (1993). Low genetic effect and age-specific family effect for symptoms of anxiety and depression in nuclear families, halfsibs and twins. *J Affect Disord, 27*, 183-195.

Tambs, K., Sundet, J.M., & Berg, K. (1985). Cotwin closeness in monozygotic and dizygotic twins: a biasing factor in IQ heritability analysis? *Acta Genet Med Gemellol (Roma), 34*, 33-39.

Tambs, K., Sundet, J.M., Eaves, L., & Berg, K. (1992). Genetic and environmental effects on type A scores in monozygotic twin families. *Behavior Genetics, 22*, 499-513.

Tambs, K., Sundet, J.M., Eaves, L., Solaas, M.H., & Berg, K. (1991). Pedigree analysis of Eysenck Personality Questionnaire (EPQ) scores in monozygotic (MZ) twin families. *Behavior Genetics, 21*, 369-382.

Tambs, K., Sundet, J.M., Magnus, P., & Berg, K. (1989). Genetic and environmental contributions to the covariance between occupational status, educational attainment, and IQ: a study of twins. *Behavior Genetics*, *19*, 209-222.

Truett, K.R., Eaves, L.J., Walters, E.E., Heath, A.C., Hewitt, J.K., Meyer, J.M., Silberg, J., Neale, M.C., Martin, N.G., & Kendler, K.S. (1994). A model system for analysis of family resemblance in extended kinships of twins. *Behavior Genetics*, *24*, 35-49.

Vandenberg, S.G. (1984). Does a special twin situation contribute to similarity for abilities in MZ and DZ twins? *Acta Genet Med Gemellol (Roma)*, *33*, 219-222.

Watkins, M.P. & Meredith, W. (1981). Spouse similarity in newlyweds with respect to specific cognitive abilities, socioeconomic status, and education. *Behavior Genetics*, *11*, 1-21.

Winokur, A., Winokur, D.F., Rickels, K., & Cox, D.S. (1984). Symptoms of emotional distress in a family planning service: stability over a four-week period. *Br J Psychiatry*, *144*, 395-399.

Zeller, R.A. & Carmines, E.G. (1980). *Measurement in the Social Sciences*. Cambridge, UK: Cambridge University Press.

CHAPTER 7

ANALYSIS OF FAMILIAL AGGREGATION IN A BINARY TRAIT BY LOGISTIC REGRESSION AND THE REGRESSIVE LOGISTIC MODEL

John L. Hopper[1], Mark A. Jenkins[1], Gregory T. Macaskill[1] and Graham G. Giles

[1]The University of Melbourne
The Faculty of Medicine Epidemiology Unit
151 Barry Street, Carlton, Victoria 3053, Australia

[2]The Anti-Cancer Council of Victoria
1 Rathdowne Street, Carlton, Victoria 3052, Australia

INTRODUCTION

This chapter is concerned with the statistical analysis of familial aggregation in a binary trait. A binary trait is one which can take two values, such as presence or absence of a particular behaviour, disease, or condition. Familial aggregation is the tendency for a trait to be more similar between individuals from the same family than between individuals from different families.

Data sets typically are derived from a number of sets of related individuals, which are given the general term "pedigrees". Pedigrees may be large kinships over a number of generations, or simply pairs (such as twins, spouses, or a parent and child). Technically, an individual on its own can be considered a pedigree. From an analytical point of view it is important to distinguish a data set derived only from pedigrees of the same size and general structure, called "regular pedigrees", from one derived from pedigrees of differing sizes and/or structures. An example of the former case would be a data set derived from twin pairs alone, or as in the application we shall consider, from sets of triples consisting of a child and both parents. These designs are considered to be regular even though some characteristics of individuals, such as their age, sex, or zygosity if twins, can differ from set to set. Regular designs may allow a less computationally burdensome statistical analysis to be conducted.

One analytical aim is to describe familial aggregation, through estimating the association between pairs of individuals of a given type. Comparisons of these associations between pairs of different genetic or environmental relationships to one another may give insight into possible genetic and environmental aetiologies. In this chapter it will be demonstrated how the established statistical methodologies of maximum likelihood theory and logistic regression can be used to perform these analyses, and how these approaches provide a framework upon which more sophisticated regressive logistic models, capable of examining simultaneously possible genetic and environmental aetiologies, can be developed.

Some of the theory will be demonstrated by application to data on smoking behaviour from parents and their offspring, based on the 1968 Tasmanian Asthma Survey, and its 1991-93 General Health Survey Follow Up. An interesting feature of this data is that the smoking behaviour is obtained on the two generations some 25 years apart, so that at the time of measurement individuals from parent and offspring generations are of approximately the same age.

BACKGROUND TO THE EXAMPLES

EXAMPLES 1 AND 2: CURRENT SMOKING IN ADULTS, AND IN THEIR PARENTS WHEN THE SAME AGE

In the 1968 Tasmanian Asthma Survey, parents of all 8,683 school children in Tasmanaia, Australia, born in 1961, were asked to complete questionnaires on the respiratory and allergic history of themselves and of their children. The parents were also asked: Do you smoke every day (or six days out of seven)?" Responses for 8,161 mothers and 8,112 fathers (95% of parents), and for 8,585 (99%) seven year olds (the cohort), were obtained. The mean age of parents was 35.2 years (standard deviation (s.d.) 6.8, range 20 to 75). The administration, methodology and questionnaires have been reported; see Gibson et al. (1969).

During 1991-93 a stratified random sample of 2,000 cohort members, half classified in 1968 by their parents as asthmatic, were followed up by mailed questionnaire. This included the question: "Do you currently smoke?" To date responses have been obtained from 1,494 subjects (75%), with mean age 31.0 years (s.d. 0.6, range 29 to 32). The response rate was independent of asthma status in 1968; see Hopper et al. (1994a) and Jenkins et al. (1994).

EXAMPLE 3: HAVING EVER SMOKED IN ADULTS, THEIR SIBLINGS AND PARENTS

During 1992-93 all women under the age of 40 living in the Melbourne Metropolitan Area with a recent diagnosis of breast cancer, as recorded on the Victorian Cancer Registry (VCR), were approached as part of a case-control-family study of breast cancer (Hopper et al., 1994b). Defined sets of relatives were administered an identical questionnaire, which included an item on smoking history. The data set relevant to this paper consists of responses to the question: "Over your lifetime, in total have you smoked more than 100 cigarettes?". This was recorded for 60 families totalling 331 individuals, comprised of 60 mothers, 60 fathers, and 211 offspring. The VCR is a population-based registry to which cancer registration is a statutory requirement, and is considered to have complete ascertainment of breast cancer. As smoking is not a recognised risk factor for breast cancer, and an 80% response rate was achieved, it was not considered necessary to adjust for ascertainment in the analysis of familial aggregation for smoking.

LOGISTIC REGRESSION

Whether or not two individuals are "alike" for a trait depends on what one might "expect" their trait value to be. For example, two pygmies may be very similar for height when compared with the rest of the human population, but could be quite different when compared with fellow pygmies. The same applies to binary traits; a pair of individuals who both have a rare condition, such as hearing impairment, might appear to be very "alike". If both individuals are 100 years old, however, the fact that they are both hard of hearing is not surprising. As another example, the concordancy between two siblings for breast cancer is more striking if they are of opposite sex, and even more so if they are both males. Therefore, associations between traits cannot be properly interpreted without first specifying the expected, or mean trait value of individuals, and this may depend on important determinants, such as age, sex, etc. For binary traits, the expected or mean value is the prevalence, and logistic regression has become a popular way of examining prevalence as a function of measured factors. The statistical process is analogous to linear regression for continuous traits.

Define a binary trait by letting $Y_i = 1$ if individual i is "affected", otherwise $Y_i = 0$. Suppose there are one or more measured covariates, $\mathbf{X} = (X_1,...,X_p)'$ which may determine the expected value of Y_i which, given the corresponding observed values $\mathbf{x}_i = (x_{1i},...,x_{pi})'$, is

$$\mu_i = E(Y_i \mid \mathbf{X}_i=\mathbf{x}_i) = P(Y_i = 1 \mid \mathbf{X}_i=\mathbf{x}_i) = \pi_i. \qquad (1)$$

This conditional mean is equivalent to a conditional probability, and should be modelled by an appropriate function of the measured covariates. Note that the mean cannot take values outside the range of 0 to 1. Moreover, $Var(Y_i \mid \mathbf{X}_i=\mathbf{x}_i) = \pi_i(1-\pi_i)$ is not independent of the mean.

For a given probability π, the corresponding "odds" are $\pi/(1-\pi)$. For any probability π, the logit, log odds or logistic transformation

$$\text{logit}(\pi) = \log[\pi/(1-\pi)] \quad (2)$$

always lies between $-\infty$ and $+\infty$, and is an increasing function of π. For this and other statistical reasons it is convenient to propose a linear logistic model for the mean of a binary trait, such as

$$\text{logit}(E(Y_i \mid \mathbf{X}_i=\mathbf{x}_i)) = \alpha_0 + \alpha_1 x_{1i} + \ldots + \alpha_p x_{pi}, \quad (3)$$

which is equivalent to

$$p_i = \exp(\alpha_0+\alpha_1 x_{1i}+\ldots+\alpha_p x_{pi})/[1+\exp(\alpha_0+\alpha_1 x_{1i}+\ldots+\alpha_p x_{pi})]. \quad (4)$$

The coefficients $\{\alpha_k\}$ can be interpreted as *log odds ratios*, a familiar concept in epidemiology. For example, suppose X_k represents an indicator or binary variable, so that x_k can equal 1 or 0, say. If two individuals, i and j , differ only on x_k ($x_{ki} = 0$, $x_{kj} = 1$, say), then by equation (3) it can be seen that $\text{logit}(E(Y_j \mid \mathbf{X}_j=\mathbf{x}_j))$ and $\text{logit}(E(Y_i \mid \mathbf{X}_i=\mathbf{x}_i))$ differ only in their a_k term. As $x_{kj}-x_{ji} = 1$, then $a_k = \text{logit}(E(Y_j \mid \mathbf{X}_j=\mathbf{x}_j)) - \text{logit}(E(Y_i \mid \mathbf{X}_i=\mathbf{x}_i))$. On substituting the definition of the logit given by equation (2), and equation (1), this becomes

$$\begin{aligned}\alpha_k &= \log\{E(Y_j \mid X_j=x_j)[1-E(Y_i \mid X_i=x_i)]/[1-E(Y_j \mid X_j=x_j)]E(Y_i \mid X_i=x_i)\} \\ &= \log(\{E(Y_j \mid X_j=x_j)/[1-E(Y_j \mid X_j=x_j)]\}/\{E(Y_i \mid X_i=x_i)/[1-E(Y_i \mid X_i=x_i)]\}) \\ &= \log(\{P(Y_j=1 \mid X_j=x_j)/[1-P(Y_j=1 \mid X_j=x_j)]\}/\{P(Y_i=1 \mid X_i=x_i)/[1-P(Y_i=1 \mid X_i=x_i)]\});\end{aligned}$$

i.e. the logarithm of the ratio of the odds that an individual is affected when $X_k = 1$, to the odds that an individual is affected when $X_k = 0$. If the probabilities of being affected, π_i and π_j by equation (1), are small (close to 0), then α_k is approximately $\log(\pi_j/\pi_i)$, so that $\exp(\alpha_k) \approx \pi_j/\pi_i$, the risk ratio. If π_i and π_j are large (close to 1), the $\exp(\alpha_k) \approx (1-\pi_i)/(1-\pi_j)$.

A powerful unifying principle in statistical inference is that of the likelihood function, and the estimation of parameters by maximising the likelihood, introduced by R.A. Fisher (1912). This method can be used in quite complicated situations, but for illustrative purposes let us consider a simple but important example.

Consider a random sample of N independent observations from a distribution parameterised by an unknown scalar, ϕ, which is to be estimated. Let $\{y_1, y_2, \ldots , y_N\}$ represent the N observations and $f(y_i,\phi)$, $i = 1,2,\ldots,N$ be the respective values of the density function at the point y_i, for a given ϕ. Being independent realisations, the density of the sample, L, is the product of the densities for each observtion; i.e. $L = f(y_1,\phi)f(y_2,\phi)\ldots f(y_N,\phi)$. The observations, $\{y_i\}$, are known, while the parameter, f, is unknown. For given $\{y_i\}$, let ϕ be the value of ϕ which maximises $L = L(\phi)$; in a sense, this is the value of ϕ which is the most likely, given the observations. It is called the *maximum likelihood estimate* of ϕ.

For many cases, maximum likelihood estimates enjoy optimal properties among all estimates. Where exact results are not available, asymptotic results can be obtained under quite general conditions. At least for large samples, ϕ has a normal distribution with mean ϕ, and a variance which can be approximated from the shape of the likelihood function, L, around the region of ϕ. If this cannot be calculated analytically, it can be approximated closely by numerical algorithms. For a number of reasons, it is more useful to work with the natural logarithm of the likelihood, LL = log(L), referred to as the log likelihood, for short.

The variance of ϕ decreases as the sample size, N, increases. The square root of this variance, s, is called the *standard error* of ϕ. Knowledge of s can be used to test hypotheses about the parameter, such as whether it differs from a given number; e.g. 1 or 0. An approximate *95% confidence interval* for ϕ is given by the interval from ϕ - 1.96s to ϕ + 1.96s, and in a sense the true value, ϕ, is reasonably certain to lie within this region. Therefore the maximum likelihood estimate can be expected to be close to the unknown true value, ϕ, the more so the larger the sample. On average, estimates derived by the method of maxiumum likelihood will be at least as close to ϕ as estimates derived by any other method.

The mathematical theory has been extensively developed. Maximum likelihood estimation can be extended to cases where the sample likelihood function, L, is expressed as a function of a vector of several unknown parameters, $\boldsymbol{\phi} = (\phi_1,\phi_2,...)'$, such as when the density, f, is a function of more than one parameter. Standard errors for each parameter can be calculated, as well as the correlation coefficient between pairs of estimates, and these can be used to test if parameters differ from one another.

A *statistical model* can therefore be derived for a given situation, in effect defined by the likelihood function in terms of a set of parameters, $\boldsymbol{\phi}$, and the observed data. The model is fitted by maximum likelihood to given data. This yields an estimate for each component of $\boldsymbol{\phi}$, and statements about how precise these estimates are can be derived. A general test procedure, called the *likelihood ratio criterion*, allows testing of whether more complicated models involving more parameters are justified. Application of this criterion can lead to selection of one (or more) parsimonious models, in which the data is described as well as possible by as few parameters as necessary. Application of the likelihood ratio criterion to logistic regression is discussed below. The reader is referred to Clayton & Hills (1993) for a self-contained and clearly written textbook on statistical modelling in epidemiology, using likelihood theory, written so as to be accessible to students with little formal training in statistics.

Returning to logistic regression, let $\{y_i\}$ be a sample of N independent observed binary values. The log likelihood of the sample is

$$LL = \sum_{i=1}^{N} \log[P(Y_i=y_i \mid \mathbf{X}_i=\mathbf{x}_i)] = \sum_{i=1}^{N} [y_i \log(\pi_i) + (1-y_i)\log(1-\pi_i)]. \quad (5)$$

Substituting the expressions given by equation (3), the log likelihood can be expressed in terms of the parameters $\{a_i\}$. This *linear logistic regression model* can then be fitted by maximum likelihood, and a computational procedure is available in most major statistical packages. (Note that the term "linear" refers to the form of the right hand side of equation (3); i.e. the logit is a linear function of the *parameters*. It is possible for the right hand side to be non-linear in a *covariate*, however. For example, the right hand side could include both age and age^2, and so represent a quadratic function of age).

The *likelihood ratio criterion* works as follows: if LL_t is the log likelihood under a model with t parameters, and LL_s is that under a nested model with s parameters, with $s < t$, then under the null hypothesis that the true model is the one given by s parameters, $-2(LL_s - LL_t)$ is approximately distributed as a $c_{t\text{-}s}^2$ variable. Therefore, if in practice twice the change in log likelihood associated with fitting a more complicated model (i.e. one which contains additional parameters) is more than might be attributed to chance, as judged by comparing the actual change to the critical values of the $c_{t\text{-}s}^2$ distribution, there is statistical justification for rejecting the simpler model, and for claiming that the additional parameters may represent real effects.

The *scaled deviance* of a (current) model is defined as $S_c = -2(LL_c - LL_f)$, where LL_c is the log likelihood under the current model, and LL_f is the log likelihood under the "full" model (in which the data is fitted exactly). Therefore, the change in scaled deviance, $S_t - S_s$, between two nested models will be approximately distributed as a $c_{t\text{-}s}^2$ variable, by the likelihood ratio criterion above. S_c on its own can be used as a rough goodness-of-fit test, by comparing with a $c_{c\text{-}f}^2$ distribution, where c is the number of parameters in the current model and f is the number of parameters in the full model. For binomial data little is known about how good the c^2 approximation is for small data sets; the approximation may be better for the *change* in deviance, rather than for the absolute deviance. For absolute deviances, valid significance values are in general not available (McCullagh & Nelder, 1983).

EXAMPLE 1: BEHAVIOUR OF AN OFFSPRING AS A FUNCTION OF THAT OF THE PARENTS

Let a family set be defined by the offspring and both parents; i.e. a regular pedigree of size 3. Let $Y_i = 1$ if the offspring reported in 1991-93 being a current smoker, and 0 otherwise. Let $X_{mi} = 1$ if the mother reported being a current smoker in 1968, else $X_{mi} = 0$, and similarly define X_{fi}. There were N = 1,397 such pedigrees in which both X_{mi} and X_{fi} were recorded. The breakdown of Y by X_m and X_f is given in Table 1.

Overall 495 offspring (35%) were current smokers. Of the 510 offspring whose mother was a smoker, 219 (43%) were current smokers, compared to 276 (31%) out of 887 whose mother was not a current smoker. Therefore the odds ratio (see equation (10)) associated with mother being a smoker, as a predictor of adult smoking, is 219x611/(291x276) = 1.67. The results of a logistic regression analysis are given in Table 2.

For males and females combined, model I shows that from fitting equation (3) with just the factor X_{mi}, the estimate of the regression coefficient, α_m, is 0.510 with standard error 0.115. Note that $e^{0.510} = 1.67$. The 95% confidence interval for the odds ratio estimate of 1.67 is from $e^{0.510-1.96x0.115} = 1.33$ to $e^{0.510+1.96x0.115} = 2.09$, and excludes unity. Similarly, model II shows that for father smoking the odds ratio is 350x405/(497x145) = 1.97, equivalent to the estimated regression coefficient, α_f, of 0.677, which has a standard error of 0.119. The 95% confidence interval is from 1.56 to 2.48.

Table 1. Current smoking status of cohort members in 1991-93 as a function of their parents current smoking status in 1968. Numbers recorded in each cell are for males, females, and both sexes combined, in vertical descent.

			Mother smokes				
			Yes		No		
			Father smokes		Father smokes		Total
			Yes	No	Yes	No	
Current	Yes	males	98	19	105	66	288
		females	79	23	68	37	207
		total	177	42	173	103	495
Smoking							
	No	males	123	32	136	186	477
		females	102	34	136	153	425
Status	total		225	66	272	339	902
Totals	males		221	51	241	252	565
		females	181	57	204	190	632
		total	402	108	445	442	1,397

Fitting equation (3) with both X_{mi} and X_{fi} as factors, the estimates of α_m and α_f become 0.355 (s.e. = 0.120) and 0.580 (s.e. = 0.124), respectively; see model III. These are equivalent to odds ratios and 95% confidence intervals of 1.43 (1.13 to 1.80) and 1.79 (1.40 to 2.28), respectively. (The change in these from their corresponding estimate in model I or II reflects mothers and fathers not being independent with respect to current smoking; see Example 2 below). The correlation between estimates was -0.30, and a test of the difference between these estimates gave an approximate c_1^2 statistic of 1.3. That is, there is no evidence that α_f is different from α_m, at the 0.05 level of significance.

Examination of the scaled deviances shows that poor fits are obtained if either of the parental smoking terms is omitted. Even when they are both included (model III), there is a suggestion ($c_1^2 = 3.9$) that the lack of fit is more than could be reasonably attributed to chance.

Table 2. Logistic regression coefficients (standard errors) from analysis of smoking status as an adult in an offspring as a function of the smoking status as an adult of mother and father, based on Table 1.

		Model I	Model II	Model III
Males and females combined				
	Mother smokes	0.510	-	0.355
		(0.115)		(0.120)
	Father smokes	-	0.677	0.580
			(0.119)	(0.124)
	Scaled deviance (d.f.)	26.23 (2)	12.55 (2)	3.85 (1)
Males				
Mother smokes	0.352	-	0.150	
		(0.155)		(0.164)
	Father smokes	-	0.698	0.652
			(0.159)	(0.166)
	Scaled deviance (d.f.)	17.38 (2)	2.49 (2)	1.65 (1)
Females				
	Mother smokes	0.725	-	0.611
		(0.174)		(0.179)
	Father smokes	-	0.655	0.514
			(0.182)	(0.188)
	Scaled deviance (d.f.)	9.99 (2)	13.98 (2)	2.30 (1)

The same models were fitted separately for males and females; see Table 2. For males, α_m and α_f were both significant when fitted alone. Model III shows, however, that whereas paternal smoking is a predictor of smoking as an adult, maternal smoking is no longer significant at the 0.05 level once paternal smoking has been taken into account. There is marginal evidence that α_m and α_f are different for males ($c_1^2 = 3.6$; p » 0.06). Furthermore, the scaled deviance for model II does not indicate that ignoring maternal smoking results in a poor fit. In contrast, for females α_m and α_f are of a similar magnitude, significant, and without both being fitted a poor fit results.

Table 3 presents further models of the combined data, examining the role of sex differences. Model IV shows that the probability an offspring was a smoker was marginally greater in males (logistic regression coefficient = 0.231, s.e. = 0.115), but fitting this extra factor had virtually no effect on the estimates of maternal and paternal smoking; cf model III, Table 2. Model V proposes that the parameter α_m depends on the sex of the offspring, so that maternal smoking has a greater influence on smoking as an adult in daughters compared to sons, while α_f is independent of the sex of the offspring. The change in scaled deviance from model IV is 3.32 (p » 0.07), and the scaled deviance itself gives no indication that this is a poor fit. Model VI, however, shows there is no justification for allowing α_f to depend on the sex of the offspring.

Table 3. Logistic regression coefficients (standard errors) from analysis of smoking status as an adult in an offspring as a function of the smoking status as an adult of mother and father, based on Table 1.

	Model IV	Model V	Model VI	Model VII	Model VIII
Males and females combined					
Mother smokes	0.361	0.168	0.150	0.572	0.553
	(0.120)	(0.160)	(0.164)	(0.253)	(0.258)
Father smokes	0.581	0.592	0.652	0.757	0.791
	(0.124)	(0.125)	(0.166)	(0.150)	(0.181)
Sex[1]	-0.231	-0.404	-0.326	-0.413	-0.363
	(0.115)	(0.150)	(0.205)	(0.151)	(0.210)
Mother.Sex[2]	-	0.427	0.461	0.418	0.436
		(0.235)	(0.242)	(0.235)	(0.241)
Father.Sex[2]	-	-	-0.138	-	-0.086
			(0.251)		(0.251)
Mother.Father[3]	-	-	-	-0.545	-0.535
				(0.268)	(0.269)
Scaled deviance (d.f.)	7.57 (4)	4.25 (3)	3.95 (2)	0.16 (2)	0.04 (1)

[1] sex is defined as 1 for females, and 0 for males

[2] the effect attributed to parent smoking in females, compared to that in males

[3] the interaction term for both mother and father smoking

Most of the fitted models assume the effect of covariates are additive on the logistic scale, without interactions. For males and females separately, as suggested by the scaled deviances corresponding to the fitted model IIIs in Table 2, there was no evidence for a significant interaction term between maternal and paternal smoking. Combining males and females, however, models VII and VIII of Table 3 suggest that an interaction term is justified; the change in scaled deviance between models V and VII is 4.1 ($p < 0.05$).

Model VII provides perhaps the most parsimonious description of the data. It predicts that for male offspring, the odds ratio associated with both mother *and* father smoking is $e^{0.572+0.757-0.545} = e^{0.784} = 2.19$, similar to that associated with mother alone smoking ($e^{0.572} = 1.77$) and with father alone smoking ($e^{0.757} = 2.13$). That is, the important risk factor for males is that *a* parent smokes; it does not appear to matter which one, or if both parents smoke. For female offspring, however, the odds ratio associated with father alone smoking is $e^{0.757} = 2.13$, with mother alone smoking it is $e^{0.572+0.418} = e^{0.990} = 2.69$, while for father *and* mother smoking it is $e^{0.757+0.572+0.418-0.545} = e^{1.202} = 3.33$. That is, compared to males, female offspring are more sensitive to the presence and extent of parental smoking.

MEASURING ASSOCIATION IN A BINARY TRAIT

There are number of measures of association in a binary trait, and different measures are more suitable to addressing different questions. The usual Pearson correlation coefficient between Y_1 and Y_2, $r = \text{Cov}(Y_1,Y_2) / [\text{Var}(Y_1)\text{Var}(Y_2)]^{1/2}$, commonly used to measure the association between continuous traits, is also a valid measure of association between binary traits. Dropping dependence on covariates for simplicity, the correlation between two binary traits Y_1 and Y_2 is

$$\begin{aligned}\rho &= [E(Y_1Y_2) - E(Y_1)E(Y_2)] / [\text{Var}(Y_1)\text{Var}(Y_2)]^{1/2} \\ &= [P(Y_1=1,Y_2=1) - \pi_1\pi_2] / [\pi_1(1-\pi_1)\pi_2(1-\pi_2)]^{1/2} \qquad (6)\end{aligned}$$

where $\pi_i = P(Y_i = 1)$ for $i = 1$ or 2 is the marginal probability that individual i is affected. Letting $D = [\pi_1(1-\pi_1)\pi_2(1-\pi_2)]^{½}$, by rearranging of equation (6) and noting that $P(Y_1=1,Y_2=0) = p_1 - P(Y_1=1,Y_2=1)$, etc. the joint probabilities can be written as

$$P(Y_1=j,Y_2=k) = P(Y_1=j)P(Y_2=k) + \delta_{jk}\rho D, \qquad (7)$$

$j, k = 0, 1$, where $\delta_{jk} = 1$ if $j = k$, else -1. The log likelihood of n independent pairs of observations, $\{(y_{1l},y_{2l})\}$, is

$$LL = \sum_{l=1}^{n} \log(P(Y_{1l}=y_{1l},Y_{2l}=y_{2l})), \qquad (8)$$

a function of ρ, π_1 and π_2. If π_1 and π_2 are constants, it can be shown that the square of the mle of ρ is X^2/n, where X^2 is the usual Pearson's chi-squared statistic for the 2x2 table of disease by individual. That is, if n_{jk} is the number of pairs for which $Y_1 = j$ and $Y_2 = k$, $j,k = 0,1$, then $X^2 = n(n_{11}n_{00}-n_{10}n_{01})^2 / [(n_{11}+n_{10})(n_{11}+n_{01})(n_{00}+n_{10})(n_{00}+n_{01})]$. This provides a simple approximate χ_1^2 test of the hypothesis $\rho = 0$. (Fisher's exact test of this hypothesis, based on summing probabilities from a central hypergeometric distribution, may be of limited use in practice as it is difficult to calculate unless the sample size is small or the strength of association extreme.)

Suppose π is a constant, independent of order. Let n_{jk} again be the number of pairs for which $Y_1 = j$ and $Y_2 = k$, $j,k = 0,1$, and write $n_d = n_{10} + n_{01}$. Let $P_c = P(Y_1{=}1 \mid Y_2{=}1)$, so that $\rho = (P_c-\pi)/(1-\pi)$. Expressing the log likelihood in terms of π and P_c, it can be shown that their maximum likelihood estimates are $(2n_{11}+n_d)/2n$ and $2n_{11}/(2n_{11}+n_d)$, respectively. Therefore the square of the mle of ρ is X^2/n, where X^2 is Pearson's chi-squared statistic for the 2x2 table of disease by individual (with $n_{10} = n_{01} = ½n_d$). The asymptotic variances of these estimates of π and P_c are cd and ad, respectively, where $a = 4n^3(n_{11}+n_d)/[n_{00}(2n_{11}+n_d)^2]$, $c = (2n_{11}+n_d)^2[(4n_{11})^{-1}+n_d^{-1}+(4n_{00})^{-1}]$, and $d = (ac-b^2)^{-1}$. For a proof of the above, see Hannah et al. (1983) Appendix C, where in addition an expression for the asymptotic variance of the mle of ρ is given. If π_1 and π_2 are not constrained to be equal, it can be shown that the mle for p_1 is the proportion of pairs for which $y_1 = 1$, and similarly for π_2.

EXAMPLE 2: CORRELATION IN CURRENT SMOKING IN ADULT SPOUSE PAIRS

The last row of Table 1 shows that, if we let Y_1 be mother a current smoker, and Y_2 be father a current smoker, then $n_{11} = 402$ is the total number of cohort members for whom both mother and father smoke, $n_{10} = 108$ corresponds to mother smokes but father does not, $n_{01} = 445$ corresponds to father smokes but mother does not, and $n_{00} = 442$ is the total number of cohort members for whom neither mother nor father smokes. Therefore the maximum likelihood estimate of the correlation in mother-father pairs for smoking r is $\{(n_{11}n_{00}-n_{10}n_{01})^2 \ / \ [(n_{11}+n_{10})(n_{11}+n_{01})(n_{00}+n_{10})(n_{00}+n_{01})]\}^{½} = \{[(402x442)-(108x445)]^2/(847x550x510x887)\}^{½} = 0.28$. The X^2 statistic is 111, indicating the correlation is highly significant.

Under the assumption that the probability an adult is a current smoker is independent of sex, the probability that one parent smokes given that the other smokes, P_c, is estimated to be $2n_{11}/(2n_{11}+n_d) = 2x402 / [2x402+(108+445)] = 0.59$ (s.e. = 0.02) and the probability that a parent smokes, p, is $(2n_{11}+n_d)/2n = [2x402+(108+445)]/2x(402+108+445+442) = 0.49$ (s.e. = 0.01). Under this assumption, then, $\rho = (P_c-\pi)/(1-\pi)$ is estimated to be 0.21 (s.e. = 0.028).

Hannah et al. (1983, 1985) demonstrate how, using a general optimisation routine to maximise likelihoods, ρ can be estimated when π is not a constant. For example, π may be fitted as a logistic function of measured covariates, as in equation (3). The flexibility of this approch has been demonstrated by an application to data on drinking status, where the proportion of drinkers is not monotone and does not follow a symmetric relationship with age. It is shown how the Box-Cox transformation can be used to obtain an appropriate scale for one or more covariates; see Hannah et al. (1983), Appendix B, and Figs 1-2. It is also possible to fit r as a function of measured covariates, such as years spent living together or living apart; see Hannah et al. (1983), Appendix A.

The odds ratio,

$$\psi = [P(Y_1=1,Y_2=1)P(Y_1=0,Y_2=0)]/[P(Y_1=1,Y_2=0)P(Y_1=0,Y_2=1)], \quad (9)$$

is a popular concept in epidemiology. If Y represents a disease state, it is approximately equal to the risk ratio (the proportional increase in the disease risk of an individual given that the other member of the pair is affected) when π_1 and π_2 are small. The natural logarithm of the odds ratio, log (ψ) is related to logistic regression coefficients, and has other desirable statistical properties.

In the simplest case where π_1 and π_2 are constants, the asymptotic maximum likelihood estimate of y is given by the observed or empirical odds ratio,

$$\psi = n_{11}n_{00}/n_{10}n_{01}, \quad (10)$$

For large samples, an approximate standard error for log(ψ), s.e., and hence confidence interval, can be calculated from the approximate formula

$$Var(\log(\psi)) » 1/n_{11} + 1/n_{10} + 1/n_{01} + 1/n_{00} = (s.e.)^2. \quad (11)$$

This, derived by the delta method, and several other methods for calculating other approximate confidence limits are given in Breslow & Day (1980); Ch. 4.

EXAMPLE 2 (CONTINUED)

The estimate of ψ, the odds ratio for an adult smoking associated with a spouse smoking, is 402x442/(108x445) = 3.70 = $e^{1.308}$. Therefore log(ψ) is estimated to be 1.308, with s.e. = $[(402)^{-1} + (108)^{-1} + (442)^{-1} + (445)^{-1}]^{½}$ = 0.127. (Note: a test of the null hypothesis, ψ = 0, is given by comparing $(1.308/0.127)^2$ = 105 with a χ_1^2 distribution; cf. the observed X^2 value, above). A 95% confidence interval ψ is from $e^{1.308-1.96*0.127}$ = 2.88 to $e^{1.308+1.96*0.127}$ = 4.74.

More generally, following Hopper et al. (1993) let $\pi_1 = P(Y_{1l}=1 \mid \mathbf{X}_{1l}=\mathbf{x}_{1l})$ and $p_2 = P(Y_{2l}=1 \mid \mathbf{X}_{2l}=\mathbf{x}_{2l})$ both be given by the same function of the covariates, described by parameters $\{\alpha_i\}$, for each pair l, l = 1,...,n. Write $\pi_{jk} = P(Y_{1l}=j, Y_{2l}=k \mid \mathbf{X}_{1l}=\mathbf{x}_{1l}, \mathbf{X}_{2l}=\mathbf{x}_{2l})$ for j, k = 0,1, so that

$$\psi = \pi_{11}\pi_{00}/\pi_{10}\pi_{01},\ \pi_1 = \pi_{10} + \pi_{11},\ \pi_2 = \pi_{01} + \pi_{00}, \text{ and } \Sigma\,\Sigma\,\pi_{jk} = 1. \quad (12)$$

Substituting the expressions given by equation (7) for each of the four π_{ij} probabilities in equation (12), and rearranging, it can be seen that

$$\rho^2 D(\psi -1) - \rho[(\psi-1)(\pi_1+\pi_2-2\pi_1\pi_2) + 1] + D(\psi -1) = 0, \tag{13}$$

Suppose the parameters ψ, $\{\alpha_i\}$, π_1 and π_2 are given. Viewing (13) as a quadratic in ρ, if it has real roots only one of them is between -1 and 1 (because the product of its roots is 1). The four joint probabilities $\{\pi_{jk}\}$ can be calculated using (12). Therefore the log likelihood (8) can be expressed in terms of ρ, $\{\alpha_i\}$, π_1 and π_2.

As for the parameter ρ above, ψ and $\log(\psi)$ can also be parameterised in terms of measured characteristics of the pair such as their (mean) age, sexes, genetic relationship to one another (e.g. zygosity if a twin pair), or cohabitational status (live together or apart; years lived together; years lived apart). Maximum likelihood estimates and asymptotic standard errors can be derived numerically. If a general optimization routine is used, a range of functions for the mean and association parameters can be explored.

Example 2 (Continued Again)

Using the optimisation program SEARCH (Lange et al., 1987), maximum likelihood estimates and standard errors were calculated for π_1 and π_2, and for various choices of ρ, ψ and $\log(\psi)$. As anticipated, π_1 was estimated to be $(n_{11}+n_{10})/n = (402+108)/1397 = 0.365$, with (binomial) standard error 0.013, and π_2 was estimated to be $(n_{00}+n_{01})/n = (445+442)/1397 = 0.635$, also with standard error 0.013. The estimate of ρ was 0.28 (s.e. = 0.024). This can be compared to its estimate under the (now patently false) assumption that $\pi_1 = \pi_2$, 0.21 with s.e. = 0.028; see Example 2 above. The estimates of ψ and $\log(\psi)$ were, of course, 3.70 and 1.308, respectively; see Example 2 (continued) above. Their numerically derived standard errors were 0.514 and 0.123, respectively, the latter slightly less than that given by the approximate formula; see Example 2 (continued).

It is of interest to examine if the association between parents in their smoking behaviour depends on, e.g. how long they have been married. As a surrogate for this, the known mean age of each set of parents was used. Using the numerical optimisation routine we estimated r as a linear function of the mean age of parents, and found an estimate of -0.0007 (s.e. = 0.0040) with a change in log likelihood of just 0.01. That is, there was no evidence that parental association differed according to mean age, but the estimated standard error suggests that an effect would only have been detectable with 80% or more power at the 0.05 level of significance if in reality ρ increased by 0.1 or more per decade. Other models, such as one proposing that ρ increases with log(mean age - 20 yrs), gave worse fits than the simpler model which assumed r was constant.

Use of Descriptive Measures of Association in Genetic Epidemiology

An important step in studying the genetic epidemiology of a trait is to establish if there is evidence for, and to quantify, familial aggregation. Obviously, estimates of measures of association between relatives such as those above can be used for this purpose. Insight into possible roles for genetic and environmental determinants of familial aggregation can be gained by comparing these estimates across different types of relatives. One classic example is by comparing the association between genetically identical or monozygous (MZ) pairs with that between fraternal or dizygous (DZ) pairs. A difference must be due either to the fact that the pairs differ in their genetic similarity, to the possibility that they may differ in their similarity in non-genetic determinants of the trait, or a combination of these explanations. It is often assumed that the latter possibility does not exist, or is trivial, in which case any difference between MZ and DZ pair associations is cited as evidence for the existence of genetic determinants of the trait.

In the example above, we have seen that there is ample evidence of familial aggregation for smoking as an adult. Parent-parent and parent-offspring associations were significant, and not insubstantial. Because parents endow half their genetic inheritance to their offspring, a significant parent-offspring association is consistent with, but certainly does not establish, a role for genetic factors in determining whether an individual is a smoker as an adult. Furthermore, ignoring for a moment the role of sex chromosomes, the genetic similarity of parent-offspring pairs is independent of their sexes. Therefore the same association between offspring and their fathers as between offspring and their mothers would also be consistent with a genetic aetiology. This was observed when the male and female offspring data were pooled.

A more critical examination of the data, however, revealed that the associations between parents and offspring were not consistent with their genetic relationship. Whereas the father-offspring association was independent of the sex of the offspring, mother-offspring association was not. Furthermore, for male offspring, the father-offspring association was greater than the mother-offspring association, which is consistent with determinants for male smoking specific to father-offspring pairs which are not shared by mother-offspring pairs. Once parent-parent and father-offspring associations were taken into account the data was adequately described, and mother-offspring association became small and not significant.

One interpretation of the above is that familial aggregation for smoking as an adult is determined *for the most part* by non-genetic factors, which are shared by parents and offspring, but to different degrees depending on the sexes of the parent and offspring. There may be a role for genetic factors, however, which explain part or all of the mother-son association, and contribute to the mother-daughter and the father-offspring associations.

The odds ratio for mother-son associations (after adjusting for paternal factors) is $e^{0.150} = 1.16$, and the upper limit of the 95% confidence interval is 1.6. These may at first sight appear to be small, but care must be taken in interpreting these odds ratios. Edwards (1960), Day & Simons (1976), Peto (1980) and more recently Khoury et al. (1988), Aalen (1991) and Hopper & Carlin (1992) have considered the stength of underlying familial mechanisms (genetic or environmental) necessary to produce a given odds ratio between relatives. In broad terms the conclusion is that the strength of these underlying factors on the trait must be at least an order of magnitude greater than the odds ratio between (first degree) relatives.

For example, suppose that the correlation in first degree relatives in a binary trait was due solely to a normally distributed polygenic susceptibility for the trait. By definition this polygenic susceptibility has a correlation of 0.5 between first degree relatives. Table 1 of Hopper & Carlin (1992) shows that, if individuals in the upper quartile of genetic susceptibility were twice as likely to be affected than those in the lower quartile (i.e. inter-quartile risk ratio = 2), then the odds ratio for disease when a relative is affected (compared with the relative being unaffected) is 1.04. If the inter-quartile risk ratio is 3, the odds ratio is 1.10, while if it is 5, the odds ratio is 1.22. Therefore, if an odds ratio of 1.16 in first degree relatives was due solely to a polygenic susceptibility, individuals in the upper quartile of genetic susceptibility must be about 4 times more likely to be affected than those in the lower quartile. That is, if a small odds ratio of 1.16 is in fact due to a polygenic susceptibility, there must be a reasonably large risk gradient across that susceptibility. If the odds ratio was 1.6 the risk ratio would be more than 10, if it was 2.0 the risk ratio would be about 20, and if it was 3.5 the risk ratio would be about 100.

Suppose instead that the association between first degree relatives is due solely to susceptibility being determined by a single genetic locus operating under a dominant mode of transmission. In this case, an odds ratio of 1.5 can only occur if the ratio of the risk in susceptible individuals to that in non-susceptible individuals is at least 6 (Easton & Peto; 1990). If the mode of transmission is recessive, the minimum risk ratio is 8.

Suppose the spouse association is due to a binary factor common to parents. That is, the factor is either present or not, and when it is there both parents have the factor. Khoury et al. (1988) showed that, to generate a spouse association equivalent to an odds ratio of $e^{1.308}$ = **3.7**, this factor must be associated with at least a 20 fold increase in the probability of being a smoker as an adult.

That is, the underlying risk factors must be substantially stronger than the odds ratio for the binary trait. Even moderate familial aggregation for a binary trait is suggestive of one or more strong risk factors which are themselves familial. These risk factors, of course, could be familial for genetic and/or environmental reasons.

THE REGRESSIVE LOGISTIC MODEL

Based on the linear logistic regression model above, a general model for familial aggregation in a binary trait can be developed. This can be extended to model transmission of an unmeasured factor (genetic or environmental) from parents to offspring, and to model factors common to groups of individuals. These *regressive logistic models* were introduced by Bonney (1986).

Let $\underline{Y} = (Y_1,...,Y_n)'$ be binary traits measured on a pedigree of size n , with the ith individual having measured covariates $\underline{X}_i = (X_{1i},...,X_{pi})'$, which could include the *genotype* of an individual at a particular locus of interest. For any ordering of pedigree members, the conditional likelihood of the observed pedigree, given the covariates $\mathbf{X} = (\underline{X}_1,...,\underline{X}_n)$, can be written using the standard laws of conditional probability as

$$P(\underline{Y} \mid \mathbf{X}) = P(Y_1 \mid \underline{X}_1)P(Y_2 \mid Y_1,\underline{X}_2)...P(Y_n \mid Y_1,...,Y_{n-1},\underline{X}_n), \qquad (14)$$

provided $P(Y_i \mid Y_1,...,Y_{i-1},\mathbf{X}) = P(Y_i \mid Y_1,...,Y_{i-1},\underline{X}_i)$ for all i = 1,...,n.

We have seen above that a logistic function can be used to model the dependence of the probability of being affected in terms of measured covariates. Extending this approach, let

$$\begin{aligned} \text{logit}(\pi_i) &= \text{logit}(E(Y_i \mid Y_1{=}y_1,\ldots,Y_{i-1}{=}y_{i-1},\underline{X}_i{=}\underline{x}_i)) \\ &= \alpha_{0i}+\gamma_1 y_1+\ldots+\gamma_{i-1}y_{i-1}+\alpha_1 x_{1i}+\ldots+\alpha_p x_{pi}. \end{aligned} \tag{15}$$

The regression coefficients, $\{\alpha_i\}$, have a similar interpretation to those in the logistic regression model.

Because it is possible for there to be a different number of Y covariates for each individual, in full generality a different baseline parameter α_{0i} would need to be allowed for each individual. Fitting a parameter for each individual is of no interest, however, so it is necessary to postulate some structure in the probabilistic model. (For simplicity it is assumed there are no missing values, either for the trait or for the covariates.)

Example 1 above has demonstrated how logistic regression parameters, such as the $\{\gamma_i\}$, can be used to represent the trait association between different members of the pedigree. Knowledge that $Y_j = 1$ for some $j < i$ increases the odds of $Y_i = 1$ by $\exp(\gamma_j)$. That is, γ_j can be interpreted as a log odds ratio for the "risk factor": Y_j is affected.

The independence of conditional probabilities in equation (14) means that, by using the logistic expression given by equation (15), the model is formally equivalent to a logistic regression model for n independent observations, and can in principle be fitted by a standard logistic regression package. In essence, this is what was achieved in the simple case of Example 1 where $n = 3$.

In general n-1 parameters are required to describe familial aggregation, so models which require fewer parameters need to be sought. Note that if $\gamma_1 = \gamma_2 = \ldots = \gamma_{n-1} = \gamma$ then on the logit scale $Y_1,\ldots,Y_{i-1}$ have equal and additive predictive value for Y_i. The first order Markov property, where the probability of a given observation conditional on prior observations involves only the immediately prior observation, implies that $P(Y_i \mid Y_1,\ldots,Y_{i-1},X_i) = P(Y_i \mid Y_{i-1},X_i)$, and extension to higher order Markov dependence is possible.

Various models can be derived that take into account biological relationships within the pedigree. For illustrative purposes, let Y_m and Y_f represent the trait values of the mother and father, and $Y_1,\ldots,Y_k$ the trait values of $k = n-2$ offspring, respectively.

If it is assumed that the offspring are dependent solely because of their common parentage, then given the outcomes of the parents of a child i , the outcomes of the siblings of i do not contribute more information on Y_i. That is,

$$\pi_i = P(Y_i \mid Y_f,Y_m,Y_1,\ldots,Y_{i-1},\underline{X}_i) = P(Y_i \mid Y_f,Y_m,\underline{X}_i). \tag{16}$$

Bonney (1986) called this type of model *Class A*, in which case

$$\text{logit}(\pi_i) = \alpha_{10} + \gamma_f y_f + \gamma_m y_m + \alpha_1 x_{1i} + \ldots + \alpha_p x_{pi}, \tag{17}$$

and it is appropriate for all offspring to have the same baseline coefficient α_{10}. The parameters γ_f and γ_m are measures of the disease association between parents and offspring, and represent the effects of factors, both genetic and environmental, common to parents and their offspring.

Dependence between siblings, in addition to that induced by their common parentage, can be added to the Class A model. If siblings are placed in birth order, models can be developed in which parents and one (or more) older siblings can account for dependence among the younger siblings (*Class B*), siblings closer in birth order are more highly correlated than siblings further apart (*Class C*), or outcomes of siblings are equally predictive, given the parental outcomes (*Class D*); see Bonney (1986).

For example, a Class B model which includes dependence on the first (oldest) sibling specifies, for i = 2,...,k, that

$$\begin{aligned} \text{logit}(\pi_i) &= \alpha_{10} + \gamma_f y_f + \gamma_m y_m + \alpha_1 x_{1i} + \ldots + \alpha_p x_{pi} && \text{if } i = 1, \\ &= \alpha_{20} + \gamma_f y_f + \gamma_m y_m + \gamma_B y_1 + \alpha_1 x_{1i} + \ldots + \alpha_p xpi && \text{if } i \neq 1. \end{aligned} \quad (18)$$

The parameter γ_B represents the dependence of siblings on the trait value of the first born child.

A Class C model,

$$\begin{aligned} \text{logit}(\pi_i) &= \alpha_{10} + \gamma_f y_f + \gamma_m y_m + \alpha_1 x_{1i} + \ldots + \alpha_p x_{pi} && \text{if } i = 1, \\ &= \alpha_{20} + \gamma_f y_f + \gamma_m y_m + \gamma_C y_{i-1} + \alpha_1 x_{1i} + \ldots + \alpha_p x_{pi} && \text{if } i \neq 1, \end{aligned} \quad (19)$$

includes a serial dependence among siblings. The parameter γ_C represents the dependence of siblings on the trait value of the previous born child.

A Class D model which includes dependence on older siblings specifies that

$$\begin{aligned} \text{logit}(\pi_i) &= \alpha_{10} + \gamma_f y_f + \gamma_m y_m + \alpha_1 x_{1i} + \ldots + \alpha_p x_{pi} && \text{if } i = 1, \\ &= \alpha_{20} + \gamma_f y_f + \gamma_m y_m + \gamma_D(y_1 + \ldots + y_{i-1}) + \alpha_1 x_{1i} + \ldots + \alpha_p x_{pi} && \text{if } i \neq 1. \end{aligned} \quad (20)$$

In this case the parameter γ_D represents the dependence of siblings on the sum of the trait values of all previous born children. This effectively assumes that the trait correlations between each pair of siblings are the same (Bonney, 1986).

The likelihood of a nuclear family can be written as

$$P(\underline{Y} \mid \mathbf{X}) = P(Y_f, Y_m \mid \underline{X}_f, \underline{X}_m) \prod_{i=1}^{n} P(Y_i \mid Y_f, Y_m, X_i), \quad (21)$$

following equation (14). The terms (16) inside the product can take the form of (17), (18), (19) or (20). The joint distribution of the parents can be expressed, as described by equation (7), with a parameter ρ_{mf} or ψ_{mf} representing the trait association between the pair, as in Example 2.

EXAMPLE 3: REGRESSIVE LOGISTIC MODELLING OF HAVING EVER SMOKED IN NUCLEAR FAMILIES

The proportion of individuals who had ever smoked (more than 100 cigarettes) was 38.3% in the 60 mothers, 71.7% in the 60 fathers, and 49.3% in the 211 offspring.

For parents, logistic regression analysis showed that the proportion who had ever smoked was significantly greater in fathers, and increased with age in both mothers and fathers at about the same rate. That is, following equation (3) the probability that a parent has ever smoked can be parameterised in terms of three regression coefficients, α_0, α_{sex} and α_{age}, say. In the modelling below, carried out using the optimisation package SEARCH (Lange et al., 1987), the estimates of these coefficients were unchanged because there are no parameters in common between the parents and their offspring. Their estimates (standard errors) were: -3.395 (1.700), 1.238 (0.394) and 0.0461 (0.0261), respectively.

Table 4. Regressive logistic coefficients (standard errors) from fitting Class A, B, C and D models to nuclear family data on smoking status (ever/never)[1]

Parameter	Class A	Class B	Class C	Class D
γ_f	0.679	0.624	0.527	0.538
	(0.354)	(0.356)	(0.365)	(0.361)
γ_m	0.704	0.669	0.682	0.657
	(0.299)	(0.302)	(0.310)	(0.303)
α_{10} (= α_{20})	-0.820	-0.905	-1.076	-0.923
	(0.314)	(0.319)	(0.331)	(0.319)
γ_B		0.597		
		(0.340)		
γ_C			1.131	
			(0.314)	
γ_D				0.288
				(0.136)
LL	-215.01	-213.45	-208.23	-212.64

[1] Estimates of α_0, α_{sex} and α_{age}, independent of type of model, given in the text.

In the offspring, however, the proportion was independent of age and sex. Therefore, for a Class A model equation (17) becomes $\mathrm{logit}(\pi_i) = \alpha_{10} + \gamma_f y_f + \gamma_m y_m$. Table 4 shows that the smoking status of an offspring depends on that of the father, with an associated odds ratio of $e^{0.679} = 1.97$ (95% confidence interval from 0.97 to 4.00), and that of the mother with an odds ratio of $e^{0.704} = 2.02$ (95% confidence interval from 1.11 to 3.63). Clearly the estimates of γ_f and γ_m are quite similar, and not statistically different from one another. When combined, the estimate of $\gamma_p = \gamma_f = \gamma_m$ was 0.693, with a standard error of 0.206 ($p < 0.001$). That is, the smoking status of the offspring depends on that of their parents.

For a Class B model equation (18) becomes $\mathrm{logit}(\pi_i) = \alpha_{10} + \gamma_f y_f + \gamma_m y_m$ for the first born offspring, and $\alpha_{20} + \gamma_f y_f + \gamma_m y_m + \gamma_B$ for younger siblings. It was found that α_{10} and α_{20} were similar, so Table 4 shows estimates of $\alpha_{10} = \alpha_{20}$. The estimates of γ_f and γ_m are again similar and the combined estimate of α_p was 0.650, with standard error 0.208 ($p < 0.002$). The estimate of γ_B of 0.597 with a standard error of 0.340 ($p < 0.08$) suggests that the smoking status of the first born sibling has an associated odds ratio 1.82, after allowing for an association with parents. Twice the difference in log likelihood between the fitted Class A and Class B models is 3.12 ($p < 0.08$).That is, there is marginal evidence that the sibling association is due to more than common parentage.

For a Class C model equation (19) becomes $\mathrm{logit}(\pi_i) = \alpha_{10} + \gamma_f y_f + \gamma_m y_m$ for the first born offspring, and $\alpha_{20} + \gamma_f y_f + \gamma_m y_m + \gamma_C$ for younger siblings. Table 4 shows that the estimate of γ_C is 1.131 with standard error 0.314 ($p < 0.001$), and twice the difference in log likelihood from the fitted model A is 13.5 ($p < 0.001$). The parental coefficient, γ_p, was estimated to be 0.616, with a standard error of 0.212 ($p < 0.005$). That is, the smoking status of the previous born child has an associated odds ratio of 3.08 (95% confidence interval from 1.67 to 5.68).

For a Class D model equation (20) becomes $\mathrm{logit}(\pi_i) = \alpha_{10} + \gamma_f y_f + \gamma_m y_m$ for the first born offspring, and $\alpha_{20} + \gamma_f y_f + \gamma_m y_m + \gamma_D$ for younger siblings. Table 4 shows that the estimate of γ_D is 0.288 with standard error 0.136 ($p < 0.04$). The likelihood ratio test gives $\chi_1^2 = 4.74$ ($p < 0.03$). The parental coefficient, γ_p, was 0.606 with standard error 0.210 ($p < 0.005$).

Therefore the dependence among siblings for smoking status is best explained by their common parentage, and by a dependence on the previously born sibling. Adjusting for the latter factor reduced the parental effect by a little over 10%. While the parental effect is consistent with genetic factors, the Markovian-like dependence within a sibship suggests that this behaviour is also under the influence of social factors.

A very important point of the regressive approach is that, by virtue of the Markov assumption, it can be used to express the likelihood of a general pedigree, not just a nuclear family, through first identifying the "ancestors" and then successively calculating the likelihood of them and their immediate progeny, as in (21); see Bonney (1986).

The likelihood of a pair of twins can also be expressed by equation (7), with association represented by a correlation, ρ_{tw}, or odds ratio, ψ_{tw}; see Hopper at al. (1990a). Note that trait associations between spouse pairs, or between twin pairs, can be modelled within the regressive logistic model framework. For example, $\log(\psi_{mf})$ can be a function of the length of marriage of the parents, and $\log(\psi_{tw})$ a function of the zygosity and sex of the twin pair, or the time they have been cohabiting or living apart.

MODELLING VERTICAL TRANSMISSION OF AN UNMEASURED FACTOR FROM PARENTS TO OFFSPRING

To model vertical transmission of unmeasured factors, i.e. from parents to offspring, of factors associated with the trait, such as genotype, behaviour, lifestyle, and culture, it is necessary to make specific assumptions about these latent factors. That is, in a probabilistic sense, how are these factors transmitted from parent to offspring, and what is the influence of these factors on the probability that an individual will develop the trait?

Suppose there exists an unobserved factor, V_i, for each individual i , that can take only one of k values. Let $\underline{V} = (V_i,...,V_n)'$ and note that, summing over all possible types, $\Sigma\, P(V_i) = 1$. The joint probability of $\underline{V}$ can be decomposed, for any ordering of the individuals, as

$$P(\underline{V}) = P(V_1,...,V_n) = P(V_1)P(V_2 | V_1)...P(V_n | V_1,...,V_{n-1})$$

$$= \prod_{i-1}^{n} P(V_i | V^{(i-1)}) \tag{22}$$

where $V^{(m)} = (V_1,...,V_m)'$. There are a total of k^n mutually exclusive vectors $\underline{V}$ possible, and assuming V is independent of X, summing over these

$$P(\underline{Y} | X) = \sum_{\underline{V}} P(\underline{V},\underline{Y} | X) = \sum_{\underline{V}} P(\underline{V})P(\underline{Y} | \underline{V},X) = \sum_{V_1}...\sum_{V_n}\prod_{I} P(V_i | V^{(i-1)})P(Y_i | V_i,Y^{(i-1)},\underline{X_i}), \tag{23}$$

provided the V-type of a relative affects the trait of an individual only through that individual's own V-type, i.e. $P(Y_i | \underline{V},Y^{(i-1)},X) = P(Y_i | V_i,Y^{(i-1)},X)$. This introduces a new "explanatory variable" for Y_i, namely V_i, which can be incorporated into the logit by introducing a new set of k-1 regression parameters $\{\gamma_{vi}\}$ depending on the number of different V-types possible. Adding this to the Class A model expressed by equation (17) gives

$$\text{logit}(\pi_i) = \alpha_{0i} + \gamma_f y_f + \gamma_m y_m + \gamma_{v1} V_{1i} + ... + \gamma_{v(k-1)} V_{i(k-1)} + \alpha_1 x_{1i} + ... + \alpha_p x_{pi}, \tag{24}$$

where $V_{ij} = 1$ if $V_i = j$, and otherwise is 0. The likelihood of the pedigree can now be expressed in terms of the regression parameters, parameter(s) describing the probability distribution of the V-types, and parameters used to model the $\{P(V_i | V^{(i-1)})\}$.

One way of modelling a factor transmitted by parents to their offspring, is to let

$$
\begin{aligned}
P(V_i \mid V^{(i-1)}) &= P(V_i) && \text{if the ancestors of i are not in the data,}\\
&= P(V_i \mid V_{sp}) && \text{if the ancestors are not in the data, but a spouse precedes i ,}\\
&= P(V_i \mid V_f, V_m) && \text{if the ancestors are in the data.}
\end{aligned} \quad (25)
$$

That is, knowledge of the V-type of the parents determines probabilistically the V-type of the offspring, and knowledge of the V-type of any other individual does not contribute any further information. To model the $\{P(V_i ô V^{(i-1)})\}$, it is therefore necessary to model $P(V_i)$, $P(V_i \mid V_s)$ and $P(V_i \mid V_f, V_m)$, for i = 1,...,k.

For example, suppose there are two V-types (k = 2). Let $P(V_i{=}1) = \pi_v$ say, and $P(V_i{=}2) = 1-\pi_v$, so that the probability distribution of the V-types is described by just one parameter, π_v. Suppose the V-types of spouses are independent, so that $P(V_i \mid V_{sp}) = P(V_i)$, and let

$$
\begin{aligned}
P(V_i{=}1 \mid V_f, V_m) &= \tau_{11} \text{ if } V_f = 1 \text{ and } V_m = 1,\\
&= \tau_{10} \text{ if } V_f = 1 \text{ and } V_m = 0,\\
&= \tau_{01} \text{ if } V_f = 0 \text{ and } V_m = 1,\\
&= \tau_{00} \text{ if } V_f = 0 \text{ and } V_m = 0.
\end{aligned} \quad (26)
$$

Then $P(V_i{=}2 \mid V_f, V_m) = 1 - P(V_i{=}1 \mid V_f, V_m)$. Each of the four paramaters $\{\tau_{ij}\}$ represents the probability that an offspring will "inherit" a V-type of 1, given the specific V-types of their parents. Clearly other variations are possible.

A SPECIAL CASE - MODELLING THE EFFECTS OF A SINGLE GENETIC LOCUS: SEGREGATION ANALYSIS

Suppose V represents an unmeasured autosomal locus with two alleles, A and a, with the genotypes AA, Aa and aa corresponding to 3 successive V-types. If the gene frequency of allele A is p_A, then under Hardy-Weinberg equilibrium

$$
\begin{aligned}
P(g_i) &= p_A^2 && \text{if } g_i = 1 \text{ (genotype = AA)},\\
&= 2p_A(1-p_A) && \text{if } g_i = 2 \text{ (genotype = Aa)},\\
&= (1-p_A)^2 && \text{if } g_i = 3 \text{ (genotype = aa)}.
\end{aligned} \quad (27)
$$

In this case only *one* parameter, p_A, is needed to describe the probability distribution of V-types. The $P(g_i \mid g_f, g_m)$ are specified by Mendelian transmission. That is, $P(1 \mid 1,1) = 1$; $P(1 \mid 1,2) = P(1 \mid 2,1) = ½ = P(2 \mid 1,2) = P(2 \mid 2,1)$; $P(1 \mid 2,2) = ¼ = P(3 \mid 2,2)$, $P(2 \mid 2,2) = ½$. Hence, under the assumption that the genotype of spouses are independent, the $\{P(V_i \mid V^{(i-1)})\}$ are described by the same parameter, π_A. Three regression parameters, including the baseline parameter, are required to model logit(π_i) as a function of the individual's genotype.

Example 3 (Continued): Segregation Analysis of Having Ever Smoked

A simple single locus model of the nuclear family data on having ever smoked can be described by the following parameters: π_A, the gene frequency of allele A; and γ_{AA}, γ_{Aa} and γ_{aa}, representing (baseline) coefficients for the three genotypes, AA, Aa and aa. In addition, as before, α_{sex} and α_{age}, representing the effects of parental sex and age, and $\alpha_{01} = \alpha_{02}$ for the offspring generation can be included, as well as any of the parameters γ_B, γ_C or γ_D.

Table 5 shows estimates from some fitted models. Because a Class C model was the best fitting model in Table 4, the Mendelian model has been fitted with and without the extra paramter γ_C.

Table 5. Regressive logistic coefficients from fitting a segregation analysis model to nuclear family data on smoking status (ever/never).

Parameter	Recessive		Dominant		Codominant	
π_A	0.653	0.594	0.275	0.133	0.653	0.579
γ_{AA}	-0.343	0.211	-1.062	7.702	-0.342	0.315
γ_{Aa}	-4.800	-4.267	-1.062	7.702	-4.754	-4.021
γ_{aa}	-4.800	-4.267	-4.936	-3.082	-4.883	-4.386
α_{sex}	1.122	1.047	0.929	0.795	1.110	0.983
α_{age}	0.0479	0.0460	0.0484	0.0349	0.0479	0.0453
$\alpha_{01} = \alpha_{02}$	0.596	0.235	0.832	0.199	0.603	0.269
γ_C		1.553		1.428		1.501
LL	-214.83	-211.17	-215.32	-211.20	-214.82	-211.15

First, note that with this relatively small nuclear family data set there is little discrimination between a recessive ($\gamma_{Aa} = \gamma_{aa}$), dominant ($\gamma_{AA} = \gamma_{Aa}$) or co-dominant ($\gamma_{AA}$, γ_{Aa} and γ_{aa} free) fit. Although the parameter estimates change considerably between the recessive and dominant models, the log likelihood differs by only 0.03. Because the likelihood surface is so flat, the maximisation package, SEARCH (Lange et al., 1987), deliberately did not produce standard errors. The parameters α_{sex}, α_{age} and γ_C, however, are not greatly influenced by mode of inheritance.

Variations on Mendelian inheritance can be modelled, following Elston & Stewart (1971). Let there be three transmission probabilities,

$$\tau_{AA} = \tau_1 = \text{P(an AA (type 1) individual transmits A to offspring)},$$
$$\tau_{Aa} = \tau_2 = \text{P(an Aa (type 2) individual transmits A to offspring)},$$
$$\tau_{aa} = \tau_3 = \text{P(an aa (type 3) individual transmits A to offspring)}, \qquad (28)$$

so that

$$P(g_i{=}AA \mid g_f{=}j, g_m{=}k) = \tau_j\tau_k,$$
$$P(g_i{=}Aa \mid g_f{=}j, g_m{=}k) = \tau_j(1-\tau_k) + \tau_k(1-\tau_j), \text{ and}$$
$$P(g_i{=}aa \mid g_f{=}j, g_m{=}k) = (1-\tau_j)(1-\tau_k). \qquad (29)$$

Mendelian transmission corresponds to the special case $\tau_{AA} = 1$, $\tau_{Aa} = ½$ and $\tau_{aa} = 0$. The $\{P(V_i \mid V^{(i-1)})\}$ are described by π_A, τ_{AA}, τ_{Aa} and τ_{aa}; see Bonney (1986).

Modelling Unmeasured Factors Common to Siblings: Horizontal Transmission

A similar approach can be used to estimate within a Class A model the increased or decreased trait association between siblings induced by factors common to the siblings, such as their shared sibship environment. Suppose for each sibling i , i = 1,..., k , there is an unobserved binary factor, H_i, which is either present ($H_i = 1$) or absent ($H_i = 0$). Let $\pi_h = P(H_i = 1)$, and suppose that if one sibling experiences the factor then so do all other siblings. Therefore

$$P(\underline{h}) = P(H_1{=}h_1,\ldots,H_k{=}h_k) = \pi h \quad \text{if } h_1{=}\ldots{=}h_k{=}1,$$
$$= 1-\pi h \quad \text{if } h1{=}\ldots{=}h_k{=}0,$$
$$= 0 \quad \text{otherwise.} \qquad (30)$$

For this sibship the conditional likelihood of $\underline{Y} = (Y_1,\ldots,Y_k)'$ is

$$P(\underline{Y} \mid Y_m, Y_f, X) = \pi_h \prod_{i=1}^{k} P(Yi \mid H_i{=}1, Y_f, Y_m, \underline{X}_i) + (1-\pi_h) \prod_{i=1}^{k} P(Y_i \mid H_i{=}0, Y_f, Y_m, \underline{X}_i), \text{ and} \qquad (31)$$

$$\text{logit}(\pi_i) = \alpha_{0i} + \gamma_f Y_f + \gamma_m Y_m + \gamma_h h_i + \alpha_1 x_{1i} + \ldots + \alpha_p x_{pi}. \qquad (32)$$

Therefore with two additional parameters, γ_h representing the log odds ratio for presence of the unobserved, hypothetical factor common to siblings, and π_h, its prevalence, an association between the trait in siblings above, or below (γ_h can take negative values), that due to parental transmission can be tested; see Hopper (1989) and Hopper et al. (1990c). Note that this is equivalent to postulating a factor, H', for which $\pi_{h'} = 1-\pi_h$, $\gamma_{h'} = -\gamma_h$, and $\alpha_{0'} = \alpha_0 + \gamma_h$.

EXAMPLE 3 (CONTINUED AGAIN)

Suppose now there is an unobserved binary factor, H, which is apparent in a proportion, π_h, of sibships, and has an associated log odds ratio γ_h. Table 6 shows the parameter estimates from including such a factor in a Class A and Class C model, and compares them with the Class A and Class C models of Table 4.

Table 6. Regressive logistic coefficients (standard errors) from fitting Class A and C models, with and without an unmeasured factor, H, common to siblings, to nuclear family data on smoking status (ever/never)[1].

Parameter	Class A		Class C	
γ_f	0.679	0.697	0.527	0.594
	(0.354)	(0.414)	(0.365)	(0.389)
γ_m	0.704	0.868	0.682	0.872
	(0.299)	(0.349)	(0.310)	(0.345)
α_{10} (= α_{20})	-0.820	-0.618	-1.076	-0.941
	(0.314)	(0.384)	(0.331)	(0.370)
γ_c			1.131	0.873
			(0.314)	(0.346)
π_h		0.123		0.077
		(0.064)		(0.060)
γ_h		-14.28		-14.63
		(N.A.)[2]		(N.A.)[2]
LL	-215.01	-209.52	-208.23	-206.32

[1] Estimates of α_0, α_{sex} and α_{age}, independent of type of model, given in the text.

[2] Standard errors were, like the parameter estimate, very large; see text.

The second column suggests that H is present in 12% of sibships, and if present it fully protects the offspring from smoking, irrespective of whether one or both of their parents smoke! The alternate interpretation is that there is a factor present in 88% of sibships; if it is not present then no siblings smoke.

The estimated standard error of γ_h is very large, and in fact meaningless. Fig. 1 shows a plot of the log likelihood surface for different values of $-\gamma_h$, with all other parameters held constant at their estimated values given in column 2 of Table 6. It can be seen that the likelihood function is flat for all values of $-\gamma_h$ greater than 6, say. A 95% confidence interval for γ_h is given by the interval $(-\infty, -1.87)$. All values of γ_h within this interval were associated with a LL within 1.92 of the maximum LL, and twice the difference in LL is less than 3.84, the critical value at $p = 0.05$ for a χ_1^2 variate.

Adding two parameters to a Class A model in order to represent the unmeasured factor H gives a change in 2LL of 11.0. It is not known exactly what distribution this statistic should have - see Discussion - but compared to a χ_2^2 distribution it has a p-value of 0.001. This model gives a better log likelihood (-209.52) than the fitted Class B and Class D models (see Table 4), although it uses one more parameter. It is not preferable to a Class C model, however.

Figure 1. Class A model with Factor H

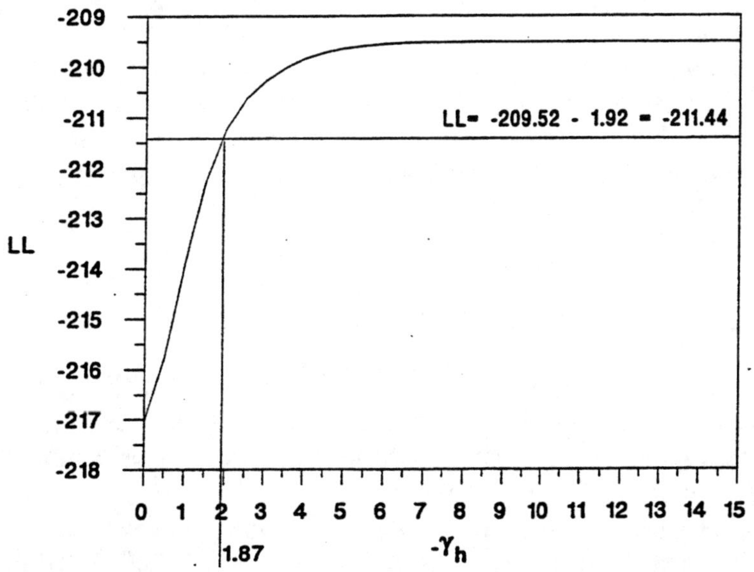

Table 6 also shows that adding these two parameters to a Class C model results in an increase in 2LL of 3.82. This is consistent with chance, if comparison is made with a χ_2^2 variable. The estimate of π_h is not strictly significant, as judged by its standard error. Fig. 2 shows that a 95% confidence interval for γ_h is $(-\infty, 1.06)$. Therefore, based on the analyses presented in Tables 4, 5 and 6, a Class C model would be chosen as giving the most parsimonious description of the data. It appears to incorporate most of the horizontal transmission within sibships that would otherwise be attributed to the factor H.

Figure 2. Class C model with Factor H

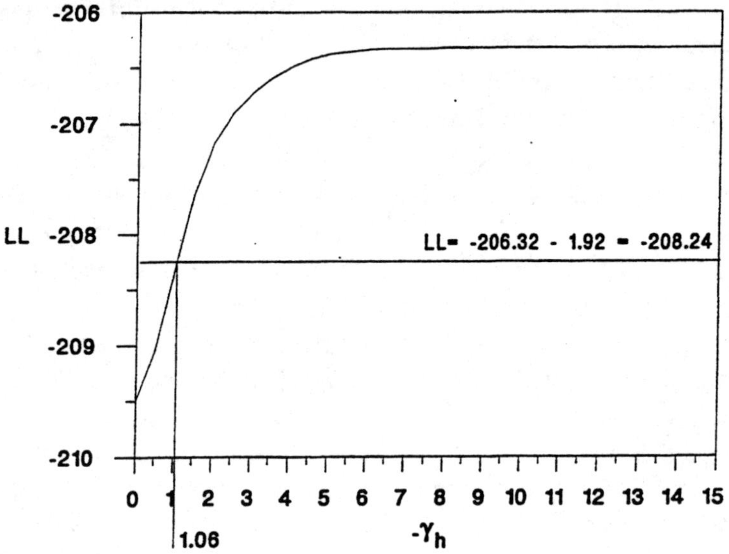

Another approach to modelling an association between groups of individuals, such as siblings, is to allow H to be an unmeasured *continuously* distributed factor. For example, it may be a normally distributed variable with zero mean, and each sibling takes the same value. Equation (31) is then replaced by an integration, which can be handled efficiently and accurately by Gauss-Hermite numerical quadrature methods; see Hopper et al. (1990a). These models have been called *compound regressive logistic models* (Bonney, 1992).

DISCUSSION

The description of these regressive logistic models in terms of explicit probabilities and logistic regression coefficients allows for straightforward interpretation. The odds ratio for different exposures, whether they be measured covariates $\underline{X}$ or familial factors such as "an affected parent", or "a particular genotype at a major locus", can be estimated as a log odds parameter and the strengths of effect compared. Moreover, the probability of being affected conditional on an exposure can be multiplied by the (predicted) prevalence of that exposure to estimate the proportion of trait associated with each exposure. This allows direct comparison of the relative importance of different risk factors; see Hopper et al. (1990c).

An estimate of the population prevalence, $P(Y_i=1 \mid \underline{X})$, can be calculated by summing over all possible exposure outcomes. If the model is correctly interpreting familial aggregation and correctly adjusting for any ascertainment (see below), it should give valid estimates of trait prevalence, even though the data has not come from a random sample of independent individuals. These prevalence estimates derived from a fitted model can be compared with estimates from, for example, population based samples, to check if the fitted model is giving realistic predictions. If population-based data are available which includes the relevant explanatory variables, it should be integrated into the analysis of family data, so as to give more accurate and precise estimates of the baseline and regression coefficients, $\alpha_0,...,\alpha_p$, associated with the covariates. It is plausible that this will improve estimates of the parameters $\{\gamma_i\}$, representing the dependence of trait status between relatives, and transmission of unmeasured familial factors.

The probabilistic description lends itself naturally to fitting and estimation by maximum likelihood. For simple designs such as nuclear familes it is not difficult to write programs to calculate pedigree likelihoods, and attach this to a maximum likelihood optimisation routine.

In general, selection between competing models can be carried out by asymptotic likelihood theory. For models involving a latent variable, such as an unobserved genetic, cultural or environmental factor occuring with a prevalence $\pi_.$ and having an associated regression coefficient $\gamma_.$, standard theory breaks down because the parameter space for logit $\pi_.$ and $\gamma_.$ is not regular. If $\gamma_. = 0$ the likelihood surface is independent of logit $\pi_.$. Therefore the correct distribution of the likelihood ratio test for comparing a model that does not contain the latent factor with one that does is unknown. A conservative approach would be to base testing on the χ^2 distribution with 2, as distinct from 1, degree of freedom.

Provided there are relatively few missing values, they can be handled in the usual way as in logistic regression. Deleting a family because one or more individuals have a missing value can result in drastic censoring of data, and may be misleading if missing values do not occur at random. For a continuous X, a missing value might be replaced by the mean or median of X. For a factor X, an extra level can be included for individuals on whom their X status is unknown, and an extra parameter estimated for this level. For the binary trait Y, if there are few missing values and the trait is rare, a reasonable stategy might be to code missing values as if the individuals were unaffected. Note that for rare traits it makes sense to code unaffected individuals as the baseline category, so that g coefficients represent the log odds ratio due to a relative being known to be affected, and due to a relative having unknown trait status.

Finally, it is important to recognise that the probabilistic description also allows a variety of ascertainment correction schemes, imperative to consider whenever sampling is not at random. The log likelihood of a pedigree, conditional on being ascertained (asc), is

$$\log P(\underline{Y} \mid asc, X) = \log P(\underline{Y} \mid X) - \log P(asc \mid X), \tag{33}$$

where the appropriate expression for $P(asc \mid \mathbf{X})$ depends on the ascertainment correction scheme invoked and can be expressed in terms of model parameters; see Hopper et al. (1990c) for an example of different ascertainment correction schemes being applied to the same data.

In summary, much is to be gained from applying logistic regression ideas to the analysis of binary pedigree data. Different models for familial aggregation can be studied, and evidence for a genetic aetiology addressed critically. Furthermore, the strength of effects and relative importance of factors can be interpreted in a straightforward probabilistic sense.

REFERENCES

Aalen, O.O. (1991). Modelling the influence of risk factors on familial aggregation of disease. *Biometrics, 47*, 933-945.

Bonney, G.E. (1986). Regressive logistic models for familial disease and other binary traits. *Biometrics, 42*, 611-625.

Bonney, G.E. (1992). Compound regressive models for family data. *Hum Hered, 42*, 28-41.

Breslow, N.E. & Day, N.E. (1980). *Statistical Methods in Cancer Research. I. The Analysis of Case-Control Studies*. Lyon: International Agency for Research in Cancer.

Clayton, D. & Hills, M. (1993). *Statistical Models in Epidemiology*. Oxford: Oxford University Press.

Day, N.E. & Simons, M.J. (1976). Disease susceptibility genes: their identification by multiple case family studies. *Tissue Antigens, 8*, 109-119.

Easton, D. & Peto, J. (1990). The contribution of inherited predisposition to cancer incidence. *Cancer Surv, 9*, 395-416.

Eaves, L.J., Eysenck, H.J. & Martin, N.G. (1989). *Genes, Culture, and Personality: An Empirical Approach*. London: Academic Press.

Edwards, J.H. (1960). The simulation of Mendelism. *Acta Geneticae, 10*, 63-70.

Elston, R.C. & Stuart, J. (1971). A general model for the genetic analysis of pedigree data. *Hum Hered, 21*, 523-542.

Fisher, R.A. (1912). On an absolute criterion for fitting frequency curves. *Messenger of Mathematics, 41*, 155-160.

Gibson, H.B., Silverstone, H., & Gandevia, B. (1969). Respiratory disorders in seven-year-old children in Tasmania: aims, methods, and administration of the survey. *Medical Journal of Australia*, 2, 201-205.

Hannah, M.C., Hopper, J.L., & Mathews, J.D. (1983). Twin concordance for a binary trait. I. Statistical methods illustrated with data on drinking status. *Acta Geneticae et Medicae Gemellologia, 32*, 127-137.

Hannah, M.C., Hopper, J.L., & Mathews, J.D. (1985). Twin concordance for a binary trait. II. Nested analysis of ever-smoking and ex-smoking traits and unnested analysis of a "committed-smoking" trait. *Am J Hum Genet, 37*, 153-165.

Hopper, J.L. (1989). Modelling sibship environment in the regressive logistic model for familial disease. *Genet Epidemiol, 6*, 235-240.

Hopper, J.L., Carlin, J.B., Macaskill, G.T., Derrick, P.L., Flander, L.B., & Giles, G.G. (1990). Incorporation of twins in the regressive logistic model for pedigree disease data. *Acta Geneticae et Medicae Gemellologia, 39*, 173-180.

Hopper, J.L., Giles, G.G., McCredie, M.R.E., & Boyle, P. (1994). Background, rationale, and protocol for a case-control-family study of breast cancer. *The Breast, in press*,

Hopper, J.L., Hannah, M.C., Macaskill, G.T., & Mathews, J.D. (1990). Twin concordance for a binary trait: III. A bivariate analysis of hay fever and asthma. *Genet Epidemiol, 7*, 277-289.

Hopper, J.L., Jenkins, M.A., Carlin, J.B., & Giles, G.G. (1995). Increase in the self-reported prevalence of asthma and hay fever in adults over the last generation: A matched parent-offspring study. *BMJ*,

Hopper, J.L., Judd, F.K., Derrick, P.L., Macaskill, G.T., & Burrows, G.D. (1990). A family study of panic disorder: reanalysis using a regressive logistic model that incorporates a sibship environment. *Genet Epidemiol, 7*, 151-161.

Hopper, J.L., White, V.M., Macaskill, G.T., Hill, D.J., & Clifford, C.A. (1992). Alcohol use, smoking habits and the Adult Eysenck Personality Questionnaire in adolescent Australian twins. *Acta Geneticae et Medicae Gemellologia, 41*, 311-324.

Jenkins, M.A., Hopper, J.L., Bowes, G., Carlin, J.B., Flander, L.B., & Giles, G.G. (1994). Factors in childhood as predictors of asthma in adult life. *BMJ, 309*, 90-93.

Khoury, M.J., Beaty, T.H., & Liang, K.Y. (1988). Can familial aggregation of disease be explained by familial aggregation of environmental risk factors? *Am J Epidemiol, 127*, 674-683.

Lange, K., Weeks, D.E. & Boehnke, M. (1987). *Programs for pedigree analysis*. Los Angeles, CA: Department of Biomathematics, University of California, Los Angeles.

McCullagh, P. & Nelder, J.A. (1983). *Generalized linear models*. London: Chapman Hall.

Peto, J. (1980). Genetic predisposition to cancer. In J. Cairns, J.L. Lyon & M. Skolnick (Eds.). *Cancer incidence in defined populations (Banbury Report no. 4)* (pp. 203-213). Cold Spring Harbor, NY: Cold Spring Harbor Laboratory

CHAPTER 8

GENETIC ANALYSIS OF COMPLEX DISEASES

Hilary Coon
University of Utah School of Medicine

GENETIC ANALYSIS OF COMPLEX DISEASES

The past decade has seen the discovery of genes for many diseases, including Duchenne muscular dystrophy (Monaco and Kunkel, 1988), cystic fibrosis (Kerem et al., 1989), neurofibromatosis (Marchuk et al., 1991; Gutman et al., 1991), and Huntington's disease (Huntington's Disease Collaborative Research Group, 1993). This chapter will review some of the techniques used to identify genes for diseases caused by a single genetic locus. Strategies will then be discussed for detecting genes underlying diseases which are more complex. These diseases are likely to be caused by the combined effects of several genes as well as multiple environmental factors.

Researchers studying the genetics of complex human behaviors must first show that these behaviors have a genetic component. Family studies, often the first step in this process, can give some indication of genetic etiology. Family members who are more closely related share more genes in common on average, so that genetic etiology is suggested when behavioral similarity becomes more pronounced as the degree of genetic relationship increases. However, family studies cannot control for shared family environment, which may also contribute to resemblance among family members.

Family studies of schizophrenia, one common behavioral disease, show an increased risk of illness as the genetic relationship to an affected individual becomes closer (Gottesman, 1991). While risk of schizophrenia in the general population is about 1%, risk to second degree relatives ranges from 2% for the uncle or aunt of a schizophrenic, to 6% for the half sibling of a schizophrenic. Risks to first degree relatives are higher, ranging from 9% for the full sibling of a schizophrenic, to 17% for the fraternal twin of a schizophrenic.

For affective disorder, the other major psychiatric illness, family studies also show an increased risk of illness as the degree of relatedness increases (Goodwin and Jamison,

1990; Nurnberger and Gershon, 1984). First degree relatives of bipolar individuals, who show episodes of both mania and depression, have a 6% to 7% risk of bipolar illness. Interestingly, these first degree relatives also have about a 12% risk of unipolar depression, an illness in which only depressive episodes occur. These familial risks are elevated compared to the 1% risk of bipolar illness and 5%-8% risk of unipolar illness in the general population.

Studies of twins offer a much more strict test of genetic etiology. Identical, or monozygotic (MZ) twins share all of their genetic material, while fraternal, or dizygotic (DZ) twins share only ½ their genes on average. Both types of twins also share environmental influences. Thus, a theory of genetic influences is supported if MZ twin pairs are concordant for the illness (both twins affected) more often than DZ twins. The strength of the genetic influence is proportional to the difference in concordance rates of MZ and DZ twins.

Results from twin studies of schizophrenia done since 1969 show concordance rates of 17% for DZ twins and 48% for MZ twins (Gottesman, 1991). These results suggest that a gene or genes predispose to schizophrenia, though the fact that identical twins are concordant only half of the time also suggests moderating environmental influences shich allow the illness to go unexpressed in some individuals, even if they may have the genetic predisposition for the disease. In a fascinating study of discordant MZ twins, Gottesman and Bertelsen (1989) found that the offspring of the unaffected twin have about the same elevated risk (17.4%) as the offspring of the schizophrenic twin (16.8%), suggesting that the unaffected twin carries a genetic predisposition to schizophrenia similar to the affected twin. Though the unaffected twin does not express the illness, the genetic risk appears to be passed on to the next generation.

For the affective disorders, about 80% of MZ twins are concordant for bipolar illness while only 13% of DZ pairs are concordant. Unipolar illness appears to have a smaller genetic component; 59% of MZ pairs are concordant and 30% of DZ pairs are concordant (Bertelsen et al., 1977; Tsuang and Faraone, 1990; Goodwin and Jamison, 1990).

Adoption studies also provide a method for determining the strength of genetic causes of illness. Because adopted children and their parents are genetically unrelated, any resemblance between them must be due to shared environmental factors. Conversely, only genes and no environment are shared between the adoptee and the biological parents. Studies which ascertain affected adoptees compare the incidence of disease in biological versus adoptive relatives. A greater incidence in the biological relatives indicates a genetic etiology. A study may also investigate adopted-away children of ill parents. A genetic hypothesis is supported if the incidence among the adoptees is elevated.

Adoption studies of schizophrenia show an increased risk of about 10% for first degree biological relatives of affected adoptees (Gottesman, 1991). The risk to first degree adoptive relatives of affected adoptees, at 1.4%, does not differ from the risk in the general population. Adoption studies of affective illness show increased risks in biological parents of adoptees with bipolar illness (7% for bipolar illness, 21% for unipolar illness) (Wender et al., 1986; Mendlewicz and Rainer, 1977; Tsuang and Faraone, 1990). Shared familial environment appears to have little effect on affective disorders, as adoptive parents of the ill adoptees showed risks close to population rates (1.8% for bipolar illness and 10.5% for unipolar illness).

In an attempt to move one step further and actually find the gene or genes which predispose to illness, researchers have turned to the molecular genetic techniques of linkage analysis. With this method, a researcher attempts to determine if a particular sequence of a known segment of DNA is inherited together with the illness. Currently, there at least 16 ongoing linkage studies of affective disorder (Nurnberger, 1993), and at least 22 for schizophrenia (NIMH report, 1993). For affective illness, several groups have reported linkage to various chromosomal regions (e.g., Egeland et al., 1987; Baron et al., 1987). Similarly, some positive linkage findings have been reported for schizophrenia (e.g., Sherrington et al., 1988). However, these results have not led to underlying genes. The lack of a breakthrough on the genetic front may not surprising, given the limitations of linkage analysis of complex behavior (Risch, 1990). We will explore these difficulties in more detail in this chapter.

There are several methods for determining if a known gene or DNA sequence is inherited together with a disease. In many cases, the lod score method provides the most powerful test of linkage; however, in the case of complex behavioral disorders such as schizophrenia and affective illness, other methods, such as the sib pair method, the Affected Pedigree Member method, or the triples method may offer better tools for determining linkage. These methods will be reviewed and applied to data on families with multiple cases of schizophrenia.

GENES AND RECOMBINATION

The genetic material which makes up chromosomes can be thought of as the blueprint of an organism. Genes provide a template for the construction of the proteins and enzymes which allow an organism to function. Because every cell has two copies of each chromosome, one inherited from the mother and one from the father, there are two copies of every gene. Genes may exist in different forms, called alleles, which ultimately determine the expression of a trait. An organism with two copies of the same allele is called a homozygote, while an organism with two different alleles is called a heterozygote. The configuration of alleles for a particular gene is called the genotype at that chromosomal location (locus).

If two traits are encoded by genes that are close together on the same chromosome, they tend to be inherited together, contrary to Mendel's law of independent assortment. Such traits are said to be genetically linked. Linkage will occur with physical proximity because the phenomenon called genetic recombination, or "crossing over", is less likely if the distance between the two genetic loci is small. During the formation of sperm or egg cells (meiosis), actual physical crossing over of chromosomes can occur. In the progenitor cells, the two copies of each chromosome are called homologues. During metaphase in cell division, the chromosomes line up in the center of the cell, with each pair of homologous chromosomes in close proximity. These paired chromosomes can break and recombine, resulting in a new arrangement of alleles from that found on the parental chromosomes. Figure 1 illustrates crossover events.

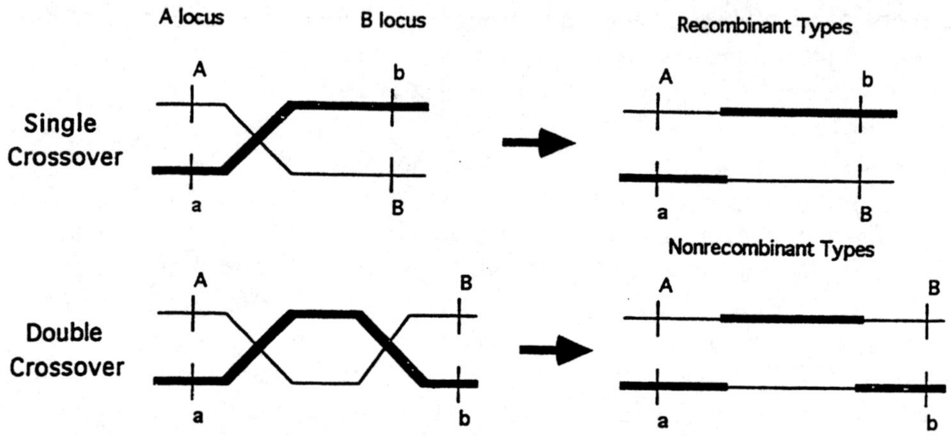

Figure 1. Crossover event.

More than one crossover can occur between two loci. An odd number of crossovers between two loci will result in a recombinant type, a configuration of alleles different from the parental chromosomes. Zero recombinations, or an even number of recombinations will result in a nonrecombinant type, a configuration of alleles identical to the parental chromosomes. When two genetic loci are far apart on a chromosome, zero, odd, and even numbers of recombinations will occur randomly, so that recombinant and nonrecombinant types will occur in the sperm or egg cells with equal frequency. In other words, the probability of a recombination (the recombination fraction, or theta) is ½ when two genetic loci are far apart. Theta is proportional to the distance between two genetic loci; it ranges from 0 for two loci that are adjacent to ½ for loci that are far apart. Theta does not measure exact distance between loci (genetic map distance) because multiple crossovers can occur between two loci, and because of a phenomenon called interference, where the occurrence of one recombination reduces the chance of another crossover nearby. A mapping function can be used to transform theta into true map distance (see Ott, 1991).

Two genetic loci are linked if theta is less than ½. Using linkage analysis, geneticists can determine if the loci are truly linked by testing if theta is significantly less than ½. In addition, the actual value of theta can be estimated to determine how close the linked loci are to one another. The following sections will describe methods for determining if two genetic loci are linked

LOD SCORE METHOD

To understand the lod score method of linkage analysis, consider the family shown in Figure 2. In this figure, circles are females, squares are males, and shading indicates illness. Spouses are joined by a horizontal line and their children are drawn below them. Thus, Figure 2 shows a family with three generations. This family could be used to test the hypothesis that a particular dominant disease is linked to a known gene which controls enzyme activity. We will assume that we can infer the genotypes of the individuals by observing their phenotypes. An individual with the disease has the genotype Dd, where D is the disease allele. An individual with the marker phenotype of high enzyme activity has the genotype Mm, where M is the allele coding for high enzyme production.

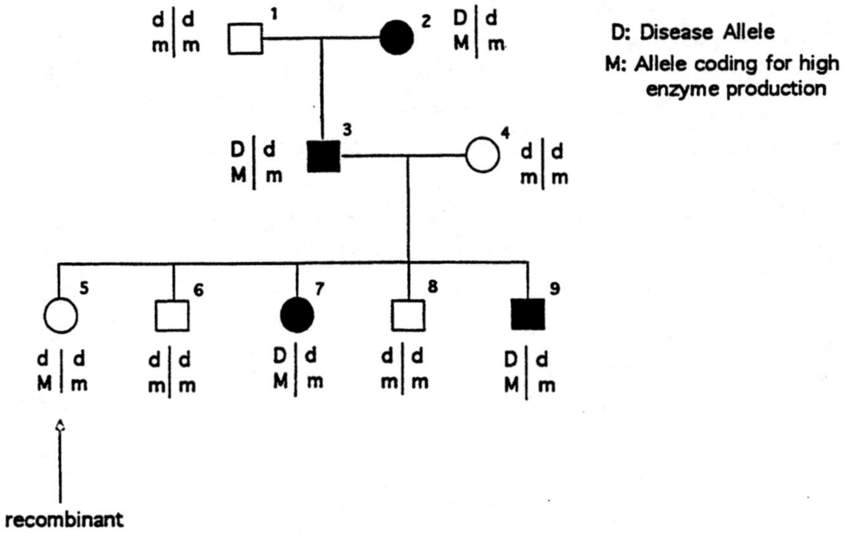

Figure 2. Example pedigree for linkage analysis

Individual 3 must have the D and M alleles on one chromosome, and d and m on the other chromosome. This information about which alleles occur together on a chromosome is called phase information. We know the phase because both the D and the M must have come from the grandmother, so they must be on the chromosome contributed by her (|$\frac{D}{M}$). All of the 5 children must receive a chromosome with the d and m alleles from their unaffected mother (|). If there were no recombination, each child would receive either the chromosome |$\frac{D}{M}$ or the chromosome |$\frac{d}{m}$ from their father; thus, the first child is a recombinant, and the others are nonrecombinants. From these data, we could estimate theta, the recombination fraction, as 1 out of 5, or 0.2. As an aside, note that if the parental genotypes at the marker locus had been mm and mm, we would not have been able to determine which children were recombinants. Homozygosity at the marker locus would have resulted in a family which geneticists call uninformative for linkage; one would not be able to distinguish between marker alleles, and therefore could not determine which allele is inherited together with the disease allele. With the advent of genetic marker loci having many alleles, spouses are rarely homozygous, and matings are almost always informative for linkage.

We can compute the probability of observing the disease status and marker status of this sibship as follows. There are two types of recombinants: |$\frac{D}{M}$ and |$\frac{d}{m}$. Because recombination occurs with frequency theta, each of the recombinant types occurs with equal frequency (θ/2). Similarly, each of the nonrecombinant types ($|\frac{D}{M}$ and $|\frac{d}{m}$) occur with frequencies (1-θ)/2. Given this information, the probability of observing the first child (the recombinant) is θ/2, and the probability of observing each of the other 4 children is (1-θ)/2. Therefore, the probability of observing all 5 children together, or the likelihood of the sibship, is $[\theta/2][(1-\theta)/2]^4$.

For any given value of θ, we now have a numerical value of the likelihood of observing this family. The crucial question is whether observing this family is more likely given a value of θ less than ½ (meaning the marker locus is linked to the disease) than given a value of θ of ½ (no linkage). To answer this question, a ratio of these two likelihoods is computed (L(θ)/L(½)). If we assume θ=0.05 (the marker and the disease locus are close together), then this likelihood ratio, or odds ratio, is $\{(0.05/2)(0.95/2)^4\}/(0.5/2)^5 = 1.296$. If we assume the marker and disease locus are more distant, at θ=0.20, then the odds ratio is $\{(0.2/2)(0.8/2)^4\}/(0.5/2)^5 = 2.621$. Assuming no linkage (θ=0.50) gives an odds ratio of $\{(0.5/2)(0.5/2)^4\}/(0.5/2)^5 = (0.5/2)^5/(0.5/2)^5 = 1.0$. Both hypotheses of linkage are more likely in this family than the hypothesis of no linkage, and it appears that moderate linkage fits the data better than tight linkage.

Table 1. Lod scores

q	Odds Ratio	Lod Score
0.00	0.00	minus infinity
0.01	0.31	-0.51
0.05	1.30	0.11
0.10	2.10	0.32
0.20	2.62	0.42
0.30	2.31	0.36
0.40	1.66	0.22
0.50	1.00	0.00

Odds ratios can be computed for each family within a sample, giving an overall likelihood ratio under a particular hypothesized value of θ. Because each family is statistically independent from each other family, the likelihood ratios from the families can be multiplied together: $\{L_1(\theta)/L_1(½)\}\{L_2(\theta)/L_2(½)\}\{L_3(\theta)/L_3(½)\}$... . To make computations easier, geneticists traditionally take the logarithm of the odds ratios so that the product becomes a sum: $\log\{L_1(\theta)/L_1(½)\} + \log\{L_2(\theta)/L_2(½)\} + \log\{L_3(\theta)/L_3(½)\} + \ldots$. The **l**og of the **od**ds ratio, gives the name **lod** score. For the data in the example, lod scores for several values of theta appear in Table 1.

Notice that when the odds ratio is less than 1.0, the lod score is negative, and when the odds ratio is greater than 1.0, the lod score is positive. In this family, θ=0 gives an odds ratio of zero and a lod score of minus infinity. These results occur because of the zero likelihood of observing a family with a recombinant under the hypothesis of 0% recombination (theta=0). The odds of observing a family with at least one recombinant when $\theta=0$ will always be zero.

A lod score of 1 indicates 10:1 odds in favor of linkage. Geneticists have considered a lod score of 3 (1000:1 odds in favor of linkage) statistically significant evidence for linkage. Note that this significance level is lower than the value of 0.05 customarily used in statistical tests. The possibility that a marker locus is on the same chromosome as a disease locus just by chance is about 5%. If the significance level is set at 0.05, then 5% of the time the researcher will conclude significant linkage exists when in fact there is no linkage. This means the chance of false positive linkage is already 5%, so the significance threshold must be lowered. A more complete discussion of the theory behind lod score analysis can be found in Ott (1991).

Computation of Lod Score

Although in theory one could compute lod scores by hand as in the above example, computer programs are generally used to calculate lod scores. A data set with many complicated pedigrees would make the analysis of data by hand impractical. Although computer programs ease the analysis, the geneticist is still required to furnish a certain amount of information about the population genetics of the disease under study, including the mode of transmission of the disease (e.g., autosomal dominant), the gene frequencies of the disease allele and of the marker alleles, and several other parameters that will be described more fully in the rest of this chapter. Usually, previous population studies provide the geneticist with some of this information, although there are often facets of the genetic model for a disease which elude definitive specification.

Complications: Incomplete Penetrance and Phenocopies

For many genetically controlled diseases, there is a chance that a person carrying the disease allele will not express the disorder. This may be due to age; perhaps the disease does not manifest itself until late in life, and the person has not yet entered the age of risk. In this way, persons in the pedigree coded as unaffected may in fact be carrying the disease allele. This reduced penetrance can arise even for a disease caused by a single gene inherited in a simple Mendelian fashion (e.g., autosomal dominant, autosomal recessive, or X-linked). How will reduced penetrance affect the linkage analysis?

Consider a family which is segregating an autosomal dominant disease (Figure 3). Assume the third child in the last generation actually has the disease allele, but is unaffected. Because we must infer the genotype from the phenotype, we would incorrectly assume the person has the normal genotype at the disease locus (dd). In this example, child 3 would appear to be a recombinant, and would lower the estimate of linkage.

We could estimate the proportion of people who are unaffected but may be carrying the disease allele, and incorporate this information into the linkage analysis. Penetrance is defined as the probability of expressing a certain phenotype given that one has a particular genotype. If we set the penetrance at 75%, a carrier of the disease allele will have a 25% chance of being unaffected. Correctly specifying the penetrance gives a more accurate estimate of the probability of linkage. Overestimating penetrance increases the error in the estimated probability of linkage; the estimate can either increase or decrease. Underestimating the penetrance lessens the information provided by unaffected individuals, which lowers the power of the analysis. Although a low estimate of penetrance may give the most prudent analysis, a true linkage may be missed.

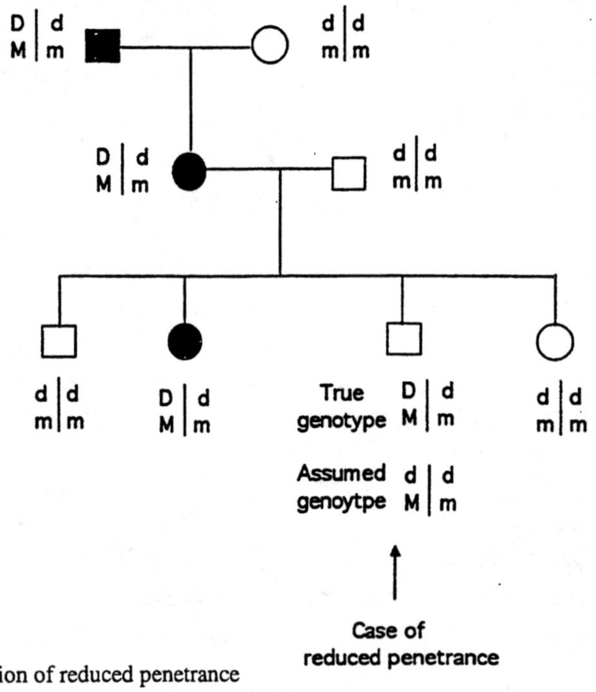

Figure 3. Illustration of reduced penetrance

When a disease is not completely penetrant, there is some uncertainty about whether unaffected individuals carry the disease allele; however, affected persons are known to carry the disease allele. Unfortunately, a researcher may also be uncertain about affected family members if some individuals appear to be affected but do not carry the disease allele. Such persons are called phenocopies, and they may be affected due to environmental toxins or trauma. In the case of a disease with phenocopies, some persons coded as affected will in fact have the normal disease genotype.

Figure 4 shows a family with a phenocopy in the last generation. Here the true genotype of this person is dd at the disease locus, but we incorrectly assume it is Dd. In this case, the person appears to be a nonrecombinant and contributes to a positive estimate of linkage, though in fact the person is a recombinant. If we assumed a 25% phenocopy rate, we would be specifying that there is a probability that ¼ of affected individuals are instead phenocopies. Correctly specifying the phenocopy rate leads to a more accurate estimate of the probability of linkage. An underestimate of the phenocopy rate will create more error in the estimate of linkage; the estimate may be either inflated or deflated. Overestimating the phenocopy rate, or mistakenly assuming some percentage of real genetic cases are phenocopies, will lessen the information provided by the affected individuals in the families, lowering the power of the analysis. An overestimate will result in a more conservative analysis, but, as with underestimates of penetrance, a true linkage may be missed.

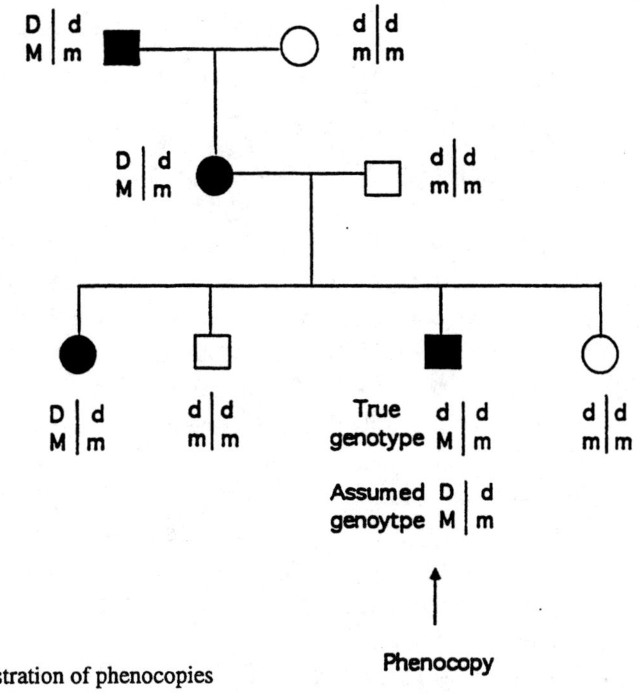

Figure 4. Illustration of phenocopies

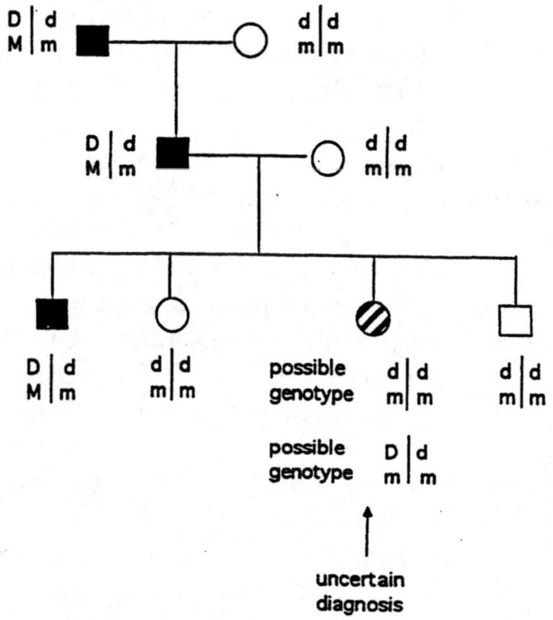

Figure 5. Illustration of diagnostic uncertainty

Linkage Analysis of Complex Diseases

Reduced penetrance and phenocopies can occur even for diseases which are controlled by a single gene inherited in a simple Mendelian fashion. For more complex diseases, however, there are several additional difficulties in performing a genetic analysis. We will use schizophrenia as an example of a complex disease, but there are many such diseases which are currently subject to genetic analysis, including Alzheimer's disease, diabetes, and cancer.

Unknown Mode of Transmission

As defined by geneticists, a complex disease must show evidence of an underlying heritable component, but families do no show the pattern of risk observed for single-gene Mendelian disorders. Families transmitting a disease caused by a single gene show high recurrence risks for first degree relatives of affected probands, ranging from 100 to 1000 times the population frequency of the illness (Risch, 1992). The risk to second degree relatives decreases by a factor of ½, and again decreases by a factor of ½ for each decrease in degree of relatedness. Complex diseases do not show this simple pattern of risk. If we take schizophrenia as an example, risks to first degree relatives range from 6 to 13 times the population frequency; risks to second degree relatives range from 2 to 6 times population frequency (Gottesman, 1991). Observed risks for schizophrenia decrease as the amount of genes shared among relatives decreases, but the numbers do not follow a simple Mendelian model. Such non-Mendelian patterns indicate that complex diseases may be caused by 1) multiple genes each having a small influence on the phenotype (the polygenic model), 2) a few genes having relatively strong effects on the phenotype perhaps together with other minor genes (the oligogenic model), or 3) one or two genes with major effects together with genes of lesser importance (the major locus model). These three possible models must also include environmental effects on the phenotype.

In order to perform linkage analysis, we must assume a simple Mendelian model of autosomal dominant, autosomal recessive, or X-linked transmission; in effect, we must assume the existence of at least one major gene with a relatively strong effect. If a disease gene exists which accounts for a large enough proportion of the phenotypic variation in the disease, linkage analysis may be able to detect this gene under a simple Mendelian model, even though other genes may contribute to the expression of the disease. If schizophrenia is polygenic, meaning no single gene accounts for a detectable proportion of the observed variation in the disease, then linkage analysis is unlikely to find any schizophrenia genes. The existence of a single major gene for schizophrenia has not been proven, though a single gene may underlie schizophrenia in a subset of families (Risch, 1990; Greenberg and Hodge, 1989; Suarez et al., 1981).

Given that the use of a single gene model may not lead us too far astray, we must then choose a mode of inheritance for that single gene. For schizophrenia, X-linkage is unlikely, as fathers transmit the illness to sons. However, it is unknown if an autosomal dominant or recessive genetic model most closely matches the true mode of transmission for schizophrenia. How do geneticists cope with this uncertainty? One approach used frequently is to test for linkage twice, once assuming the disease allele is dominant, and once assuming it is recessive. Performing two tests increases the chances of a false positive result; therefore, a correction for multiple tests must be done. In practice, most researchers test multiple models for complex diseases, as the true model is unknown. There are several adjustments that can be made to lod scores to correct for multiple models. One of the most conservative is to simply subtract $\log_{10}t$ (t=number of tests) (Kidd and Ott, 1984). For 2 tests (one dominant model and one recessive model), lod scores would be reduced by 0.30.

In addition to multiple models, researchers usually test multiple DNA markers for linkage. Corrections may not need to be made for tests of multiple markers because as each marker is eliminated, the probability of detecting true linkage among the remaining markers is slightly greater. This increase in the prior probability of linkage roughly balances the increased chance of observing a false positive for up to 100 markers (Ott, 1991).

With complex diseases, the effect of multiple tests and multiple markers is not straightforward, and a threshold lod score of 3 may not be appropriate. The most practical solution to this dilemma is to determine an empirical significance value of the result using computer simulation methods.

Although misspecification of the genetic model for a complex disease is likely, it may not preclude success of linkage approach. Studies have shown that misspecification of the genetic model has more effect on the estimates of q than on the estimates of the actual lod score (Clerget-Darpoux, 1991). Major genes have been found for several complex diseases, including Alzheimer's disease (Goate et al., 1991; Pericak-Vance et al., 1991; Shellenburg et al., 1992), breast cancer (Hall et al., 1990), and diabetes (Julier et al., 1991; Bell et al., 1991; Froguel et al., 1992)

DIAGNOSTIC UNCERTAINTY

In addition to uncertainty of the mode of genetic transmission, complex diseases are also typically characterized by uncertainty in the definition of the phenotype itself. This diagnostic uncertainty means that some individuals who are truly unaffected may be mistakenly coded as affected, and vice versa. Figure 5 shows a pedigree with an individual of uncertain affection status. The M allele is associated with the disease in the other affected members of this family. If the third child in the last generation is truly affected, she must be a recombinant, and evidence of linkage is correctly reduced. On the other hand, if she were misdiagnosed as affected when she is truly unaffected, she is not a recombinant although she would appear to be one. In this case, evidence of linkage would be incorrectly reduced.

Several strategies can be employed to minimize the effects of diagnostic uncertainty. First, diagnoses can be verified by clinicians blind to the study in an attempt to eliminate false positive and false negative diagnoses. Follow-up diagnoses can be made to determine any changes in affection status. Second, a diagnostic hierarchy can be established such that persons who are definitely affected are coded differently from uncertain or borderline cases. Initial scanning for linkage can be done using only the strictly defined cases. If a possible linkage is found, borderline cases can then be considered. Although including the borderline cases in the group of affected individuals may allow the detection of a disease gene, there is also more risk that these cases represent phenocopies.

Family Size

Another problem specific to the genetic analysis of schizophrenia is the dearth of large families transmitting the illness. Large families with multiple affected cases carry more information for genetic linkage, and may be more likely to be segregating a dominant form of the disease (Greenberg, 1992). However, persons with schizophrenia are less likely to marry and have children. Some relatively large pedigrees have been found, but many families that have been studied are small, with only 2 or 3 affected individuals. About 63% of schizophrenics do not have a family history of the illness (Gottesman, 1991), lessening possibility of finding additional large families transmitting schizophrenia.

Genetic Heterogeneity

In 1988, a team of researchers identified a possible gene for schizophrenia on chromosome 5 using linkage analysis of 5 Icelandic and 2 British pedigrees (Sherrington et al., 1988). At the same time, a separate research team reported schizophrenia linkage evidence from one large Swedish pedigree which excluded the same chromosomal region (Kennedy et al., 1988). Since these reports, numerous other researchers have also found no evidence for linkage to this region of chromosome 5 (see McGuffin et al., 1990, for review). One possible explanation for these discrepant results could be genetic heterogeneity; a gene on chromosome 5 may in fact be a cause of schizophrenia in the families studied by Sherrington et al., but may not cause schizophrenia in other families. The limited number of large families segregating schizophrenia compounds the problems of genetic heterogeneity by creating difficulties for the verification of a positive linkage finding. If multiple genes cause schizophrenia, it is likely that only a subset of families will show linkage to a marker for one of these hypothetical genes. Verification using linkage techniques may be difficult, and researchers may have to turn to direct mutation scanning techniques in genes in the area of linkage known to affect brain function.

In fact, even one occurrence of positive results such as the findings of Sherrington et al. (1988) may be difficult to come by. If only a subgroup of families *within* a study show evidence for linkage while other families are negative, an overall lod score may miss the suggestion of a positive finding. A researcher can formally test for heterogeneity within a sample using the HOMOG program (Ott, 1991). Unfortunately, this method is not ideally suited for the relatively small pedigrees transmitting schizophrenia.

To circumvent these problems, many researchers have placed their hopes in diagnostic centers and collaborative efforts, combining their data so that their work might be used to the best advantage. Researchers also do not rely on overall lod scores, instead checking lod scores for individual families. If lod scores are surprisingly large in a subset of families, the region may deserve further investigation with additional DNA markers. Another alternative is to search for identifying characteristics to place families into subgroups which might reflect underlying differences in genetic causes. Although diagnostic characteristics and drug response have not yet yielded definite subtypes of schizophrenia, research efforts are still being devoted to this issue.

NONPARAMETRIC METHODS

Finally, many researchers are turning to methods of detecting linkage which do not require model specification (nonparametric methods). Because the mode of transmission, gene frequency, penetrance, and phenocopy rates are unknown, the use of these methods may be more prudent in the search for genes which predispose individuals to genetically complex diseases. The simplest of these techniques is the sib pair method, first outlined by Penrose (1935).

The S.A.G.E. computer program package (S.A.G.E., 1992), is available for sib pair analysis. The program works by using marker data to estimate the proportion of genes shared through a common ancestor (identical by descent, or IBD). The program then performs three tests to determine: 1) if the proportion of genes IBD for concordant affected-affected pairs is greater than ½; 2) if the proportions of alleles shared IBD for unaffected-unaffected pairs is greater than ½; and 3) if the proportion of shared alleles for discordant pairs is less than ½. Linkage is indicated if one observes this pattern of deviations from the expectation of sharing ½ of the alleles IBD. In addition, the squared sib pair difference for affection status is computed, where affection status is quantified as 1 for affected and 0 for unaffected. This squared sib pair difference is regressed on the proportion of genes IBD at each marker locus, to determine if the proportion of shared alleles is a significant predictor of sibling similarity for disease status. In other words, are the shared alleles contributing to the illness? This regression test will detect linkage if there is a significant difference in the proportion of genes IBD for concordant sib pairs (either affected-affected pairs or unaffected-unaffected pairs) versus discordant pairs.

Although this sib pair regression method offers more statistical power by including information from unaffected individuals, results should be viewed with caution for complex diseases because some unaffected persons may be gene carriers. Unlike the lod score method, the sib pair method has no mechanism for specifying reduced penetrance. Thus, the test of whether affected-affected sib pairs share significantly more than ½ of their alleles IBD may be the safest test for complex diseases where reduced penetrance is probable.

The sib pair method has the advantages of simplicity and no assumed underlying genetic model. However, the sib pair technique is not as powerful as the lod score method, and requires pedigrees with multiply affected subjects within a sibship. Although the sib pair method is often useful in the smaller pedigrees available for analysis for an illness such as schizophrenia, if ill members of the pedigree have more distant genetic relationships, another nonparametric technique which may be preferable is the Affected Pedigree Member (APM) method (Weeks and Lange, 1988). This method also uses only the affected members of the pedigree, and is based on the probability that pairs of relatives within a pedigree share alleles. The program gives a positive result either if alleles are shared through a common ancestor (IBD) or if alleles are shared, but originate from both the maternal and paternal sides of the pedigree (identical by state, or IBS); thus, positive findings may indicate either linkage or an association between the disease and the marker. However, with DNA markers that are highly polymorphic, affected relatives will likely share alleles due to identity by descent, and the APM method will approximate an IBD method.

The test statistic from the APM method is adjusted for marker allele frequency using three correction functions: $f(p) = 1$, $f(p) = 1/\sqrt{p}$, and $f(p) = 1/p$, where p is the marker allele frequency. The function $f(p) = 1$ does not weight the statistic for the marker allele frequencies, though these frequencies are used in the basic calculation of the probability of allele sharing among the affected relatives. This function may give inaccurate results if the marker allele frequencies are far from equal. Particularly with highly polymorphic markers, alleles can have quite small frequencies. If relatives share a rare allele, the finding should be weighted more heavily than if they share a common allele. The lack of such weighting can lead to either false positive or false negative results. The other two functions weight the statistic such that rare alleles shared among relatives carry more significance. The intermediate function ($f(p) = 1/\sqrt{p}$) more often provides a normal distribution of test statistics, and is the recommended correction function (Dan Weeks, personal communication). Empirical significance levels of the statistics generated by the program can be derived using the SIMULF program (Weeks and Lange, 1988), which uses randomly generated genotypes and provides a distribution of statistics given the marker allele frequencies. This simulation technique alleviates the problems of non-normal distributions which may result under the weighting functions.

Although highly polymorphic markers give a result that approximates an IBD method (i.e., one is more likely to be testing for linkage rather than association), this advantage is offset by the sensitivity of the APM method to marker allele frequency estimates (Babron et al., 1993). To circumvent this problem, APM analyses must be performed using multiple methods of marker allele frequency estimation. While markers should be tested using published allele frequencies (when available), frequencies should also be estimated from the disease families. This estimation technique ensures that the frequencies match the population from which the disease families were drawn. The analysis using these estimates should provide conservative results; if a rare allele does in fact occur together with schizophrenia in the families, its frequency estimated from the family members will be elevated, ant the sharing of this rare allele will carry less statistical weight.

The triples method is a new nonparametric technique which may also prove useful in the analysis of complex genetic diseases. This method determines the probability of the pattern of alleles shared identical by descent in any triple of second-degree relatives (e.g., affected cousins). For example, for non-inbred cousins, only one or zero alleles can be shared IBD. Thus, the resulting patterns for a triple of cousins could be: 0,0,0; 0,0,1; 0,1,0; 0,1,1; 1,0,0; 1,0,1; 1,1,0; or 1,1,1. If the marker is linked to the disease gene, the cousins will share marker alleles more often than if the marker were unlinked. The use of second degree relatives decreases the chance of a false positive result, as the sharing of alleles for these more distantly related relatives will happen less often by chance.

EXAMPLE

To illustrate the concepts of linkage analysis of complex illness, the following sections contain an example of an analysis of 3 DNA markers on chromosome 22 in a sample of nine schizophrenia families (for an expanded analysis, see Coon et al., 1994a). We chose these markers after lod scores suggesting linkage were found in a preliminary genome-wide survey for schizophrenia susceptibility genes (Coon et al., 1994b). Two genetic markers which both mapped to the distal end of the long arm of chromosome 22, showed a suggestion of linkage, though no lod scores were greater than 3, the threshold for statistical significance. However, these loci did not have many alleles, and were therefore not maximally informative for linkage. To follow up the findings, we used three DNA markers from the distal end of chromosome 22: D22S274, D22S276, and D22S279 (Weissenbach et al., 1992). These markers are stretches of DNA with varying numbers of repeats of a two base pair sequence. Such dinucleotide repeat markers have many alleles, and are highly informative for linkage. D22S274 is the most distal marker on chromosome 22q. This marker has nine alleles and a heterozygosity value of 0.81 (a measure of the chance that matings will be informative for linkage); thus, the marker is likely to be quite informative. D22S276 lies five centiMorgans (cM) from D22S274, and D22S279 lies 2 cM from D22S276. D22S276 has eight alleles and a heterozygosity of 0.70, and D22S279 has five alleles and a heterozygosity of 0.65.

The sample used consists of nine multigenerational families, which were identified by screening for probands among patients hospitalized for schizophrenia. Following informed consent, family members were interviewed using a semi-structured interview, the Schedule for Affective Disorders and Schizophrenia - Lifetime Version (Endicott and Spitzer, 1979). When available, physicians' records and clinical records were obtained. All information on each subject was collected with identifying names deleted and given to two psychiatrists who made consensual diagnoses based on Research Diagnostic Criteria.

The pedigrees have three to six cases of schizophrenia. Of the 120 subjects in the nine pedigrees, 32 are affected cases; 24 of these have chronic schizophrenia, five have subchronic schizophrenia, and three have chronic schizoaffective disorder. Only these strictly defined cases of schizophrenia are included in analyses of DNA markers; no cases of schizotypal personality or other psychiatric disorders are considered affected. All subjects are Caucasians of European ancestry, ascertained from one geographic region, the Intermountain West. Families from this area are typically multigenerational, closely knit families with low levels of drug and alcohol abuse.

High molecular weight genomic DNA was isolated from consenting subjects. Genotypes were read independently by two technicians who were blind to the diagnostic status of the family members.

Lod score analyses between schizophrenia and these DNA markers were first carried out using the LINKAGE software package (Lathrop et al. 1984). Using a maximum penetrance of 0.6, we assumed first autosomal dominant then autosomal recessive transmission. Disease allele frequency was set at 0.01 for the dominant model and at 0.12 for the recessive model. Reduced penetrance was made to be dependent on age of onset by fixing penetrances at 0.05 (onset < 15), 0.5 (onset 15-30), and 0.6 (onset > 30). We allowed for phenocopies by setting the penetrances for the normal genotype at 0.0005 (age < 15), 0.001 (age 15-30), and 0.002 (age > 30). These model parameters follow the assumptions of a population disease prevalence ranging from 0.8% to 1.0%, and a phenocopy rate ranging from 11% to 15%.

Table 2. Linkage Analysis of Chromosome 22 DNA Markers.

	Lod Score						
Marker	0.00	0.01	0.05	0.10	0.20	0.30	0.40
D22S274	-4.10 (0.51)	-2.86 (0.81)	-1.06 (1.40)	-0.19 **(1.58)**	**0.37** (1.33)	0.36 (0.78)	0.17 0.25)
D22S276	-0.50 (1.59)	0.17 (1.76)	0.90 (2.07)	**1.04** **(2.09)**	0.81 (1.63)	0.42 (0.92)	0.10 (0.27)
D22S279	-1.31 (-1.17)	-0.82 (-0.29)	0.04 (0.75)	0.40 **(1.11)**	**0.50** (1.05)	0.33 (0.65)	0.12 (0.21)

Note. Maximum penetrance was set at 0.6. Results for the recessive model appear in parentheses. Maximum lod scores are highlighted.

Table 2 presents the results of linkage analyses with the chromosome 22 markers. We report lod scores at various values of the recombination fraction θ. Recall that a value of q=0.0 means that we assume there are no recombinants in the family. Because the lod score is the log of the odds ratio, the score of 2.09 for D22S276 indicates that the odds are about 100 to 1 in favor of linkage, with a recombination frequency between this marker and a putative disease gene of 0.10. These results assume recessive inheritance of schizophrenia. Assuming dominant inheritance gives only about 10 to 1 odds in favor of linkage, again with a recombination frequency of 0.10. Lod scores for the other two loci also give suggestions of linkage, but all scores fail to meet the significance threshold of a lod score of 3.0 (1000 to 1 odds in favor of linkage).

One possible explanation of these findings is that some families in our sample are linked while others are not (genetic heterogeneity). We tested this hypothesis using the HOMOG program (Ott 1991), which determines if linkage can be detected in a subset of families with the remainder of families unlinked. For the three chromosome 22 markers, these tests for linkage in a subset of families were not significant (p values ranged from 0.17 to 0.50).

Because these findings were inconclusive, we tested the data further using nonparametric methods. If we incorrectly specified the genetic model, false negative findings could occur. Table 3 shows sib pair results for the 3 chromosome 22 DNA markers in the 9 families described above. Although findings again suggest linkage, the regression results were significant only for D22S276 when an adjustment was made for multiple tests. The individual tests for each of the three types of sib pairs showed that affected-affected pairs shared significantly more than ½ of their alleles IBD for D22S276.

Table 3. Sib Pair Analysis of Chromosome 22 DNA Markers.

Marker	Concordant vs discordant pairs p value	Mean estimated proportion of marker alleles shared identical by descent		
		0 sibs affected (n=57)	1 sib affected (n=81)	2 sibs affected (n=23)
D22S274	0.043	0.51	0.47	0.67
D22S276	0.016[a]	0.52	0.45	0.69[b]
D22S279	0.020	0.54	0.45	0.58

[a] Significant after adjustment for multiple tests.

[b] Significantly greater than ½ after adjustment for multiple tests.

Table 4. Affected-Pedigree-Member Method Analysis of Chromosome 22 DNA Markers.

Marker	f(p)=1	prob	f(p)=1/√p	prob	f(p)=1/p	prob
D22S274	2.16 (1.20)	0.04 (0.12)	3.18 (1.86)	0.0007 (0.04)	0.69 (-0.08)	0.31 (0.54)
D22S276	-0.07 (0.49)	0.54 (0.30)	5.02 (1.80)	0.0002 (0.04)	5.36 (0.84)	0.0002 (0.05)
D22S279	0.11 (-0.08)	0.47 (0.51)	-0.49 (-0.36)	0.36 (0.61)	-0.56 (-0.18)	0.72 (0.55)

Note. Results are adjusted for marker allele frequency (p) using the function f(p) = 1/√p. Empirical probabilities were calculated using the SIMULF program: 10,000 replicates were generated for each marker, and the resulting distribution of scores allowed the calculation of the probability of the observed score. Estimated marker allele frequencies were computed from CEPH families; results using marker allele estimates taken from the disease families are shown in parentheses.

Table 4 summarizes the findings using the Affected-Pedigree-Member method for the chromosome 22 DNA markers. The table shows the three corrections for marker allele frequency (no correction: f(p)=1; moderate correction: f(p)=1/Öp; strong correction: f(p)=1/p). Although significance was found for several markers using one of these correction functions, none of the markers showed significance for all three methods. A robust finding would show significance independent of the marker allele correction; therefore, although the results suggest linkage, they do not meet strict standards for significance. In addition, these results are for marker allele frequencies estimated from reference families. Note how the statistics drop dramatically when marker allele frequencies estimated from the disease families are used (results in parentheses). Again, however, this estimation technique gives a conservative lower bound for the APM method.

The three types of analysis of the chromosome 22 DNA markers show a suggestion of linkage, though results are not conclusive. Next steps will include testing more DNA markers in this chromosomal region and turning to other researchers to confirm or refute this suggestion of a finding (see Coon et al., 1994a). In addition, a hint of linkage such as this will generate a search for genes in the region which are known to function in some way in the brain. Sequences of these candidate genes in schizophrenics could then be scanned for possible mutations.

ENDOPHENOTYPES

One final strategy, which may be the best new hope in the search for susceptibility genes for schizophrenia, is the use of physiological characteristics correlated with the disease, or endophenotypes (Matthysse, 1990; Matthysse et al., 1992; Holzman et al., 1988). Such endophenotypes are simpler traits to define, often have higher penetrances, and frequently show patterns of inheritance more likely to be due to effects of a major gene. Because psychiatric diagnosis may not identify all individuals who are carriers of a genetic predisposition for schizophrenia, an endophenotype may supply a definition for both affected and carrier individuals (Gottesman and Bertelsen, 1989). A gene which is found to underlie a physiological characteristic correlated with schizophrenia may in turn be involved in the etiology of schizophrenia itself.

As an example of this strategy, linkage analyses have been performed on the 9 families describe above using a previously characterized neurophysiological measure, gating of evoked auditory response (Freedman et al., 1987). An auditory response measure is an appropriate endophenotype for schizophrenia because of evidence for sensory disturbances in schizophrenics (Venables, 1964; Patterson et al., 1986). Individuals with the disease often show an inability to filter out background stimuli, which may be a result of a failure in a neurological inhibitory pathway (Adler et al., 1982; Freedman et al., 1987). This auditory filtering process can be observed in the electroencephalograhpic (EEG) measurement of evoked response to repeated auditory signals. In normal individuals, the EEG response to the repetition of a sound is observed to be substantially lower than the response to the initial sound. In effect, the first sound conditions neuronal pathways so that sensitivity to a repetition of the same sound is lessened. In most schizophrenics, this conditioning is absent, and evoked response is phenomenon is measured using the amplitude of the EEG wave occurring 50 ms after each stimulus sound. The amplitudes of these P50 waves are measured, and the ratio of the second to the first wave amplitude computed. This P50 ratio is closer to zero in individuals with normal P50 because response to the second stimulus has been suppressed in these individuals.

For the analysis of the P50 ratio, values from 0% to 40% were considered normal and those greater than 50% were considered abnormal; 3 subjects who had ratios between 40% and 50% were coded as unknown. This classification was based on previous findings of a 40% threshold value below which 95% of a control population scored (Siegel et al., 1984). As reported by Siegel et al. (1984), one cluster of scores was formed by normal control individuals with no family history of schizophrenia, and another distinct cluster was formed by schizophrenics. Within first degree relatives of schizophrenics, these two clusters were repeated, with approximately half of the unaffected relatives grouping around 0% (normal) and the others falling above 40%. Some of the variation observed within these clusters will be due to measurement error; therefore, a region of uncertainty between the 40% threshold and 50% was used. In our sample, 2 schizophrenics had P50 values between 40% and 50%; the remainder had P50 values over 50%. These distributional properties of P50 are consistent with effects of a major gene.

Although the mode of inheritance for P50 has not been tested, previous data are consistent with autosomal dominant transmission (Siegel et al., 1984). In our sample, 7 of the 8 marriages where P50 is known for both parents have one normal and one abnormal parent (these marriages are from separate pedigrees). Of the 34 offspring of these marriages, 18 have abnormal P50s, resulting in a ratio not significantly different from 0.5 ($X^2 = 0.12$, $p > 0.7$). Within each family, this ratio of abnormal to normal P50 ranges from 0.25 to 0.83, with 4 of the families giving ratios of 0.5. Further, our data are not compatible with X-linked inheritance, as male to male transmission is observed in 4 out of the 9 pedigrees.

Based on the above data, we tested an autosomal dominant model with relatively high penetrance (0.80). Gene frequency was set at 0.05, and we used a penetrance of 0.01 for the normal genotype, resulting in a 10% phenocopy rate. These parameters give a population prevalence of 0.02 for abnormal P50 which conforms to estimates from control populations (Siegel et al., 1984).

Using this model, the 3 chromosome 22 markers were tested (Table 5). Although the analyses using the schizophrenia phenotype suggested linkage, these results are strongly negative. We may be incorrect in assuming that abnormal P50 and schizophrenia share underlying genetic causes, though the observed cosegregation of the two traits suggests otherwise. Alternatively, a gene causing abnormal P50 may be only one of many genes conferring susceptibility to schizophrenia. The suggestion of linkage on chromosome 22 may be indicating a gene having some effect on schizophrenia but not causing abnormal P50. Of course, the most prudent explanation is that our positive results with the schizophrenia phenotype are due to chance. Further exploration of this chromosomal region using the P50 endophenotype will be required to settle these questions.

Table 5. Linkage Analysis of Chromosome 22 DNA Markers using the P50 Endophenotype.

	Lod Score						
Marker	0.00	0.01	0.05	0.10	0.20	0.30	0.40
D22S274	-5.01	-4.34	-2.81	-1.71	-0.54	-0.07	0.05
D22S276	-6.06	-5.50	-4.00	-2.74	-1.23	-0.48	-0.12
D22S279	-5.99	-5.45	-4.11	-3.00	-1.51	-0.60	-0.13

Note. Maximum penetrance was set at 0.8.

CONCLUSION

Although this chapter has focused on applications to schizophrenia, many complex human behaviors are currently under investigation. These behaviors include mental retardation, autism, Alzheimer's disease and dementia, Tourette's syndrome, attention deficit hyperactivity disorder, panic disorder, antisocial behavior and conduct disorder, alcoholism and substance abuse, anxiety disorder, and sexual orientation. Although progress has been slow to date in the molecular genetic study of these behaviors, the addition of new methodological techniques to the arsenal of analytical tools should speed the discovery of genes predisposing to these traits. New methods currently being researched include refinements on the nonparametric methods mentioned in this chapter, methods for detecting associations to disease loci, regression techniques which allow the addition of environmental effects to the traditional genetic models, new more powerful tests for genetic heterogeneity, and methods to test models including more than one behavior and/or more than one disease locus.

Finding the genetic causes of schizophrenia and other complex genetic diseases using standard linkage techniques has proven to be a remarkably difficult task. More DNA markers are becoming available; because these markers have many alleles, they will almost always be informative for linkage. These better genetic markers, combined with careful analysis using traditional techniques, analyses using new approaches, collaborative efforts, and the exploration of endophenotypes such as P50 may help surmount some of the difficulties involved in the genetic analysis of complex behaviors.

REFERENCES

Babron, M-C., Martinez, M., Bonaiti-Pellie, C., & Clerget-Darpoux, F. (1993). Linkage detection by the Affected-Pedigree-Member method: What is really tested? *Genetic Epidemiology, 10*, 389-394.

Bell, G.I., Xiang, K-S., Newman, M.V., Wu, S-H., Wright, L.G., Fajans, S.S., Speilman, R.S., & Cox, N.J. (1991). Gene for non-insulin-dependent diabetes mellitus (maturity-onset diabetes of the young subtype) is linked to DNA polymorphism on human chromosome 20q. *Proceedings of the National Academy of Sciences USA, 88*, 1484-1488.

Bertelsen, A., Harvald, B., & Hauge, M. (1977). A Danish twin study of manic depressive disorders. *British Journal of Psychiatry, 130*, 330-351.

Clerget-Darpoux, F. (1991). The uses and abuses of linkage analysis in neuropsychiatric disorder. In P. McGuffin & R. Murray (Eds.). *The new genetics of mental illness* (pp. 44-57). Oxford: Butterworth-Heinemann.

Coon, H., Holik, J., Hoff, M., Reimherr, F., Wender, P., Myles-Worsley, M., Waldo, M., Freedman, R., Byerley, W. (1994a). Analysis of chromosome 22 markers in 9 schizophrenia pedigrees. *Neuropsychiatric Genetics, 54*, 72-79.

Coon, H., Jensen, S., Holik, J., Hoff, M., Myles-Worsley, M., Reimherr, F., Wender, P., Waldo, M., Freedman, R., Leppert, M., & Byerley, W. (1994b). A genomic scan for genes predisposing to schizophrenia. *Neuropsychiatric Genetics, 54*, 59-71.

Endicott, J. & Spitzer, R.L. (1979). A diagnostic interview: The Schedule for Affective Disorders and Schizophrenia. *Archives of General Psychiatry, 35*, 837.

Freedman, R., Adler, L.E., Gerhardt, G.A., Waldo, M., Baker, N., Rose, G.M., & Drebing, C. (1987). Neurobiological studies of sensory gating in schizophrenia. *Schizophrenia Bulletin, 13*, 669-677.

Freedman, R., Adler, L.E., Waldo, M.C., Pachtman, E., & Franks, R.D. (1983). Neurophysiological evidence for a defect in inhibitory pathways in schizophrenia: Comparison of medicated and drug-free patients. *Biological Psychiatry, 18*, 537-551.

Froguel, P.H., Vaxillaire, M., Sun, F., Velho, G., Zouali, H., Butel, M.O., Lesage, S., Vionnet, N., Clement, K., Fougerousse, F., Tanizawa, Y., Weissenbach, J., Beckmann, J.L., Lathrop, G.M., Passa, P.H.A-P, & Cohen, D. (1992). Close linkage of glucokinase locus on chromosome 7p to early-onset non-insulin-dependent diabetes mellitus. *Nature, 356*, 162-164.

Goate, A., Chartier-Harlin, M-C., Mullan, M., Brown, J., Crawford, F., Fidani, L., Giuffra, L., Haynes, A., Irving, N., James, L., Mant, R., Newton, P., Rooke, K., Roques, P., Talbot, C., Pericak-Vance, M., Roses, A., Williamson, R., Rossor, M., Owen, M., & Hardy, J. (1991). Segregation of a missense mutation in the amyloid precursor protein gene with familial Alzheimer's disease. *Nature, 349*, 704-706.

Goodwin, F.K. & Jamison, K.R. (1990). *Manic-Depressive Illness*. New York: Oxford University Press.

Gottesman, I.I. (1991). *Schizophrenia genesis: the origins of madness*. New York: WH Freeman.

Gottesman, I.I. & Bertelsen, A. (1994). Confirming unexpressed genotypes for schizophrenia: Risks in the offspring of Fisher's Danish identical and fraternal discordant twins. *Archives of General Psychiatry, 46*, 867-872.

Greenberg, D.A. (1992). There is more than one way to collect data for linkage analysis. *Archives of General Psychiatry, 49*, 745-750.

Greenberg, D.A. & Hodge, S.E. (1989). Linkage analysis under "random" and "genetic" reduced penetrance. *Genetic Epidemiology, 6*, 259-264.

Gutman, D.H., Wood, D.L., & Collins, F.S. (1991). Identification of the neurofibromatosis type 1 gene product. *Proceedings of the National Academy of Sciences USA, 88*, 9658-9662.

Hall, J.M., Lee, M.K., Newman, B., Morrow, J.E., Anderson, L.A., Huey, B., & King, M-C. (1990). Linkage of early-onset familial breast cancer to chromosome 17q21. *Science, 250*, 1684-1689.

Holzman, P.S., Kringlen, E., Matthysse, S., Flanagan, S.D., Lipton, R.B., Cramer, G., Levin, S., Lange, K., & Levy, D.L. (1988). A single dominant gene can account for eye tracking dysfunctions and schizophrenia in offspring of discordant twins. *Archives of General Psychiatry, 45*, 641-647.

Huntington's Disease Collaborative Research Group , (1993). A novel gene containing a trinucleotide repeat that is expanded and unstable on Huntington's disease chromosomes. *Cell, 72*, 971-983.

Julier, C., Hyer, R.N., Davies, J., Merlin, F., Soularue, P., Briant, L., Cathelineau, G., Deschamps, L., Rotter, J.I., Froguel, P., Boitard, C., Bell, J.I., & Lathrop, G.M. (1991). Insulin-IGF2 region on chromosome 11p encodes a gene implicated in HLA-DR4-dependent diabetes susceptibility. *Nature, 354*, 155-159.

Kennedy, J.L., Giuffra, L.A., Moises, H.W., Cavalli-Sforza, L.L., Pakstis, A.J., Kidd, J.R., Castiglione, C.M., Sjogren, B., Wetterberg, L., & Kidd, K.K. (1988). Evidence against linkage of schizophrenia to markers on chromosome 5 in a northern Swedish pedigree. *Nature, 336*, 16-170.

Kerem, B., Rommens, J.M., Buchanan, J.A., Markiewicz, D., Cox, T.K., Chakravarti, A., Buchwald, M., & Tsui, L-C. (1989). Identification of the cystic fibrosis gene: Genetic analysis. *Science, 245*, 1073-1080.

Kidd, K.K. & Ott, J. (1984). Power and sample size in linkage studies. Human Gene Mapping 7: Seventh International Workshop on Human Gene Mapping. *Cytogenetics and Cell Genetics, 37*, 510-511.

Lathrop, G.M., Lalouel, J-M., Julier, C., & Ott, J. (1984). Strategies for multilocus linkage analysis in humans. *Proceedings of the National Academy of Sciences USA, 81*, 3443-3446.

Marchuk, D.A., Saulino, A.M., Tavakkol, R., Swaroop, M., Wallace, M.R., Andersen, L.B., Mitchell, A.L., Gutmann, D.H., Boguski, M., & Collins, F.S. (1991). cDNA cloning of the type 1 neurofibromatosis gene: Completed sequence of the NF1 gene product. *Genomics, 11*, 931-940.

Matthysse, S. (1990). Genetic linkage and complex diseases: A comment. *Genetic Epidemiology, 7*, 29-31.

Matthysse, S., Levy, D.L., Kinney, D., Deutsch, C., Lajonchere, C., Yurgelun-Todd, D., Woods, B., & Holzman, P.S. (1992). Gene expression in mental illness: A navigation chart to future progress. *Journal of Psychiatry Research, 26*, 461-473.

McGuffin, P., Sargeant, M., Hett, G., Tidmarsh, S., Watley, S., & Marchbanks, R.M. (1990). Exclusion of a schizophrenia susceptibility gene from the chromosome 5q11-q13 region. New data and a reanalysis of previous reports. *American Journal of Human Genetics, 47*, 524-535.

Mendlewicz, J. & Rainer, J.D. (1977). Adoption study supporting genetic transmission in manic-depressive illness. *Nature, 268*, 327-329.

Monaco, A.P. & Kunkel, L.M. (1988). Cloning of the Duchenne/Becker muscular dystrophy locus. *Advances in human genetics, 17*, 61-98.

NIMH Activities , (1993). Schizophrenia-related grants, fiscal year 1991. *Schizophrenia Bulletin, 19*, 171-194.

Nurnberger, J.I. (1993). Status report on linkage studies of affective disorders. *Psychiatric Genetics, 3*, 207-214.

Nurnberger, J.I. & Gershon, E.S. (1984). Genetics of affective disorders. In R. Post & J. Ballenger (Eds.). *Neurobiology of Mood Disorders* (pp. 76-101). Baltimore: Williams and Wilkins

Ott, J. (1991). *Analysis of human genetic linkage*. Baltimore: The Johns Hopkins University Press.

Patterson, T., Spohn, H.E., Bogia, D.P., & Hayes, K. (1986). Thought disorder in schizophrenia. *Schizophrenia Bulletin, 12*, 460-472.

Penrose, L.S. (1935). The detection of autosomal linkage in data which consist of pairs of brothers and sisters of unspecified parentage. *Annals of Eugenics, 6*, 133-138.

Pericak-Vance, M.A., Bebout, J.L., Gaskell, P.C., Yamaoka, L.H., Hung, W-Y., Alberts, M.J., Walker, A.P., Bartlett, R.J., Haynes, C.A., Welsh, K.A., Earl, N.L., Heyman, A., Clark, C.M., & Roses, A.D. (1991). Linkage studies in familial Alzheimer disease: evidence for chromosome 19 linkage. *American Journal of Human Genetics, 48*, 1034-1050.

Risch, N. (1990). Genetic linkage and complex diseases, with special reference to psychiatric disorders. *Genetic Epidemiology, 7*, 3-16.

Risch, N. (1992). Genetic linkage: interpreting lod scores. *Science, 255*, 803-804.

S.A.G.E., (1992). *Statistical Analysis for Genetic Epidemiology, Release 2-1. Computer program package available from the Department of Biometry and Genetics, LSU Medical Center, New Orleans.*

Shellenburg, G.D., Bird, T.D., Wijsman, E.M., Orr, H.T., Anderson, L., Nemens, E., White, J.A., Bonnycastle, L., Weber, J.L., Alonso, M.E., Potter, H., Heston, L.L., & Martin, G.M. (1992). Genetic linkage evidence for a familial Alzheimer's disease locus on chromosome 14. *Science. 258*. 668-671.

Sherrington, R., Brynjolfsson, J., Petursson, H., Potter, M., Dudleston, K., Barraclough, B., Wasmuth, J., Dobbs, M., & Gurling, H. (1988). Localization of a susceptibility locus for schizophrenia on chromosome 5. *Nature, 336*, 164-167.

Siegel, C., Waldo, M., Mizner, G., Adler, L.E., & Freedman, R. (1984). Deficits in sensory gating in schizophrenic patients and their relatives. *Archives of General Psychiatry, 41*, 607-612.

Suarez, B., O'Rourke, D., & vanEerdewegh, P. (1981). Power of the affected sib pair method to detect disease susceptibility loci of small effect: an application to multiple sclerosis. *American Journal of Medical Genetics, 14*, 309-326.

Tsuang, M.T. & Faraone, S.V. (1990). *The genetics of mood disorders*. Baltimore: Johns Hopkins University Press.

Venables, P. (1964). Input dysfunction in schizophrenia. In B.A. Maher (Ed.). *Progress in Experimental Personality Research* (pp. 1-47). New York: Academic Press

Weeks, D.E. & Lange, K. (1988). The Affected-Pedigree-Member method of linkage analysis. *American Journal of Human Genetics, 42*, 315-326.

Weissenbach, J., Gyapay, G., Dib, C., Vignal, A., Morissette, J., Millasseau, P., Vaysseix, G., & Lathrop, M. (1992). A second-generation linkage map of the human genome. *Nature, 359*, 794-801.

Wender, P.H., Kety, S.S., Rosenthal, D., Schulsinger, F., Ortmann, J., & Lunde, I. (1986). Psychiatric disorders in the biological and adoptive families of adoptive individuals with affective disorders. *Archives of General Psychiatry, 43*, 923-929.

CHAPTER 9

WHOLE-GENOME SCREENING USING THIRD-DEGREE RELATIVES: POWER TO DETECT EXCESS ALLELE-SHARING AT A PREDISPOSING LOCUS

Elena L. Grigorenko[1] *and Alexei A. Chikanian*[2]

[1]Department of Genetics and Psychology, Yale University and Psychological Institute, Russian Academy of Education

[2]Department of Physics, Yale University and Lebedev Physical Institute, Russian Academy of Science

WHOLE-GENOME SCREENING USING THIRD-DEGREE RELATIVES: POWER TO DETECT EXCESS ALLELE-SHARING AT A PREDISPOSING LOCUS

It has been noted by many researchers that today's human genetics is challenged by an inconvenient reality that has become apparent in the past decade: many traits of interest do not follow simple Mendelian monogenic inheritance (e.g., Ott, 1990; Lander & Schork, 1994; Risch, 1990a). Such traits are referred to as complex traits. Complex inheritance may result from (1) incomplete penetrance, (2) the presence of phenocopies, (3) genetic heterogeneity or (4) polygenic inheritance. Thus, it appears that there is not a perfect correspondence between genotype at a single locus and phenotype. That is, individuals carrying a susceptibility allele may have a higher relative risk of disease, but some gene carriers may be unaffected and some noncarriers may be affected. Examples of "complex" traits are susceptibilities to heart disease, hypertension, diabetes, cancer, and behavioral disorders.

The genetic dissection of complex traits (i.e., localizing a gene without knowing the biological function of its product) imposes challenging problems that attract many investigators. There are now high expectations for finding solutions to unsolved puzzles and a variety of budding experimental and analytical methods. As more closely spaced and highly informative markers have been cataloged in the human genome (Gyapay, Morissette, Vignal et al., 1994; Murray, Buetow, Weber et al., 1994; Nakamura et al., 1987; Weber & May, 1989), there has also been an increasingly strong demand to develop applied sampling and analytic strategies that will be suitable for dealing with highly polymorphic markers and will enable investigators to locate individual loci affecting complex traits. Lander and Schork (1994) classified modern techniques for the genetic dissection of complex traits into four categories: linkage analysis, allele-sharing methods, association studies in human populations, and genetic analysis of large crosses in model organisms such as the mouse and rat (targeted to mapping polygenic traits, including quantitative trait loci). These analytic strategies have been described in detail in a number of excellent reviews (e.g., Festing, 1979; Ott, 1991; Weir, 1990), which included some enlightening examples and discussions of remaining statistical/analytical needs.

One of these needs is related to the fact that, although the pace of molecular technology is resulting in ever more efficient way of typing individuals for a large numbers of markers, yet the possibility of efficient typing of many large extended families, or large unselected samples of individuals, or many pairs of relatives in a search for relatively-low-contributing-loci expressed in complex traits remains unlikely. Questions have also been raised as to the desirability of these strategies from a statistical perspective. Thus, there is a need to develop new strategies of nonrandom sampling that could enhance the power of detecting linkage or association between marker loci and a trait locus. A number of authors (Carey & Wiliamson, 1991; Eaves & Meyer, 1994; Risch & Zhang, 1995) have suggested different sampling strategies based on characteristics of the phenotype of interest. Another related issue is the need to develop tactics that allow one to utilize non-randomly selected family material for performing an efficient, inexpensive, and reliable whole-genome screening. These strategies would not necessarily provide precise estimation of recombination fractions, but rather, result in the identification of regions of the genome in which there is a significant genetic component important in the etiology of the disease.

Among the most promising analytical strategies that can be implemented in the framework of genome-wide screening, is the methodology related to analyzing the patterns of allele-sharing. Allele-sharing methods are based on the notion that a set of affected relatives will tend to have inherited the same allele at a predisposing locus more often than expected under random Mendelian segregation. These nonparametric analytic strategies have the advantage of not requiring a correctly specified model of inheritance. In parametric genetic analyses misspecification of the genetic model may lead, not only to loss of power, but also to false exclusion (Martinez & Goldin, 1991; Clerget-Darpoux & Bonaiti-Pellie, 1992). There are a number of additional advantages of these methods. As compared to parametric approaches, allele-sharing strategies are (1) less bound to structures of pedigrees; (2) less laborious in terms of marker-typing; and (3) less computationally demanding. There are different subtypes of allele-sharing strategies, some of them consider only affected relatives of certain degree of relatedness (e.g., the affected sib-pair method), whereas others incorporate into the analyses all types of affected relatives in pedigrees (e.g., the affected-pedigree-member method). Thus, due to the advantages described above, allele-sharing methods seem to offer a particularly robust and cost-efficient approach for the genetic dissection of complex traits (Kruglyak & Lander, 1995).

Recently there have been several attempts to investigate applicability of allele-sharing methods to whole-genome searches (e.g., Brown, Gorin, & Weeks, 1994). The affected-pedigree-member methods was also applied to whole-genome search in a recent study on manic depression (Berrettini, Ferraro, Goldin, et al., 1994). The affected sib-pair method was used as a screening tool in the search for the potential involvement of single-loci and multiple-loci in hypertension (Wilson, Elston, Tran, & Siervogel, 1991) and diabetes (Davies et al., 1994). Nikali et al (1995) assumed a founder mutation in a Finnish population, and conducted a primary screening of the genome using samples from just four affected individuals in two consanguineous pedigrees. The identification of a shared chromosomal region in these four patients provided the first evidence that infantile-onset spinocerebellar ataxia gene locus is on the long arm (q) of the chromosome 10. This finding was confirmed by conventional linkage analysis in the complete family material.

Weber (1994) has started a genome-wide screening using an affected-distant-relative method for genetic studies of Tourette Syndrome. The affected-distant-relative paradigm requires less intense clinical effort in evaluating phenotypes and is more cost-efficient in genotyping. DNA from the affected individuals is analyzed using dense sets of polymorphic markers distributed evenly throughout all the chromosomes. The longest chromosomal segments (or patches) shared among the affected individuals are likely to encompass the disease genes.

In this paper we explore applicability of the method of allele-sharing by distant affected relatives to whole-genome screening (Weber, 1994). This report summarizes simulation studies which explore the informativeness and power of the distant-affected-relatives method as a prelude to extensive genotyping for marker and candidate loci.

MODEL

The principal concern of the simulations was the evaluation of the power of the allele-sharing method in a small nonrandomly selected set of distant relatives for the whole-genome search. Prior to providing a description of the results of the simulation study, the underlying assumptions of the allele-sharing methods are summarized briefly.

UNDERLYING ASSUMPTIONS

General Characteristics of the Allele-Sharing Methods. The task, that is implicitly built into allele-sharing methods, is to show that the inheritance pattern of a chromosomal region is not consistent with random Mendelian segregation, that is, that affected relatives inherit identical copies of the chromosome more often than expected by chance (Lander & Schork, 1994). Allele-sharing methods assume that robust phenotype definitions are implemented in such analyses. If the phenotypes are not well defined, allele-sharing methods could prompt a high percentage of false positives. In this case, more general and resource intensive family studies are recommended.

Since allele-sharing methods assume no model for the inheritance of the trait, researchers view them as more robust than linkage analysis. The expectation is that affected relatives will show excess allele-sharing even in the presence of such complicating factors as incomplete penetrance, phenocopies, genetic heterogeneity, and high-frequency disease alleles. However, allele-sharing methods have been shown to be less powerful than a correctly specified linkage model (Lander & Schork, 1994).

The underlying purpose of allele-sharing methods is to evaluate how often a particular copy of a chromosomal region is shared identical-by-descent (IBD) among affected relatives. This identity implies that the region is inherited from a common ancestor within the pedigree. The observed frequency of IBD sharing at a locus is compared with random expectation under the assumption of no linkage between the trait and the locus. Lander and Schork (1994) define an identity-by-descent affected-pedigree-member (IBD-APM) statistic as follows:

$$T(s) = \sum_{ij} x_{ij}(s)$$

where $x_{ij}(s)$ is the number of copies shared IBD at position s along a chromosome, and where the sum is taken over all distinct pairs (i,j) of affected relatives in a pedigree. The results can be summed over multiple families and a weighted statistic, $T(s)$, could be obtained. Under the assumption of random segregation, $T(s)$ tends to be normally distributed with a mean μ and variance σ calculated based on the kinship coefficients of the affected relatives. Deviation from random segregation is detected when the statistic $(T - \mu)/\sigma$ exceeds a critical threshold, that is specific for different pairs of relatives (Lander & Schork, 1994).

Sib Pairs. The earliest and the simplest form of the allele-sharing methods is affected sib pair analysis. The idea of using data on siblings to study the inheritance of a trait originated with Penrose (Penrose, 1935). Formal theory of the affected sibling method was developed in the 1970s by Haseman and Elston (1972), Day and Simons (1976), Thomson and Bodmer (1977), Green and Woodrow (1977), and Suarez, Rice and Reich (1978). The sib-pair approach utilizes the IBD sharing pattern. This pattern measures how many alleles individuals share as copies of a single gene from common ancestors. For example, sibling pairs can share two, one, or zero alleles at a given locus identically by descent from their parents. If a disease susceptibility gene is linked to a marker gene, affected siblings will share marker alleles IBD more often then would be expected in the absence of a linked disease susceptibility gene. Excess allele-sharing can be measured with a simple χ^2 test (Holmans, 1993; Knapp, Seuchter, & Baur, 1994a; Owerbach & Gabbay, 1994; Suarez, Rice, & Reich, 1978; Weitkamp, 1981). In addition, several parametric and likelihood ratio test statistics have been proposed (cf Tierney, 1994; Shan & Green, 1994). The sib-pair method has been applied to a number of diseases that have been found to be associated with the highly polymorphic human leukocyte antigen (HLA) complex, including insulin dependent diabetes mellitus, multiple sclerosis and schizophrenia (Bishop, Falk & MacCluer, 1986; Seboun, Robinson, Doolittle et al., 1989; Tiwari & Terasaki, 1985), as well as in studies of homorosexuality (Hamer, Hu, Magnuson, et a., 1993).

Allele-sharing methods can also be utilized in the analyses of quantitative traits. Initially suggested by Penrose (1983), this approach was developed by Hill (1975) and Haseman and Elston (1972). The underlying logic of the method is that phenotype similarity between two relatives should be correlated with the number of alleles shared at a predisposing locus. This approach was generalized to other relatives (Amos & Elston, 1989; Olson & Wijsman, 1993), pedigrees (George & Elston, 1987), and multivariate phenotypes (Eaves & Meyer, 1994; Plomin, McClearn, Smith et al., 1994). Recently, there has been a resurgence of interest in the theoretical and analytical aspects of studying allele-sharing in its application to quantitative traits (e.g., Amos, 1994; Eaves, 1994; Fulker & Cardon, 1994; Fulker, Cherny & Cardon, 1995; Goldgar, 1990; Goldgar & Oniki, 1992; Guo, 1994; Vogler, this volume).

Identical-by-Descent Versus Identical-by-State. It is often not known if two relatives inherited a chromosomal region IBD, that is, if one chromosomal region is a copy of the other or if they are both copies of the same ancestral gene. Rather, it is observed that two relatives have the same alleles at genetic markers in the region, but the ancestral origin of these regions is not known without typing additional, and perhaps many, other relatives. Shared alleles of unknown ancestry are called identical-by-state (IBS). It has been shown for loci that are not highly polymorphic, that methods using the IBS relation have unacceptably low power (Risch, 1990c). However, it is usually safe to infer IBD status from IBS observations when a dense collection of highly polymorphic markers has been examined. A number of approaches have been developed to cope with the important practical difficulty of deriving IBD from IBS. The first was described by Lander and Schork (1994), who inferred IBD sharing on the basis of the marker data either via the expected value of the IBD statistic or via the probability distribution over the IBD statistic (expected IBD-APM methods) (Amos, Dawson, Elston, 1990; Kruglyak & Lander, 1995). The second method uses a statistic based explicitly on IBS sharing (IBS-APM method) (Bishop & Williamson, 1990; Lange, 1986; Weeks & Lange, 1988; 1992). Additionally, Risch (1990c) suggested a likelihood-ratio method. The main difference between this likelihood-ratio method and the IBS method of Bishop and Williamson (1990) is that it takes into account not only how many markers an affected pair share but also which ones they are (e.g., rare or common). The IBS-APM is the least computationally intensive and the easiest to apply. This method has been successfully used for complex diseases such as Alzheimer's, breast cancer, and melanoma (Cannon-Albright et al., 1992; Haile et al., 1990; Hall et al.; Pericak-Vance, et al., 1991; St George-Hyslop et al., 1990).

Methodological Complications of the Allele-Sharing Methods. It has been shown that the allele-sharing methods are not robust to a misspecification of the marker allele frequencies (Babron, et al., 1993; Martinez et al., 1992). This methodological difficulty is often referred to as a major drawback due to the difficulty of accurately detrmining allele frequencies for highly polymorphic markers. Underestimating or misspecifying the frequency of a marker allele may lead to the occurrence of false positive results, since counts of co-occurrence of frequent alleles, if not controlled for population frequency, might be higher than amount of matches between rare alleles. It is necessary to obtain valid estimates of the marker allele frequencies appropriate for the sample under study. The preferred strategy is to utilize a population estimate of allele frequencies, obtained for the population from which families enrolled in the study derive. Another approach proposed by Boehnke (1991) allows the estimation of these frequencies from the data themselves using all the family information. However, the latter estimates may be biased, in particular if affected individuals are preferentially typed.

Allele-sharing methods contrast the null hypothesis of no association between marker variation and disease against a very general alternative hypothesis. The price for lack of specificity in the alternative hypothesis and the genetic model is loss of statistical power, and the onus is on devising ways to increase this power (Ward, 1993).

It has been suggested that the inclusion in the allele-sharing analyses of specific types of close relatives (e.g., sib pairs) may lead to false positive evidence for linkage or can markedly reduce the power of allele-sharing methods to detect linkage (Goldin & Weeks, 1993). However, the use of more distant relatives decreases the chance of a false positive result, as the sharing of alleles for these relatives will happen less often by chance (Bishop & Williamson, 1990; Cantor & Rotter, 1987; Coon, in press; Risch, 1990c). In addition, looking at large clusters of distantly-related individuals (triads, quadruples, etc.) should also minimize the probability of the Type I error.

Affected-Distant-Relatives Method. One attempt to overcome loss of statistical power while applying allele-sharing methods was implemented in the development of the affected-distant-relatives method (Weber, 1994). This method is based on the assumption that it is more striking for affected distant relatives to share a rare marker allele than a common marker allele. This paper deals with a subtype of this method, a technique, that involves third-degree relatives ("cousins method" - CM). The CM is a nonparametric approach that shares key features with the affected sibling method, but requires fewer individuals. The aim of this method is to provide an efficient initial screening rather than to definitely establish linkage. The CM capitalizes on the vast number of highly informative markers identified in recent years and is based on the expected pattern of marker alleles IBD closely linked to a disease-related locus. The CM uses third-degree relatives definitely affected with the same disorder. For non-inbred cousins, only one allele can be shared IBD. If a marker is closely linked to a disease gene, third-degree relatives will share marker alleles more often than if the marker were unlinked. As in any nonparametric method, the CM statistic measures increased marker similarity, and thus makes no assumptions concerning how the disease is inherited.

It has been suggested that larger sets of affected relatives (e.g., triads), rather than pairs, are more powerful for studying complex disease whenever heterogeneity, sporadic cases, and incomplete penetrance are suspected (Sribney & Swift, 1992; Whittemore & Halpern, 1994). In addition, it is computationally simpler to deal with sets of relatives of the same degree (e.g., second- or third-degree relatives), so all sets in the analysis are assumed to have, on average, a certain amount of shared genetic material. Practical considerations suggest that the search for triples of affected third-degree relatives should not be limited to first cousins, but rather include mixed relative types such as two first cousins and a great-grant parent (Grigorenko, Chikanian, & Pauls, 1995). Then, for each triplet, one can construct the test statistic that is based on the joint IBD distribution produced by combinations of marker alleles IBD in three-way comparisons among the three cousins.

The CM retains the substitution of IBS relations, suggested by Weeks and Lange (1988). This assumption imposes a serious limitation on the CM. Bishop and Williamson (1990) showed that the distortion in the IBD distribution, inferred on the basis of the IBS distribution, is the greatest for the more distant relationships. However, this distortion depends on the allele frequency of the trait; for example, rare trait studies may demonstrate that distant relationships are sufficiently powerful even though many of those sets of relatives are not IBD at the trait locus.

Below, we will consider a collection of affected relative triplets of two fixed types (R) - three first cousins or two first cousins and a great-grandparent. Each triple shares either 0 or 1 allele identical by descent (IBD) at any locus s, and the allele-sharing proportion $x(s)$ is defined to be the proportion of affected relative triplets that share an allele IBD at s. Letting $\mu(s)$ denote the expected value of the allele-sharing statistic under random Mendelian segregation, and substituting IBD by IBS, the condition $x(s) > \mu(s)$ suggests that there is a predisposition locus at or near s.

In this study, the effect of a variety of factors on the power to detect disease-related allele-sharing by studying triplets of third-degree relatives was estimated by means of computer simulation. These factors are (1) mode of inheritance; (2) marker polymorphism; (3) the distance between marker and disease; (4) penetrance and phenocopy rate; (6) heterogeneity, and (6) misspecified marker allele frequencies.

MODEL SPECIFICATION

Monte Carlo simulations were conducted to explore the feasibility of using affected third-degree relatives to identify the most likely regions for the location of susceptibility genes. Twenty-four sets of parameter values (Table 1) were used to simulate data sets for further analyses. Models 1-12 were simulated twice for two different linked markers (a 13-allele marker and a 2-allele marker). For each of the 36 data sets, the starting population was approximately 24,000 individuals. The simulations were carried out under the simplifying assumption of no genetic interference, i.e., that crossovers occur independently of one another. The Haldane mapping function relating the recombination fraction to segment length was used. Simulations were performed assuming (1) random mating, (2) population expansion, determined by the Poisson branching process, (3) linkage equilibrium at the population level, (4) gender-dependent recombination values, (5) variable penetrance, (6) sporadic cases. For each combination of parameters, 100 replicates were simulated and analyzed. The hypothetical chromosome region, modeling the current state of knowledge on chromosome 15, is summarized in Figure 1a-b.

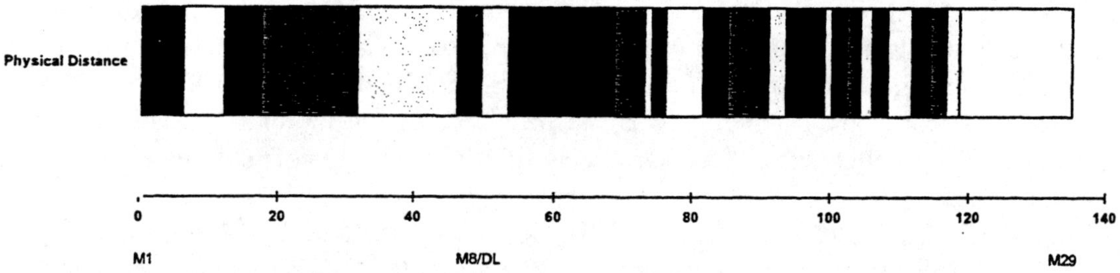

Figure 1a. Diagram of the region of the genome simulated for third-degree affected relatives studies.

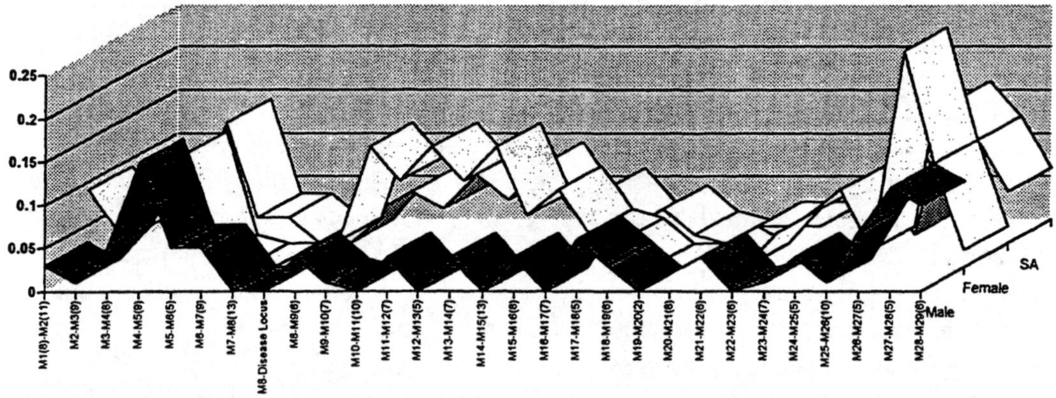

Figure 1b. A schematic representation of the gender-specific (Male and Female) and sex-averaged (SA) recombination fractions at the simulated region.

A disease marker was simulated within a region of the genome marked by the 29 fully informative markers M1-M29. The simulations were based on theoretical estimates of optimal marker spacing and information that have been derived previously (Goldin & Geshon, 1988; Bishop & Williamson, 1990; Risch, 1990b, 1990c). When studying genetic heterogeneity, two markers were separated by a total of 100 CM and were assumed to be independent.

Disease Models. For the purposes of the simulation we designed relatively simple models. These models, however, may provide insight into more complex models. We assumed that there are two alleles at the trait locus, one of which (D) is a disease-predisposing allele. The genetic model for the trait was defined by the allele frequency, p_D, and the matrix of penetrances, f_{DD}, f_{Dd}, and f_{dd}, one for each genotype at the trait locus. The genetic models are shown in Table 1.

Table 1. Genetic models used in the simulations

Model	Allele frequencies			Penetrances (%)						Phenocopies (%)		Data set	
	OLM	TLM		f_{DD}		f_{Dd}		f_{dd}					
	p_D	p_{D1}	p_{D2}	M	F	M	F	M	F	M	F	OLM[1]	TLM
(1) Dominant	.01	.01	.01	90	80	90	80	0	0	2	1	1	13
				90	80	90	80	0	0	20	10	2	14
				60	40	60	40	0	0	2	1	3	15
				60	40	60	40	0	0	20	10	4	16
				30	20	30	20	0	0	2	1	5	17
				30	20	30	20	0	0	20	10	6	18
(2) Recessive	.20	.20	.20	90	80	0	0	0	0	2	1	7	19
				90	80	0	0	0	0	20	10	8	20
				60	40	0	0	0	0	2	1	9	21
				60	40	0	0	0	0	20	10	10	22
				30	20	0	0	0	0	2	1	11	23
				30	20	0	0	0	0	20	10	12	24

D = Disease-predisposing allele; F = Female, M = Male; OLM = One-locus model; TLM = Two-locus model

[1] Models 1-12 were simulated twice for two different linked markers: - a 2-allele marker and a 13-allele marker.

Ascertainment Rules. Two types of affected third-degree relatives were ascertained - relatives in the same generation (cousins, intra-generation triplets) and relatives in two different generations (triplets including great-grandparents, inter-generation triplets). We decided to ascertain both type of triplets due to the consideration that triplets of affected third-degree relatives tend to be rather rare, so it is practically unlikely to ascertain large samples of such triplets. In addition, we have shown (Grigorenko, Chikanian, & Pauls, 1995), that inter-generation triplets can be found significantly more often than intra-generation triplets. Finally, it has been demonstrated that the inclusion of grandparents and great-grandparents in the analyses of affected relatives tends to increase the power of detecting linkage (Bishop & Williamson, 1990; Risch, 1990c).

All triples were independently ascertained. Pedigrees of triple sets were examined to ensure that, for at least three generations, the sets were unrelated.

In this study we used three different ascertainment schemes (A, B, C). Under the scheme A, the program was asked to select randomly 3 inter-generation and 4 intra-generation triplets. The scheme B randomly included 10 inter-generation and 20 intra-generation triplets. Finally, the ascertainment scheme C was a complete ascertainment scheme, according to which all independent triplets of affected third-degree relatives, detected in the population, were ascertained. The number of triplets ascertained under scheme C varied with the simulated genetic model (from a couple dozen in the frame of low-penetrant recessive disorder with a low rate of sporadic cases to a few hundred in fully penetrated conditions).

Heterogeneity. Data were simulated under one-locus and two-locus disease models. For the sake of simplicity, we used the same penetrance matrix and the same amount of sporadic cases for both models.

Marker Polymorphism. Ideally, allele-sharing approaches are best applied to typing information obtained on polymorphic markers that have equally frequent alleles. However, recent studies show that virtually none of highly polymorphic markers tend to have alleles at equal frequencies. On the contrary, alleles of highly polymorphic markers demonstrate an unequal frequency distribution with 1-3 alleles of high frequency and the remaining alleles being relatively rare. These observations were implemented in the simulation. Some of the markers had equal, and some--variable allele frequencies among alleles at every site. The hypothetical chromosome region was designed by combining the information on chromosome 15 from the CHLC map (Murray, Buetow, Weber et al., 1994), NIH/CEPH linkage map (NIH/CEHP Collaborative Mapping Group, 1992), and Genothon (Gyapay, Morissette, Vignal et al., 1994). Twenty eight highly polymorphic markers and one RFLP were placed on a simulated chromosome accordingly to the combined information from these sources (see Figure 1 a-b). Due to observed variation of allele frequencies at simulated markers, the weighting function was introduce to correct for unequal frequencies. The weighting function $f(p) = 1/p$, where p is an estimated allele frequency, was implemented in all analyses. In addition, the effects of three different weighting functions ($f(p) = 1$, $f(p) = 1/sqrt(p)$, and $f(p) = 1/p$) on false-positive determination of allele-sharing were studied (Weeks & Harby, 1995).

Under a one-locus model, we simulated a 13-allele marker linked at $\theta_M = 0.00$, $\theta_F = 0.00$ proximal to the disease locus, and, a 2-allele marker, also linked at $\theta = 0.00$ (equal for males and females).

Under a two-locus model, in addition to the previously described disease locus, another disease locus was simulated 100 CM distant. This locus was located on a different arm of the simulated chromosome. A 2-allele proximal adjacent locus was linked to the disease locus at $\theta = 0.00$ (equal for males and females).

Distance Between Marker and Disease. We evaluated the power to detect excess allele-sharing as a function of the distance (recombination fraction) between the marker and the trait locus. These analyses were carried out within the context of dominant Model 1 (see Table 1) and a 13-allele linked marker. The recombination fraction was varied between 0.00 and 0.30.

Misspecification of Marker Allele Frequencies. It has been shown that allele-sharing methods are sensitive to misspecification of marker allele frequencies (Babron, et al., 1993; Martinez et al., 1992; Van Eerdewegh, et al., 1993). Incorrect specification of the frequency of a marker allele may result in false-positive evidence for linkage, since co-occurences of rare marker alleles may provide a more significant contribution to the summary statistic than co-occurences of frequent alleles (Weeks & Harby, 1995).

In this study we investigated the effects of misspecifying allele frequencies in two ways. First, implementing the weighting factor $f(p) = 1/p$, we used the allele frequency estimates that were obtained via sampling at random 50 unrelated individuals from the simulated populations. This procedure allowed us to examine sampling error that is present in any allele-frequency determining procedure. Further, we explored the effect of misspecifying a frequency of a randomly selected allele at a marker unlinked to the trait marker within the context of Models 1-6 (Table 1). This allows us to determine how often a false-positive signal is registered (e.g., linkage is falsely concluded) if the allele frequencies are specified incorrectly.

THE SIMULATIONS

The implemented model was of a multi-locus multi-allele system with random union of gametes in a randomly mating population with discrete generations. Population growth was determined by Poisson branching process. In every family sibship size was a random number, distributed by Poisson {Z} with mean Z. In the simulations presented below, the mean sibship size in populations was set as Z=2.5. In order to create enough relative connections, the simulation was carried for five consecutive generations. Initial population size was set at 24,500 individuals as generation 1 was created, which resulted in the sample size of ~38,000 individuals at the fifth generation.

From each parent, a chromosome, forming a gamete, was chosen at random. Cross-overs to the complementary strand between each pair of loci were simulated as a function of the recombination fraction between loci. The generated recombination fractions were sex-specific and corresponded to CHLC map of chromosome 15 (Murray, Buetow, Weber, et al., 1994). For each simulated offspring, phenotypic status was generated by sampling at random in correspondence to the parameters in Table 1.

The simulation program was written in FORTRAN and used the random number generators from the CERN library (CERNLIB, 1995).

DATA ANALYSIS

Utilizing the previously developed methods of estimating of excess of allele-sharing in affected relatives, we used the following statistic for the analysis of the simulated data sets

$$T(s) = \sum_{ijk} x_{ijk}(s)$$

where $x_{ikj}(s)$ is the number of copies shared IBS by triplet members at position s along a chromosome, and where the sum is taken over all independent triplets (i,j,k) of affected relatives, ascertained in the sample. The outcome of the application of the CM is a matrix m x a (where m is a number of markers and a is a number of alleles at every studied marker), containing the counts of a given allele (a) shared IBS at a given marker (m). Instead of correcting for allele-frequencies in the analytic formula, we used the weighting function $f(p) = 1/p$, where p represents allele frequencies, estimated on the basis of the simulated population, prior to obtaining a summary statistic $t(s)$. The summary statistic $t(s)$ denotes the amount of allele-sharing at position s along a simulated chromosome.

Having corrected for allele frequencies in the population, we constructed a one-way table consisting of the distribution of occurrences of allele-sharing among all members of each independent triple. The probability of occurrence of allele-sharing in the site s that is linked to the disease locus is assumed to be β times as large as the probability of occurrence of allele-sharing in the locus s, unlinked to the disorder, where β is independent of linkage to the disease ($\theta = 0.5$). Thus the expected number $\theta(s)$ of allelic co-occurrence among third-degree relatives satisfies a model of the form

$$log\, \theta\, (s) = \alpha + \beta(s).$$

Traditionally, researchers perform a set of comparisons for every locus contrasting the observed number of allelic co-occurrences among affected relatives to the expected proportions of allele-sharing proportions, that are specific for various types of relatives (e.g., Tierney, 1994). In this paper we apply a different method, utilizing the techniques for contingency tables analyses (Argesti, 1990; Wickens, 1989). Instead of comparing observed to expected allelic co-occurrences, we fit a set of models, screening the studied genome interval for the maximum value of $\beta(s)$. The assumption, underlying this method is that, when a multilocus system corrected for allele frequencies is subjected to model fitting, the majority of loci should be unlinked to the disorder. That is, these loci provide expectations under the null hypothesis. Thus, comparing to theoretically expected values of IBS at every given locus is not necessary and can be substituted with a series of model fitting analyses, the goal of which is to determine the maximum values of $\beta(s)$. In addition, the suggested technique allows fitting multilocus models. As has been shown, (Knapp, Seuchter, Baur, 1994b; Schork, Boehnke, Terwilliger, Ott, 1993; Weeks & Lange, 1992), the multilocus allele-sharing statistic is more significant than the single-locus statistic. For highly polymorphic markers, most of the information for linkage comes from the two flanking markers closest to the disease-predisposing locus. Thus, in most practical genetic analyses the two-locus version of allele-sharing statistics is advocated. Knapp et al (1994b) showed that, even though the probability of Type I error is still high, a straightforward extension of the well-known affected-sib-pair mean test for two marker loci can be much more powerful than single-marker-locus analysis. This motivated us to perform an analogous comparison with the CM.

The tests based on multiple affected relatives in pedigrees (e.g., Weeks & Lange, 1988; Knapp et al., 1994a; Bishop & Williamson, 1990), though more widely applicable than the technique we introduce, requires special software. In addition, a special modifications of pairwise comparisons between affected pedigree members would need to be made in order to compare larger sets of affected relatives. Our simpler technique can be implemented in any powerful statistical software (BMDP, SAS, SPSS). The method allows the identification of the locus at which excess allele-sharing between affected third-degree relatives is observed and provide odds ratios of this locus being the putative candidate for linkage over other screened loci.

RESULTS AND DISCUSSION

Allele-sharing methods are an appealing design for linkage analysis for a trait that is familial and perhaps even genetic, but no mode of inheritance is clearly apparent when pedigrees are examined. It has been convincingly shown that the power to detect linkage using IBS allele-sharing methods depends on the relationship of the affected pedigree members, the polymorphism of the marker, the recombination distance between a trait locus and the closest marker, and the mode of inheritance of the trait (Bishop & Williamson, 1990). Our simulation study was designed (1) to systematically appraise the magnitude of these effects in the context of the CM and (2) to study the power of the CM in the presence of such complex phenomena as incomplete penetrance and sporadic cases.

Due to CPU time constraints imposed by the extensive computation required for each simulation we carried out only 100 replicates for each simulated data set. Power results are presented in a number of different forms. First, we report proportions of true and false positives for the data, simulated under single-disease-locus models after the indices of allele-sharing for each triple were corrected for population allele frequencies and the data were subjected to model fitting. Second, we display the results of model-fitting to the corrected for allele-frequencies data. Since the number of replicates is small, all power results are reported relative to a theoretical false-positive rate of 0.01. The format of presentation matches the one suggested by Weeks and Harby (1995).

EFFECTS OF MODEL OF INHERITANCE, SAMPLE SIZE, AND MARKER POLYMORPHISM

Implementing the Monte-Carlo approach, we have evaluated the power of the CM to detect excessive allele-sharing among relatives, which can be interpreted as a first indication of linkage to a single-locus disease. At the first stage of analysis we calculated the proportions of true and false positive allele-sharing for 12 different single-disease-locus models by dividing the observed allele-sharing by estimated allele frequency. Figure 2 displays our results under the 6 dominant models (data sets 1-6, Table 1) as a function of sample size and number of ascertained triplets. The first and most obvious observation is that a high rate of sporadic cases is more detrimental for the CM power to detect true allele-sharing than reduced penetrance. It is also evident that power increases as sample and marker informativeness increases. We simulated the disease allele in the dominant model at a rare frequency ($p = 0.01$), therefore, it is unlikely that it would appear in a given family more than once. Thus, the high phenocopy rates "dilute" the true excess allele-sharing and decrease the power to detect the putative locus.

Figure 3 presents the results under 6 recessive models (data sets 7-12, Table 1). Again, the results are shown as a function of sample size, and marker polymorphism. As is apparent, 13-allele markers are more powerful than 2-allele markers. For recessive models power appears to be higher in cases with a high amount of sporadic cases (models 8, 10, & 12) when compared to corresponding dominant models.

It has been shown that power of various allele-sharing tests depends on the mode of inheritance (Knapp et al., 1994a). In our study, the absolute power to detect disease-related excess allele-sharing seems to be higher for recessive models, while the magnitude of obtained effects (e.g., parameter estimates, determining the inclusion of exclusion of the studied area) appears to be higher for dominant models. In other words, the values of averaged *β(s)* are higher for dominant models. When detecting excess allele-sharing at loci linked to the disease loci, that results in higher odds ratios. At the loci unlinked to the disease, that results in lower odds ratios, allowing the establishment of a probability of exclusion of linkage to a particular chromosomal region. Thus, dominant models provide higher confidence intervals for the detected effects (Figures 4 and 5). Even so, the power to detect true allele-sharing appears to be higher for recessive models in general, and especially for situations with higher phenocopy rates.

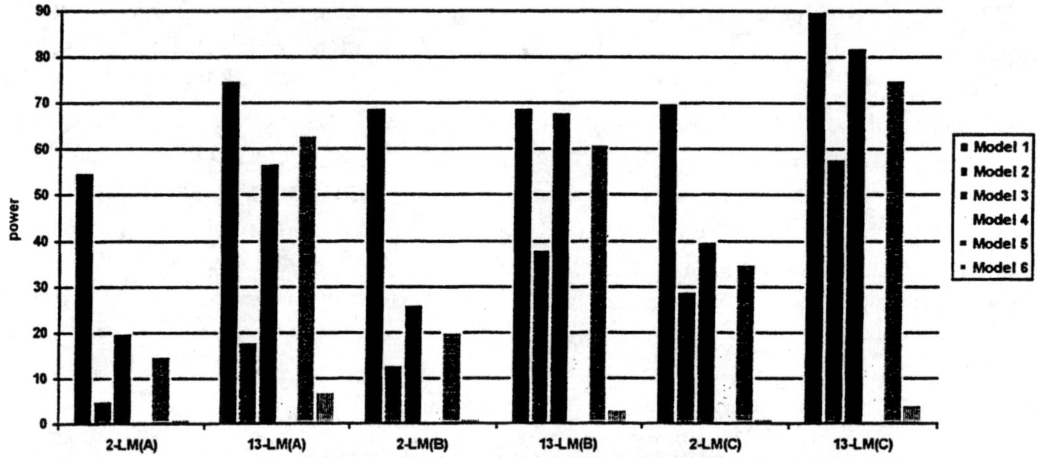

Figure 2. Power (%) under the sampling schemes A (7 triplets), B (30 triplets), and C (all independent triplets): dominant models 1-6 from Table 1 with a 2- and 13-allele marker, linked at θ= 0.0.

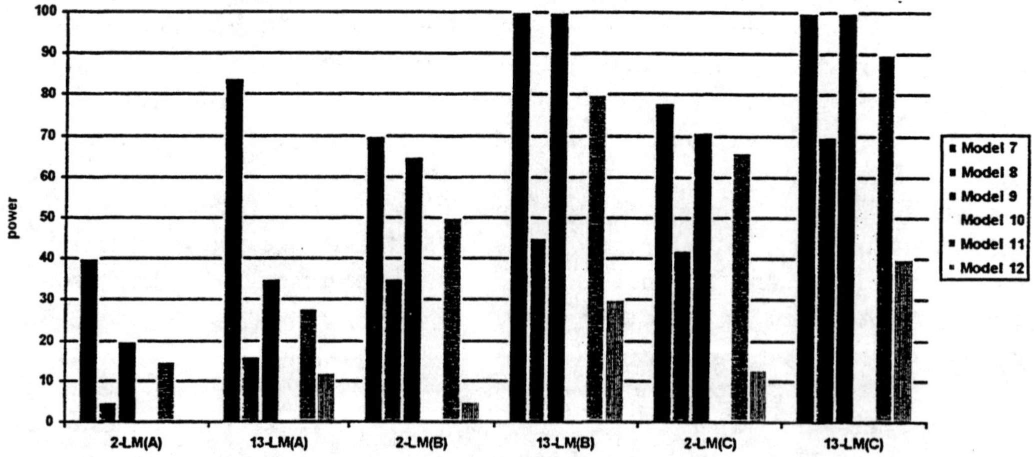

Figure 3. Power (%) under the sampling schemes A (7 triplets), B (30 triplets), and C (all independent triplets): recessive models 7-12 from Table 1 with a 2- and 13-allele marker, linked at θ= 0.0.

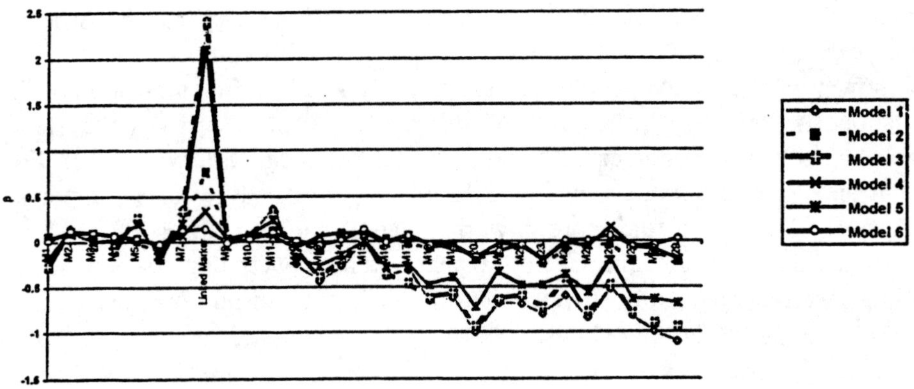

Figure 4. Plotted averaged β-s for dominant models 1-6 with a 13-allele marker, linked at θ= 0.0 (sampling scheme B, 30 triplets).

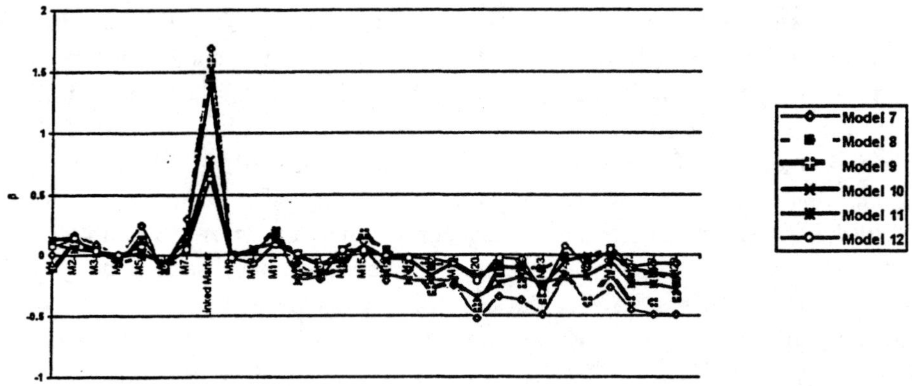

Figure 5. Plotted averaged β-s for recessive models 7-12 with a 13-allele marker, linked at θ= 0.0 (sampling scheme B, 30 triplets).

This may be because of the higher frequency of recessive disease alleles (p = 0.20), which could have entered some triplets independently. Thus, some of the sporadic cases may have the disease allele simply due to its high population frequency. In contrast, a dominant disease will segregate through generations at low frequency. That is why it is important to include intra-generation triplets in the analyses, so they will bring more power to detect excess allele-sharing. These findings are consistent with the conclusions of Weeks and Hardy (1995), who suggest that large pedigree structures are more advantageous for studying dominant disease because they contain more affecteds, and therefore, more power to detect linkage. A similar observation was made by Bishop and Williamson (1990), who state that "for a partially penetrant recessive trait, affected sib pairs are the most informative, while for a trait with dominance, other relationships may be considerably more informative" (p. 262). Therefore, the examination of segregation patterns in pedigrees to indicate whether the trait has evidence of dominant or recessive inheritance is necessary prior to conducting whole-genome screening. The choice of different types of relatives for analysis of excess allele-sharing may be differentially suitable given different modes of inheritance (Bishop & Williamson, 1990).

The power to detect excess allele-sharing, which could be a first indication of linkage, depends on the polymorphism level of the marker, or PIC value (PIC stands for polymorphic information content, Botstein, While, Skolnick, & Davis, 1980). For both dominant and recessive models, use of a 13-allele marker (PIC = 0.89), appears to be approximately 1.5- to 2.5-fold more powerful than a 2-allele marker (PIC = 0.41), see Figures 2 and 3. It is interesting to note that for some models (e.g., dominant model 1 with high penetrance and low phenocopies) the difference in power between a 2-allele and a 13-allele marker is only about 15%, while for some other models (e.g., recessive model 9) the difference is almost 3 times as great. The increase in power is roughly proportional to the PIC ratio (0.89/0.41=2.2).

The effect of PIC on the power to detect linkage has been thoroughly examined (cf. Weeks & Harby, 1995). Boehnke (1990), conducting a simulation study of traditional linkage analysis of a quantitative trait, found that the PIC predicts the effect of marker information on the required sample size. Risch (1990c), investigating the effects of marker polymorphism within the context of affected-relative-pair IBD-based statistics, showed a strong link between the PIC value and power, when marker information was collected only on pairs of affected relatives with missing information on the intervening relatives. Thus, in order to obtain lod scores whose magnitude is at least half of the possible expected, the PIC must be greater than 0.80 for sib pairs and greater than 0.90 for second- and third-degree relative pairs (Risch, 1990c). Bishop and Williamson (1990), studying properties of IBS statistics, compared that maximum possible power for a grandparent-grandchild pair when using an 8-allele marker with equally frequent alleles (PIC = 0.861) and a 4-allele marker with equally frequent alleles (PIC = 0.703), and showed that the power is twice as high (50% versus 25%) for the 8-allele marker. Similarly, Weeks and Harby (1995) showed that the power to detect linkage using the APM method depends on the polymorphism level of the marker. As with all linkage studies, the polymorphism of the marker is important, with higher polymorphism implying greater information, and, respectively, greater power to detect linkage.

However, as has been observed, the effects of a low PIC can be attenuated by using a system of several closely linked markers in a multilocus analysis (Risch, 1990b; Weeks & Lange, 1992). In addition, multilocus analysis tends to increase the absolute power to detect disease-related excess allele-sharing.

In addition to fitting one-locus model, we analyzed the power of two-locus and three-locus model via applying them to the data set 2 (dominant model with medium level of penetrance and high level of phenocopies). That is, models were fitted that reflected excess allele-sharing at the linked marker as well as at one (two-locus model) and two (three-locus model) adjacent markers. The results of these analyses are presented in Figure 6. Our findings are similar to those of Knapp et al. (1994b), Holmans (1993), and Schork et al. (1993) in that there is a pronounced superiority of appropriate multi-marker tests with respect to power. The multi-locus analyses improve absolute power to detect excess disease-relevant allele-sharing and increase the magnitude of specific effects, improving the confidence interval for statistics of allele-sharing at the region of probable linkage.

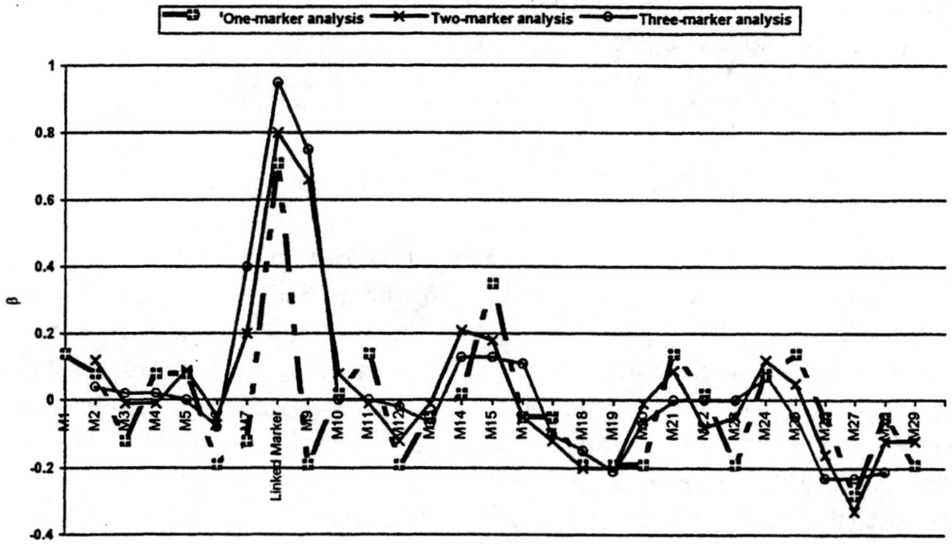

Figure 6. Plotted averaged β-s for the dominant models 2 with a 13-allele marker (scheme A, 7 triplets) using one-, two-, and three-maker analysis.

EFFECT OF PHENOCOPY RATE AND PENETRANCE

Another factor that has considerable impact on the power of CM is the frequency of phenocopies. The power to detect excess allele-sharing decreases as the phenocopy rate increases, as displayed in Figures 2-5, and even low-phenocopy rate can have disproportionately large effects. The power with no phenocopies and high and medium penetrance rate (Models 1, 3 and 7, 9, Figures 2 and 3) is approximately the same for all three sampling schemes, when a 13-allele marker used. Power drops to 25% and below, when the rates of sporadic cases go up and the level of penetrance goes down. In general, higher phenocopy rate and lower penetrance is more detrimental to the power of detecting excess allele-sharing in dominant models. Weeks and Hardy (1995) demonstrated in their simulation study that power drops to 50% of its maximum at a phenocopy rate of approximately 0.07. Bishop and Williamson (1990), in the context of a model of a partially penetrant dominant trait, studied the descrease in informativeness of affected pairs of grandparents and grandchild as a function of the phenocopy rate. They demonstrated that the information content strongly dependes on the rate of sporadic cases: power drops by 61% when the phenocopy rate is 50% (disease allele frequency is equal 0.01) and by 56% when the phenocopy rate is 18% (disease allele frequency was equal 0.10). The study of Bishop and Williamson (1990) considered pairs of affecteds as the sampling unit; however, they suggested that the power of a larger set of related affected individuals should be less sensitive to the phenocopy rate than is the power of an affected pair. It appears that a larger set of related affected individuals is less likely to contain phenocopies and more likely to have a penetrant trait than a pair of affecteds (Weeks & Hardy, 1995). It is clear that triplets of affected relatives will still be impacted by phenocopies, though significantly less then pairs of relatives. Triplets ascertained from highly loaded families for the trait are more likely to carry the trait allele and hence be more informative for detecting disease-relevant excess allele-sharing.

EFFECT OF DISTANCE BETWEEN MAKER AND DISEASE

Figure 7 depicts the power loss when the recombination fraction between the marker and the disease increases for the dominant models 1-6 with a 13-allele marker (scheme B). Power declines as θ increases, with the power dropping to 50% of its maximum at a recombination fraction of about 0.15 for models 1, 3, and 5 (low phenocopies rate) and at θ = 0.07 for models 2, 4, and 6 (high sporadic cases). Weeks and Harby (1995) showed that the power of APM drops approximately 50% at a recombination fraction of about 0.12. Similar results have been obtained before, with about 50% of power loss when θ exceeds 10% (Bishop & Williamson, 1990; Ott, 1991; Risch, 1990b), suggesting that the rate of power drop for the CM is similar to that of other parametric and nonparametric techniques.

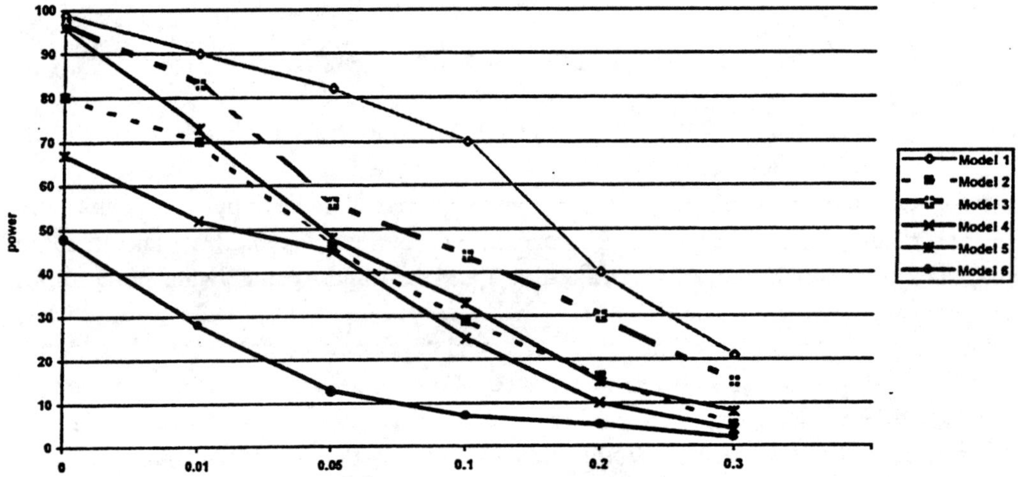

Figure 7. Effects on power of distance between the marker and the disease: dominant models 1-6 with a 13-allele marker (sampling scheme B, 30 triplets).

Figure 8 displays changes in averaged β*(s)*, shown for dominant model 1, when the linked marker is moved away from the disease locus. It is noticeable, that the increase in the recombination fraction between the marker and the disease not only influences the absolute power to detect the disease-related excess allele-sharing, but also the power to evaluate the probability of exclusion of linkage at the distant from the disease locus markers.

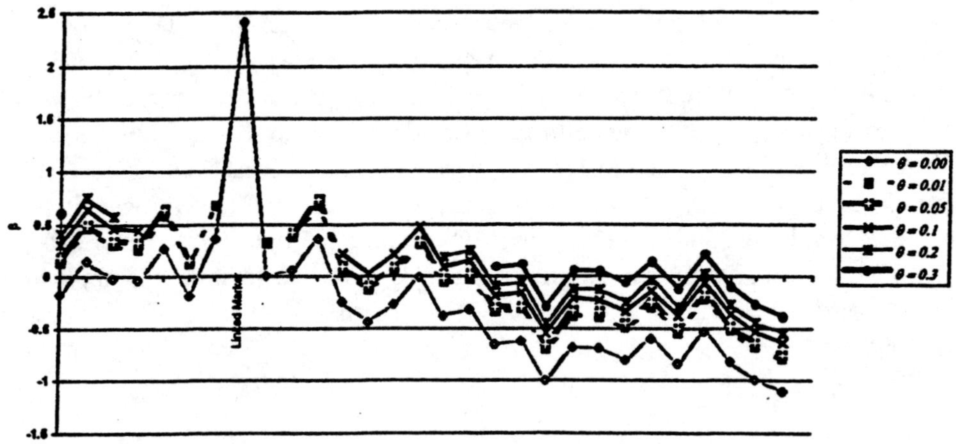

Figure 8. Plotted averaged β-s for dominant model 1 with a 13-allel marker, linked at θ = 0.00, 0.01, 0.05, 0.1, 0.2, & 0.3 (sampling scheme B, 30 triplets).

In general, the power of CM decreases rapidly with the recombination distance. For a genomic search, this implies that markers should be approximately a recombination distance of 5-10 CM, so that the trait is never greater than a 0.1--0.2 recombination distance from the closest marker.

EFFECT OF HETEROGENEITY

Studying the effect of heterogeneity, we assumed both disease loci to contribute equally to the prevalence of the trait. The effect of heterogeneity is shown in Figure 9. The comparisons are presented for Models 1-12 and 13-24 for ascertainment scheme B. Power drops to 30% of its maximum when about 50% of the triplets are unlinked. The power

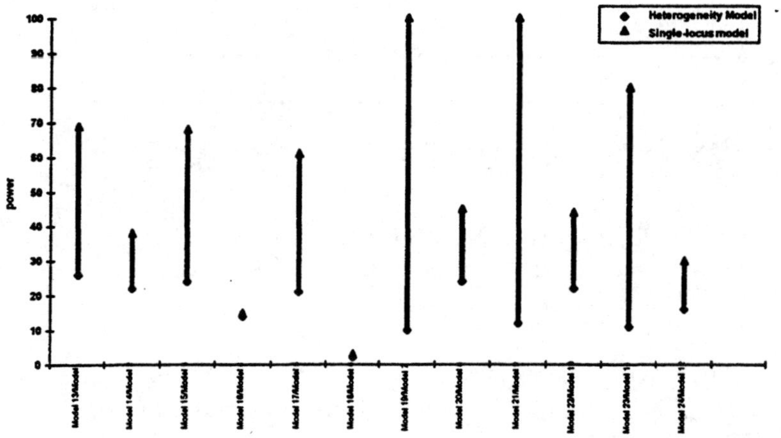

Figure 9. Effects of heterogeneity: power in models 1-24 with a 13-allele marker (sampling scheme B, 30 triplets).

decreases dramatically regardless of the other parameters of the models (e.g., phenocopies and penetrance). By comparison, in the admixture approach using lod scores, the number of families necessary to detect linkage at $q = 0.001$ increases twofold, when 20 to 30% of families are unlinked (Cavalli-Sforza & King, 1986, as cited in Weeks & Hardy, 1995). This suggests that the CM performs as well as other allele-sharing methods and may also be less sensitive to nonallelic heterogeneity when compared to traditional lod score approach (Weeks & Harby, 1995). However, more analytical work is needed in order to investigate this issue and to study the applicability of nonparametric methods to test heterogeneity (Matise & Weeks, 1993; Weeks & Harby, 1995).

EFFECT OF ALLELE FREQUENCY MISSPECIFICATION

As researchers have shown (Ott, 1992; Weeks & Harby, 1995), misspecification of allele frequency might be quite detrimental for both parametric and nonparametric approaches. However, while parametric methods are more sensitive to under-estimation, nonparametric methods have been shown to be influenced by both under- and over-estimation of allele frequency. This specificity of nonparametric methods led to the development of different weighting functions.

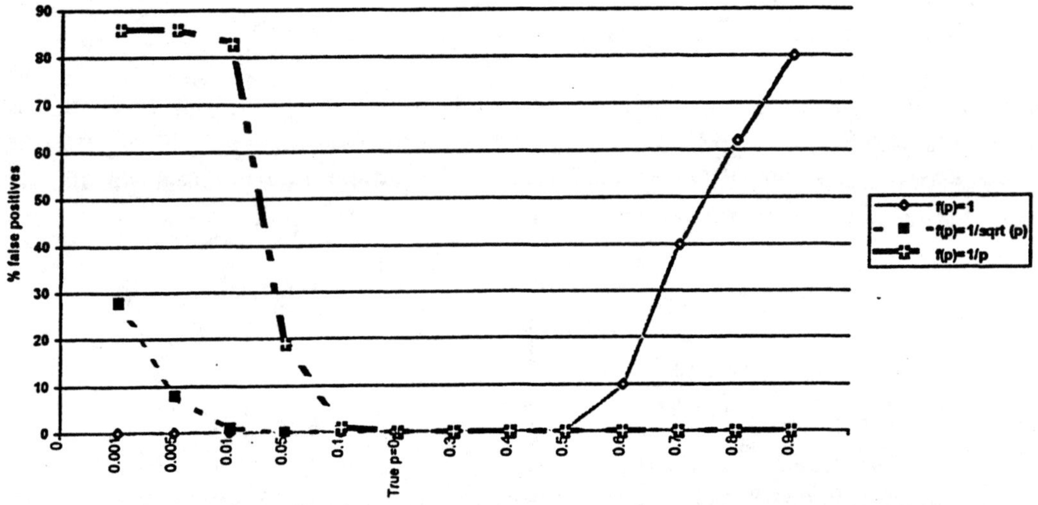

Figure 10. Effect of allele frequency misspecification on false-positive determination of excess allele-sharing: dominant model 4 (sampling scheme B, 30 triplets) with an unlinked 5-allele marker.

Figure 10 displays the effects of misspecifying the marker allele frequency under the dominant model 4 with an unlinked 5-allele marker. Because the marker is highly polymorphic, at the true allele frequency, no false-positives are obtained. By contrast, when the analysis involves an incorrect allele frequency, false positive results may appear. According to Figure 7, the misspecification of the allele frequency in the interval 0.1-0.5 is safe, assuming that the marker is highly polymorphic; otherwise, the weighting functions are necessary. It appears that the weighting function $f(p)=1/sqrt(p)$ is the best, since it has the lowest maximum false-positive rate. However, incorrectly lowering frequencies may result in 30% rate of false-positive findings. This figure shows that true disease-related excess allele-sharing may be missed under the function $f(p)=1$, with overestimated allele frequencies, and under the function $f(p)=1/p$, with underestimated allele frequencies.

CONCLUSIONS

In this research, we have examined the utility of the CM and have shown that several factors have a major influence on the power. These factors are the mode of inheritance of the trait, the polymorphism of the marker, the recombination distance between a trait locus and the closest marker, the penetrance, and the phenocopy rate. The overall conclusion of these analyses is that, if a trait is homogeneous and sufficient triplets are available, then the correct identification of disease-relevant excess allele-sharing is feasible. However, complex interactions of the above factors prohibit simplistic statements about efficiency of the CM in general.

There are a number of possible limitations to our findings. First, the major disadvantage of all simulation studies is that the results are restricted to the particular circumstances simulated (Elston, 1989). Although we believe that we evaluated the power of the CM for a fairly wide variety of disease models, this principal objection can never be overcome by any simulation study (Knapp, Seuchter, & Baur, 1994). Second, our approach is that CM analyses are by necessity exploratory, since there is a high dependency on allele frequency and, therefore, a high probability for false positives. If a significant result is obtained, then confirmation will be required. As we stated earlier, the CM, as well as any other allele-sharing method, is not intended for fine mapping of a disease locus. This is probably best done by standard parametric methods once a putative disease-predisposing locus is identified and the initial finding is replicated. Third, the current limitations for affected-distant-relatives studies are related to the lack of sufficient numbers of sets of distantly-related relative, all of whom are affected. Our theoretical estimates of the ability to identify or exclude regions of genome on the basis of statistically significant excess/not-excess allele-sharing by means of a set of highly polymorphic markers will not, necessarily, correspond to practical availability of sets of distant affected relatives.

With all the mentioned above restrictions in mind, we nonetheless conclude that affected-third-degree-relative analysis provides an appropriate method for detection or exclusion of excess allele-sharing, enriching the arsenal of existing methodological tools for genetic dissection of complex traits. Together, these findings imply that, under appropriate conditions, whole-genome screening for excess allele-sharing sets of distant affected relatives could be achievable and fruitful. Affected-distant relatives are well suited for screening multiple markers to identify promising linkage relationships, which may later be confirmed using other markers, by standard lod-score analysis, or both.

REFERENCES

Agresti, A. (1990). *Categorical data analysis*. New York: Wiley.

Amos, C.I. (1994). Robust variance-components approach for assessing genetic linkage in pedigrees. *American Journal of Human Genetics, 54*, 535-543.

Amos, C.I. & Elston, R.C. (1989). Robust methods for the detection of genetic linkage for quantitative data from pedigrees. *Genetic Epidemiology, 6*, 349-360.

Anonymous, (1992). A comprehensive genetic linkage map of the human genome. NIH/CEPH Collaborative Mapping Group. *Science, 258*, 148-162.

Babron, M.C., Martinez, M., Bonaiti-Pellie, C., & Clerget-Darpoux, F. (1993). Linkage detection by the Affected-Pedigree-Member method: what is really tested? *Genetic Epidemiology, 10*, 389-394.

Berrettini, W.H., Ferraro, T.N., Goldin, L.R., Weeks, D.E., Detera-Wadleigh, S., Nurnberger, J.I.,Jr., & Gershon, E.S. (1994). Chromosome 18 DNA markers and manic-depressive illness: evidence for a susceptibility gene. *Proceedings of the National Acadamy of Sciences USA, 91*, 5918-5921.

Bishop, D.T. & Williamson, J.A. (1990). The power of identity-by-state methods for linkage analysis. *American Journal of Human Genetics, 46*, 254-265.

Boehnke, M. (1990). Sample-size guidelines for linkage analysis of a dominant locus for a quantitative trait by the method of lod scores. *American Journal of Human Genetics, 47*, 218-227.

Botstein, D., White, R.L., Skolnick, M., & Davis, R.W. (1980). Construction of a genetic linkage map in man using restriction fragment length polymorphisms. *American Journal of Human Genetics, 32*, 314-331.

Brown, D.L., Gorin, M.B., & Weeks, D.E. (1994). Efficient strategies for genomic searching using the affected-pedigree-member method of linkage analysis. *American Journal of Human Genetics, 54*, 544-552.

Cannon-Albright, L.A., Goldgar, D.E., Meyer, L.J., Lewis, C.M., Anderson, D.E., Fountain, J.W., Hegi, M.E., Wiseman, R.W., Petty, E.M., Bale, A.E., & et al, (1992). Assignment of a locus for familial melanoma, MLM, to chromosome 9p13-p22. *Science, 258*, 1148-1152.

Cantor, R.M. & Rotter, J.L. (1987). Marker concordance in pairs of distant relatives: a new method of linkage analysis for common diseases. *American Journal of Human Genetics, 41*, A252.

Carey, G. & Williamson, J. (1991). Linkage analysis of quantitative traits: increased power by using selected samples. *American Journal of Human Genetics, 49*, 786-796.

Cavalli-Sforza, L.L. & King, M.C. (1986). Detecting linkage for genetically heterogeneous diseases and detecting heterogeneity with linkage data. *American Journal of Human Genetics*, *38*, 599-616.

Clerget-Darpoux, F. & Bonaiti-Pellie, C. (1992). Strategies based on marker information for the study of human diseases. *Annals of Human Genetics*, *56*, 145-153.

Day, N.E. & Simons, M.J. (1976). Disease susceptibility genes--their identification by multiple case family studies. *Tissue Antigens*, *8*, 109-119.

Eaves, L. & Meyer, J. (1994). Locating human quantitative trait loci: Guidlines for the selection of sibling pairs of genotyping. *Behavior Genetics*, *24*, 443-455.

Eaves, L.J. (1994). Effect of genetic architecture on the power of human linkage studies to resolve the contribution of quantitative trait loci. *Heredity*, *72*, 175-192.

Elston, R.C. (1989). Man bites dog? The validity of maximizing lod scores to determine mode of inheritance. *American Journal of Medical Genetics*, *34*, 487-488.

Festing, M. (1979). *Inbred strains in biomedical research.* New York: Oxford University Press.

Fulker, D.W. & Cardon, L.R. (1994). A sib-pair approach to interval mapping of quantitative trait loci. *American Journal of Human Genetics*, *54*, 1092-1103.

Fulker, D.W., Cherny, S.S., & Cardon, L.R. (1995). Multipoint interval mapping of quantitative trait loci, using sib pairs. *American Journal of Human Genetics*, *56*, 1224-1233.

George, V.T. & Elston, R.C. (1987). Testing the association between polymorphic markers and quantitative traits in pedigrees. *Genetic Epidemiology*, *4*, 193-201.

Goldgar, D.E. (1990). Multipoint analysis of human quantitative genetic variation. *American Journal of Human Genetics*, *47*, 957-967.

Goldgar, D.E. & Oniki, R.S. (1992). Comparison of a multipoint identity-by-descent method with parametric multipoint linkage analysis for mapping quantitative traits. *American Journal of Human Genetics*, *50*, 598-606.

Goldin, L.R. & Weeks, D.E. (1993). Two-locus models of disease: comparison of likelihood and nonparametric linkage methods. *American Journal of Human Genetics*, *53*, 908-915.

Green, J.R. & Woodrow, J.C. (1977). Sibling Method for Detecting HLA-linked genes in disease. *Tissue Antigens*, *9*, 31-35.

Guo, S.W. (1994). Computation of identity-by-descent proportions shared by two siblings. *American Journal of Human Genetics*, *54*, 1104-1109.

Gyapay, G., Morissette, J., Vignal, A., Dib, C., Fizames, C., Millasseau, P., Marc, S., Bernardi, G., Lathrop, M., & Weissenbach, J. (1994). The 1993-94 Genethon human genetic linkage map. *Nature Genetics*, *7*, 246-339.

Haile, R.W., Goldstein, A.M., Weeks, D.E., Sparkes, R.S., & Paganini-Hill, A. (1990). Genetic epidemiology of bilateral breast cancer: a linkage analysis using the affected-pedigree-member method. *Genetic Epidemiology*, *7*, 47-55.

Hall, J.M., Lee, M.K., Newman, B., Morrow, J.E., Anderson, L.A., Huey, B., & King, M.C. (1990). Linkage of early-onset familial breast cancer to chromosome 17q21. *Science, 250*, 1684-1689.

Hamer, D.H., Hu, S., MagL, , Hu, N., & Pattatucci, A.M. (1993). A linkage between DNA markers on the X chromosome and male sexual orientation. *Science, 261*, 321-327.

Haseman, J.K. & Elkston, R.C. (1972). The investigation of linkage between a quantitative trait and a marker locus. *Behavior Genetics, 2*, 3-19.

Hill, A.P. (1975). Quantitative linkage: a statistical procedure for its detection and estimation. *Annals of Human Genetics, 38*, 439-449.

Holmans, P. (1993). Asymptotic properties of affected-sib-pair linkage analysis. *American Journal of Human Genetics, 52*, 362-374.

Hyer, R.N., Julier, C., Buckley, J.D., Trucco, M., Rotter, J., Spielman, R., Barnett, A., Bain, S., Boitard, C., Deschamps, I., & et al, (1991). High-resolution linkage mapping for susceptibility genes in human polygenic disease: insulin-dependent diabetes mellitus and chromosome 11q. *American Journal of Human Genetics, 48*, 243-257.

Knapp, M., Seuchter, S.A., & Baur, M.P. (1993). The effect of misspecifying allele frequencies in incompletely typed families. *Genetic Epidemiology, 10*, 413-418.

Knapp, M., Seuchter, S.A., & Baur, M.P. (1994a). Two-locus disease models with two marker loci: the power of affected-sib-pair tests. *American Journal of Human Genetics, 55*, 1030-1041.

Knapp, M., Seuchter, S.A., & Baur, M.P. (1994b). Linkage analysis in nuclear families. *Human Heredity, 44*, 37-51.

Kruglyak, L. & Lander, E.S. (1995). High-resolution genetic mapping of complex traits. *American Journal of Human Genetics, 56*, 1212-1223.

Lander, E.S. & Schork, N.J. (1994). Genetic dissection of complex traits. *Science, 265*, 2037-2048.

Lange, K. (1986). The affected sib-pair method using identity by state relations. *American Journal of Human Genetics, 39*, 148-150.

Martinez, M. & Goldin, I.R. (1991). Detection of linkage for heterogenous disorders by using multipoint linkage analysis. *American Journal of Human Genetics, 49*, 1300-1305.

Matise, T.C. & Weeks, D.E. (1993). Detecting heterogeneity with the affected-pedigree-member (APM) method. *Genetic Epidemiology, 10*, 401-406.

Murray, J.C., Buetow, K.H., Weber, J.L., Ludwigsen, S., Scherpbier-Heddema, T., Manion, F., Quillen, J., Sheffield, V.C., Sunden, S., Duyk, G.M., & et al, (1994). A comprehensive human linkage map with centimorgan density. Cooperative Human Linkage Center (CHLC). *Science, 265*, 2049-2054.

Nakamura, Y., Leppert, M., O'Connell, P., Wolff, R., Holm, T., Culver, M., Martin, C., Fujimoto, E., Hoff, M., Kumlin, E., & et al, (1987). Variable number of tandem repeat (VNTR) markers for human gene mapping. *Science, 235*, 1616-1622.

Nikali, K., Suomalainen, A., Terwilliger, J., Koskinen, T., Weissenbach, J., & Peltonen, L. (1995). Random search for shared chromosomal regions in four affected individuals: The assignment of a new hereditary ataxia locus. *American Journal of Human Genetics, 56*, 1088-1095.

Olson, J.M. & Wijsman, E.M. (1993). Linkage between quantitative trait and marker loci: methods using all relative pairs. *Genetic Epidemiology, 10*, 87-102.

Ott, J. (1990). Cutting a Gordian knot in the linkage analysis of complex human traits. *American Journal of Human Genetics, 46*, 219-221.

Ott, J. (1991). *Analysis of human genetic linkage*. Baltimore: The Johns Hopkins University Press.

Owerbach, D. & Gabbay, K.H. (1994). Linkage of the VNTR/insulin-gene and type I diabetes mellitus: increased gene sharing in affected sibling pairs. *American Journal of Human Genetics, 54*, 909-912.

Penrose, L.S. (1935). The detection of autosomal linkage in data which consists of pairs of brothers and sisters of unspecified parentage. *American Journal of Human Genetics, 6*, 133.

Penrose, L.S. (1983). Genetic linkage in graded human characters. *Annals of Eugenics, 9*, 133-138.

Pericak-Vance, M.A., Bebout, J.L., Gaskell, P.C., Jr., Yamaoka, L.H., Hung, W.Y., Alberts, M.J., Walker, A.P., Bartlett, R.J., Haynes, C.A., Welsh, K.A., & et al, (1991). Linkage studies in familial Alzheimer disease: evidence for chromosome 19 linkage. *American Journal of Human Genetics, 48*, 1034-1050.

Plomin, R., McClearn, G.E., Smith, D.L., Vignetti, S., Chorney, M.J., Chorney, K., Venditti, C.P., Kasarda, S., Thompson, L.A., Detterman, D.K., & et al, (1994). DNA markers associated with high versus low IQ: the IQ Quantitative Trait Loci (QTL) Project. *Behavior Genetics, 24*, 107-118.

Risch, N. (1990a). Genetic linkage and complex diseases, with special reference to psychiatric disorders. *Genetic Epidemiology, 7*, 3-16.

Risch, N. (1990b). Linkage strategies for genetically complex traits. III. The effect of marker polymorphism on analysis of affected relative pairs. *American Journal of Human Genetics, 46*, 242-253.

Risch, N. (1990c). Linkage strategies for genetically complex traits. II. The power of affected relative pairs. *American Journal of Human Genetics, 46*, 229-241.

Schork, N.J., Boehnke, M., Terwilliger, J.D., & Ott, J. (1993). Two-trait-locus linkage analysis: a powerful strategy for mapping complex genetic traits. *American Journal of Human Genetics, 53*, 1127-1136.

Seboun, E., Robinson, M.A., Doolittle, T.H., Ciulla, T.A., Kindt, T.J., & Hauser, S.L. (1989). A susceptibility locus for multiple sclerosis is linked to the T cell receptor beta chain complex. *Cell, 57*, 1095-1100.

Sribney, W.M. & Swift, M. (1992). Power of sib-pair and sib-trio linkage analysis with assortative mating and multiple disease loci. *American Journal of Human Genetics, 51*, 773-784.

Suarez, B.K., Rice, J., & Reich, T. (1978). The generalized sib pair IBD distribution: its use in the detection of linkage. *Annals of Human Genetics*, *42*, 87-94.

Thomson, G. & Bodmer, W. (1977). The genetic analyysis of HLA and disease. In A. Svejgaard & J. Dausset (Eds.). *HLA and disease* (pp. 56-63). Copenhagen: Munksgaard

Tierney, C. (1994). *Power of affected sibling method tests for linkage*. (un pub)

Tiwari, J.L. & Terasaki, P.I. (1985). *HLA and disease associations*. New York: Springer Verlag.

Van Eerdewegh, P., Hampe, C.L., Suarez, B.K., & Reich, T. (1993). Alzheimer's disease: A piscatorial trek. *Genetic Epidemiology*, *10*, 395-400.

Ward, P.J. (1993). Some developments on the affected-pedigree-member method of linkage analysis. *American Journal of Human Genetics*, *52*, 1200-1215.

Weber, J.L. (1994). Patch mapping of Tourette syndrome genes. In *TSA permanent research fund recipients:1994* Bayside, New York: Tourette Syndrome Association

Weber, J.L. & May, P.E. (1989). Abundant class of human DNA polymorphisms which can be typed using the polymerase chain reaction. *American Journal of Human Genetics*, *44*, 388-396.

Weeks, D.E. & Harby, L.D. (1995). The affected-pedigree-member method: power to detect linkage. *Human Heredity*, *45*, 13-24.

Weeks, D.E. & Lange, K. (1988). The affected-pedigree-member method of linkage analysis. *American Journal of Human Genetics*, *42*, 315-326.

Weeks, D.E. & Lange, K. (1992). A multilocus extension of the affected-pedigree-member method of linkage analysis. *American Journal of Human Genetics*, *50*, 859-868.

Weir, B.S. (1990). *Genetic data analysis*. Sunderland,MA: Sinauer.

Weitkamp, L.R., Stancer, H.C., Persad, E., Flood, C., & Guttormsen, S. (1981). Depressive disorders and HLA: a gene on chromosome 6 that can affect behavior. *New England Journal of Medicine*, *305*, 1301-1306.

Whittemore, A.S. & Halpern, J. (1994). A class of tests for linkage using affected pedigree members. *Biometrics*, *50*, 118-127.

Wickens, T.D. (1989). *Multiway contingency tables analysis for the social sciences*. Hillsdale, NJ: Lawrence Erlbaum.

Wilson, A.F., Elston, R.C., Tran, L.D., & Siervogel, R.M. (1991). Use of the robust sib-pair method to screen for single-locus, multiple-locus, and pleiotropic effects: application to traits related to hypertension. *American Journal of Human Genetics*, *48*. 862-872.

CHAPTER 10

PARTITIONING THE GENETIC VARIATION OF QUANTITATIVE TRAITS USING MARKER LOCI

George P. Vogler

Pennsylvania State University, Pennsylvania, USA

INTRODUCTION

Just as linkage analysis using individual highly polymorphic markers to map monogenic traits was prominently at the interface between molecular genetics and genetic epidemiology in the 1980s, multipoint mapping of QTLs for complex quantitative traits presents one of the most exciting challenges in human genetics today. Recent technical advances in molecular genetics allow for the use of highly polymorphic markers that are spaced closely throughout the genome. At the same time, genetic epidemiological studies have demonstrated convincingly that there are genetic influences on complex quantitative phenotypes such as risk factors for cardiovascular disease and dimensions of personality. Traditional quantitative genetic models such as path analysis or segregation analysis provide descriptive information regarding familial resemblance but cannot be used effectively to identify or localize specific genes. Linkage analysis can detect genes for monogenic traits, but this technique is designed to detect single major genes rather than genes that contribute to phenotypic variation for multifactorial traits that may be polygenic or oligogenic in origin.

The methodology of molecular genetics has advanced to a point where it is now practical to consider systematic exploratory approaches for the detection of quantitative trait loci which depend on the availability of a large number of highly polymorphic genetic markers distributed throughout the genome. Currently, libraries of microsatellite markers are available and automated marker detection technology exists so that it is possible to obtain markers on large samples that are distributed as closely as every 2 cM throughout the genome. Practical considerations, such as cost, make it appealing to consider multi-phase screening strategies in which a sparser map, such as 5 cM, 10 cM, or even 20 cM, is used for an initial screening. The use of a very fine, high-resolution map can then be reserved for a second phase to examine in fine detail those regions which show potential in the initial screening. The tradeoff between the increased statistical power and the increased likelihood of obtaining numerous false positive results when a high-density set of markers is used is a problem which is receiving much attention at present.

Many methods for locating genes for quantitative traits were developed for application to plant or animal experimental paradigms consisting of selection experiments or crosses (Lander & Botstein, 1989; Paterson et al., 1988, 1990; Gora-Maslak et al., 1991; Plomin & McClearn, 1993). Recently, work has begun to focus more intently on methods to localize genes that influence quantitative traits in humans. Yet most of the models that have been developed for the study of human quantitative traits are designed to look for the linkage or association between a single trait locus and individual markers (e.g., Haseman & Elston, 1972; Hill, 1975; Penrose, 1983; George & Elston, 1987; Amos & Elston, 1989; Wilson et al., 1990; Fulker et al., 1991; Haley & Knott, 1992; Olson & Wijsman, 1993; Plomin, 1993). Some approaches advocate the use of selected samples to increase power (Carey & Williamson, 1991). The reliance on information from the tail of the distribution to make inference regarding the population (e.g., Rao et al., 1988) is appropriate only if the genetic mechanism affecting phenotypic variation in the extreme tails of a distribution is the same as the mechanism which results in most of the phenotypic variation in the normal range, but this approach can provide substantial improvement in power to detect QTLs when this assumption is met (Schork, 1993).

There have been a few recent attempts to develop interval mapping methods that use marker information to localize genes for complex multifactorial traits in humans to specific chromosomal regions (Goldgar, 1990; Fulker & Cardon, 1994). These methods extend the Haseman & Elston (1972) approach, which relates sib-pair trait differences to the estimated sibling genetic correlation at a single marker locus, to multiple markers that span a chromosomal region. Goldgar's (1990) approach uses identity-by-descent status at the multiple markers to determine the proportion of the chromosomal region that is shared identical by descent. The size of the region can be very small (flanked by two close markers) for a very focused study, or very large (up to an entire chromosome) for an exploratory investigation. This methodology allows detection of genetic variation due to (1) a single major locus in the region; (2) multiple loci in the region; and (3) multiple unlinked loci that are not located within the region.

The model to partition human quantitative genetic variation into effects due to loci on specific chromosomal regions is sufficiently general to allow for the detection of (1) true polygenic variation that clusters to a particular region, (2) oligogenic variation in which a single locus in a particular region can be identified but which does not account for all (or even a large) portion of the genetic variation, and (3) variation due to a major locus.

This method was developed for use with sib pairs (Goldgar, 1990; Goldgar & Oniki, 1992). Power calculations have been published for a sampling scheme that represents large unusual data sets such as the CEPH or Venezuelan pedigrees (40 families with 8 offspring in each), but which is not available in practice in most studies.

In this report, the multipoint mapping methodology is incorporated into the classical twin study design. DZ twins provide the same information as sib pairs using identity-by-descent status of markers to quantify variability among pairs in the proportion of a chromosomal region shared identical by descent. The proportion of a chromosomal region shared by a pair of twins can be related to observed phenotypic similarity within pairs. The addition of MZ twin information on the phenotype permits the same contrast as in the classical twin design. The combination of the DZ marker information with the MZ and DZ phenotypic information permits decomposition of phenotypic variation into genetic effects due to loci on the chromosomal region flanked by a pair of markers; genetic effects due to loci that are not in the region under investigation; environmental effects that are shared by twins; and environmental effects that are not shared by twin pairs. The replacement of sibs in Goldgar's (1990) design with DZ twins, and the inclusion of MZ twin phenotypic data, permits an increase in resolution of genetic variation that is analogous to the additional information provided by the classical twin study compared to that provided by a sib correlation.

METHODS

THE QTL MODEL

The multipoint mapping method as developed by Goldgar (1990) permits one to make inferences about R, the proportion of genetic material between a pair of chromosomal markers that is shared IBD by a sib pair, based on observed IBD and recombination status of the pair of markers and assuming some model of recombination, such as the well-known Haldane model, in the interval between markers. The expected values of R is calculated as a straightforward algebraic function of l, the known distance between markers in Morgans, and q, the recombination fraction between the two markers, which itself is a function l. Under the conventional Haldane model of recombination, upon which most genetic maps are based, $q = 1/2\ (1-\exp(-2l))$. The expected values and variances of R are derived and presented by Goldgar (1990) for the 7 possible combinations of IBD and recombination status: (1) IBD at both markers; no recombination in either twin; (2) IBD at both markers; recombination in both twins; (3) non-IBD at both markers; no recombination in either twin; (4) non-IBD at both markers; recombination in both twins; (5) IBD at one marker and non-IBD at the second marker; recombination in one twin; (6) IBD at one marker and uninformative at the second marker; (7) non-IBD at one marker and uninformative at the second marker. The 7 equations for the expected values are each simply functions of the known parameter l. The value of R is calculated for each twin pair based on their observed pattern of IBD and recombination status if this is known from the availability of parental marker information.

THE TWIN MODEL

The classical twin design contrasts the resemblance between MZ twins, who share 100% of their genetic material with the resemblance between DZ, twins who share on the average 50% of their genetic material. Shared environmental effects are assumed to contribute equally to both zygosities and unshared environmental effects also can contribute to phenotypic variation. The classical twin design has been well-developed and extended to many complex situations beyond the simple univariate phenotype analysis (Eaves, 1977; Martin & Eaves, 1977; Neale & Cardon, 1992). Numerous studies have evaluated the assumptions of the model critically and found them to be generally valid (Loehlin & Nichols, 1976; Plomin et al., 1976; Matheny et al., 1976; Kendler et al., 1993; but see also Clifford et al., 1984; Kaprio et al., 1990). Numerous applications of the methodology have provided consistent and replicable results for many phenotypes (see, for example, Neale & Cardon, 1992).

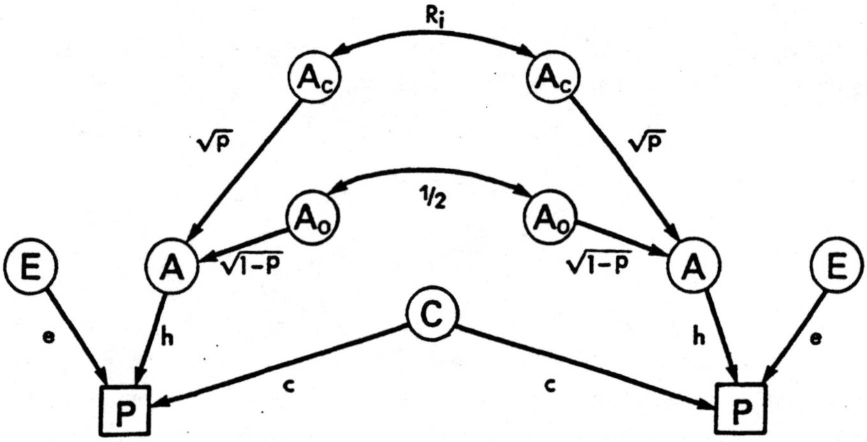

Figure 1. *Path diagram of QTL model in DZ twins.* Phenotypic measurements (P) are shown for two twins, with a common environment (C), unique environment (E), and additive genetic factors (A) that are divided between those within the chromosomal region under investigation (A_C) and those distributed throughout the remainder of the genome (A_O). Parameters of the model include : h^2, the genetic heritability; c^2, the cultural heritability or proportion of phenotypic variance that is due to shared environmental effects; e^2, the proportion of phenotypic variance that is due to nonshared environmental effects; and p, the proportion of additive genetic variance that is due to genes located in the chrosomal region flanked by a pair of markers, with $1 - p$ being the proportion of additive genetic variance that is due to genes located in the remainder of the genome.

The methodology of the proposed study takes advantage of the fact that DZ twins have the same genetic relationship as siblings and therefore can be used in the QTL model to localize loci and to distribute the genetic variation into that attributable to loci within a particular interval and that attributable to loci in the remainder of the genome. The detection of QTLs is made possible because even though DZ twins share on the average 50% of their genetic material, the proportion of genetic material in a particular chromosomal region between two markers varies around a mean of 50% and is quantifiable for each twin pair based on IBD and recombination status for that particular twin pair. The advantage of using twins over sib pairs is that the addition of phenotypic information (not marker data) on MZ twins permits the phenotypic variation to be divided into genetic, shared environmental, and nonshared environmental variation, whereas with sib pairs the genetic and shared environmental variation are completely confounded, thus reducing the power of the method to detect QTLs.

The path model for the twin QTL model is presented in Figure 1. Incorporation of the calculated proportion of a chromosomal region shared by a DZ twin pair into the basic twin model is shown in the Figure by the use of the parameter R_i. This parameter is calculated in advance from the marker data using IBD status and recombination status (if available), and the known distance (l) between markers using Goldgar's (1990) equations. This parameter R_i is fixed at the calculated value for each twin pair i. The proportion of the additive genetic variance attributable to genes within the chromosomal region is denoted by p, which is an estimable but unstable parameter. Rejection of the null hypothesis that $p = 0$ is taken as evidence of the presence of genes in the region of interest. For MZ twins, the path between A for the two twins reduces to unity regardless of p, so the MZ twins have no information for estimating p but do contribute to the estimation of genetic heritability (h) and shared environmental effects (c).

Expected values of the twin correlations are given below (in practice, covariances rather than correlations are analyzed using maximum likelihood but for ease of exposition the standardized case is presented here):

$$r_{MZ} = h^2 + c^2$$
$$r_{DZ} = h^2\,[1/2\,(1 - p) + R_i p] + c^2$$

where R_i is the the proportion of genes shared IBD by twin pair i between a pair of markers. This parameter is calculated for a twin pair based on the observed recombination and marker IBD status using Goldgar's equations. It is fixed for each twin pair at the calculated value. In addition to the twin correlations, we have the additional information that the standardized phenotypic variance of unity if a function of the parameters:

$$1 = h^2 + c^2 + e^2 .$$

In the statistical analysis of twin data, covariances are used rather than correlations. There is a phenotypic vector **x** and an expected mean vector **m**, and a covariance matrix **S** with the expected value of the phenotypic variance for twins of the appropriate zygosity as the two diagonal elements, and the expected value of the covariance between members of a twin pair off the diagonal. For MZ twins, the covariance is $V_A + V_C$, where V_A is the additive genetic variance and V_C is the shared environmental variance. For DZ twins, the covariance is $[1/2\,(1 - p) + R_i p]V_A + V_C$. The log likelihood of observing the twin phenotypic values in the ith twin pair is given by the log of the multivariate normal density:

$$\ln L_i = -1/2\,[\ln|\mathbf{S}_i| + (\mathbf{x}_i - \mathbf{m})'\mathbf{S}_i^{-1}(\mathbf{x}_i - \mathbf{m})] + \text{constant}$$

where $|\mathbf{S}_i|$ is the determinant of $\mathbf{S}_i$, and $\mathbf{S}_i^{-1}$ is the inverse of $\mathbf{S}_i$. For n twin pairs, the likelihood is summed over all pairs. This procedure is used for DZ twins, for whom R_i is calculated separately for each pair pased on marker information. For MZ twins, a more efficient procedure using the Wishart distribution is used to avoid multiple inversions of S. A numerical optimization routine is used to maximize the log likelihood. The likelihood is evaluated once with p as an estimated parameter and once under the hypothesis that $p = 0$ (i.e., there are no QTLs in the chromosomal region flanked by the markers).

POWER

Simulations were conducted to explore in a general manner the feasibility of using the QTL twin methodology. The simulation procedure assumed that two fully informative markers are available on each member of a DZ twin pair. Recombination status between a pair of markers is assessed for each twin, along with identity-by-descent status for a twin pair at each marker. In the absence of observed recombination, a twin pair can be identical by descent (IBD) or non-IBD at both marker loci. Similarly, if recombination is evident, a twin pair can be IBD or non-IBD at both markers. Finally, there are a several ways that a twin pair can be IBD at one locus but not IBD at the second locus.

The expected value of R, the proportion of the chromosomal region flanked by the two markers that is shared IBD by a DZ twin pairs can be calculated for each marker outcome combination. R is a function of l, the known length of the region in Morgans between two markers, and q, the recombination fraction. If the Haldane mapping function is used, q is merely a function of l.

Data were simulated under several models, chosen to be comparable with Goldgar's (1990) sibling simulations: a model with a chromosomal segment of 100 cM length with 3 marker loci 50 cM apart and 5 trait loci; 100 cM segment length and 5 trait loci but with 5 marker loci 25 cM apart; 5 trait loci and 5 marker loci distributed over a total length of 200 cM; and a single locus with an effect equal to the total effect of the 5 trait loci. Ten additional trait loci were randomly distributed outside of the chromosomal region. The models are represented graphically in Figure 2.

Assumptions of the simulation model include an allele frequency of 0.5 for each locus; each locus is modelled to be in Hardy-Weinberg equilibrium; linkage equilibrium is assumed at the population level; random mating occurs; all markers are completely informative with phase known; and the Haldane mapping function relating the recombination fraction to segment length is used.

Data were simulated under each model for low (0.25) and high (0.75) heritability; low (0.25), intermediate (0.50), and high (0.75) proportion of the additive genetic variance attributable to loci within the chromosomal region; and low (0.25) and high (0.75) proportion of the environmental variance attributable to common environment.

For each combination of parameters, 100 samples of 500 MZ and 500 DZ pairs were simulated and analyzed by maximum likelihood twice: once with p as an estimated parameter and once under the null hypothesis of $p = 0$. The proportion of times the null hypothesis of no effect of genes in the chromosomal region is rejected is reported as an indication of the power of the model.

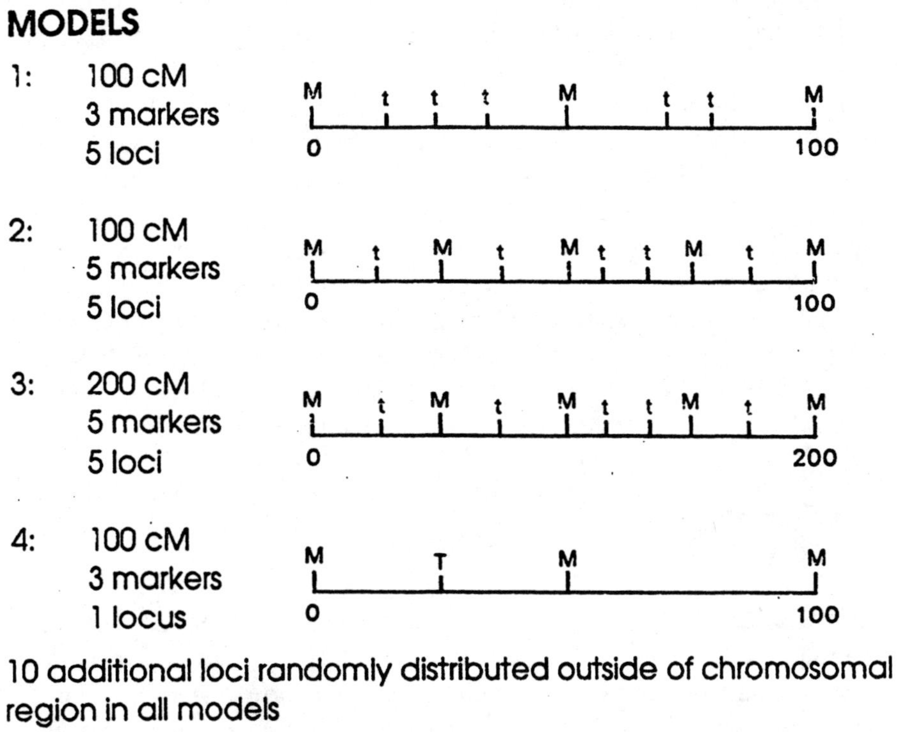

Figure 2. *Graphic representation of the chromosomal marker models used in the simulation study.* Each figure represents a chromosomal segment of the length presented. There are either 3 or 5 markers evenly spaced on the segment, with either 5 trait loci of small effect (t) or 1 trait locus of major effect (T) located on the segment (after Goldgar, 1990).

To simulate data for a DZ twin pair, fully informative marker loci are simulated for the parents of the twins, along with the trait loci appropriate for the model of trait loci within the chromosomal region flanked by the markers. Twins are simulated by selecting at random one of the two parental chromosomes. Recombination between pairs of adjacent loci is incorporated by using Haldane's mapping function to determine the recombination fraction between loci as a function of distance. Ten additional loci are generated in the parents for the traits and one is transmitted at random to each twin. Shared environmental variance is added to both members of the twin pair by generating a normal random deviate with a mean of 0 and a variance of c^2. Specific environmental variance is added to each member of the twin pair individually by generating additional normal random deviates with a mean of 0 and a variance of $(1 - h^2 - c^2)$. The phenotypic value for each twin is calculated by adding the genotypic effects for each locus, the shared environmental effect, and the specific environmental effect.

A similar procedure was used to calculate the phenotypic value for the MZ twins, with the following differences. Trait loci were generated only once for each twin pair and the values were applied to both members of the pair. Thus, the phenotypic values were calculated as the sum of the locus effects and shared environmental effects, which were the same for both twins, plus the specific environmental effect, which was generated separately for each twin. There was no need to generate marker loci for the MZ twins.

The results of the simulation study are summarized in Table 1. The first two columns contrast the 1-locus and 5-locus 3-marker with a chromosomal distance of 100 cM. The power is consistently higher for the single locus model wherever there is any power worth noting, but only by a modest amount. In contrasting the 3- and 5-marker models with 5 trait loci and a length of 100 cM (second and third columns), there may be some increase in power with more markers, although it is not consistent and appears to occur primarily in intermediate ranges of parameter values. The last two columns contrast a length of 100 cM and 200 cM with 5 markers and 5 trait loci. Not surprisingly, wherever there was substantial power there is a marked loss with increasing region length. Finally, it is noteworthy that power is good only when heritability is high.

DISCUSSION

The massive effort that is underway in human gene mapping has made systematic identification and processing of closely spaced highly polymorphic markers a reality. This technological advance provides great opportunity for the investigation of complex quantitative traits at the level of individual loci. Whereas linkage analysis is now a mainstream tool for identifying genes for monogenic traits, methods for detecting genes for complex traits in humans are only now beginning to mature. Many of the most perplexing problems in human genetics involve complex traits that have been shown repeatedly to have genetic effects that do not follow a simple Mendelian pattern. These include risk factors for cardiovascular disease, psychiatric disorders, and behavioral traits. Gene mapping of loci for quantitative traits using methods similar to that used in this investigation show great promise for bringing the investigation of quantitative traits to the next generation.

There are currently several initiatives under way to apply this methodology to a number of quantitative risk factors for cardiovascular disease. Although progress has been made in recent years to reduce the enormous consequences of cardiovascular disease on health and mortality, it remains one of the most pressing health problems in most western societies (American Heart Association, 1991). The complexities of the causes and effects of cardiovascular disease makes it very difficult to control and treat. Numerous studies have examined risk factors for cardiovascular disease. Some risk factors, like diet, exercise habits, and other behaviors can be controlled; others are less easily modified.

Numerous quantitative risk factors have been shown to have genetic influences using twin, family, and population-based (including case-control) studies. The large number of genetic studies of risk factors for cardiovascular disease have demonstrated rather conclusively that there are genetic influences that frequently account for half or more of the phenotypic variance. Total cholesterol, high-density-lipoprotein cholesterol (HDL), low-density-lipoprotein cholesterol (LDL), and triglycerides may all have about 50% of their variation attributable to genetic factors (e.g., Namboodiri et al., 1984; Iselius, 1988; Perusse et al., 1989; Vogler et al., 1991). ApoA2 and ApoB generally show moderate to strong genetic effects; ApoA1 shows great variability in heritability estimates (Lamon-Fava et al., 1991); and genetic effects on quantitative ApoE, other than the ApoE polymorphism itself, have not been conclusively characterized. Genetic effects on systolic and diastolic blood pressure have been studied extensively (e.g., Burns & Lauer, 1986; Rice et al., 1989; Tambs et al., 1992). Although heritability estimates vary widely, there appears to be about 50% of the variability attributable to genetic effects (Hopkins & Williams, 1986).

The traditional models for the genetics of quantitative traits have served their purpose well in identifying those risk factors for cardiovascular disease which have a genetic component. The traditional conclusion drawn from such models is that a certain percentage of the phenotypic variation is genetic, often polygenic, and the remainder of the variation is due to other factors that are presumably environmental in origin. The methodology outlined in this paper will take the genetic analysis of these risk factors to the next generation by merging quantitative genetic analysis with the techniques available from molecular genetics. This approach can be used to search for chromosomal locations for genes that influence the complex quantitative risk factors.

The simulation results indicate that power is affected substantially by the heritability and the distance between flanking markers. The most promising phenotypes that can be studied using this method are those that exhibit high heritability. The incorporation of the QTL method into the twin design provides a level of refinement to the estimation of the genetic variation that is a substantial improvement on the resolution of sib pairs. Heritability estimates from twin studies frequently are higher than heritability estimates from family studies. Whether this effect is due to greater resolution power of the twin design or failure of the underlying assumptions of either the twin design or the family design is an open question. If the gene mapping methodology is able to assign a large portion of the genetic variation to loci in specific regions, this will be important confirmation of the validity of heritability estimates from twin studies.

An additional effect on power is that the heterogeneity of available markers will be high, but not completely informative as in the simulation study. Currently available microsatellite markers that are distributed throughout the genome have heterozygosity of at least 65%, with most of the markers having 75% or greater heterozygosity. The reduction of power that the lack of informativeness in real applications can be somewhat offset by the use of a substantially denser marker system than was simulated.

Any genome-wide scanning method that uses a large number of markers presents a particularly acute potential for spurious positive findings resulting from multiple comparisons (Type I error). This is true for methods such as exploratory linkage analysis in which linkage to a large number of markers is evaluated. Exploratory searches for QTLs using the method outlined in this paper presents a similar problem when a large number of markers, and consequently a large number of possible chromosomal intervals that can be evaluated for QTL effects, are used. The number of comparisons that are made can be controlled by adjusting the size of the region to be searched to be larger in order to make the number of comparisons smaller. One strategy is to start the search by looking at very large areas, such as arms of longer metacentric chromosomes (which have the centromere located near the center of the chromosome), or entire chromosomes for shorter metacentric chromosomes or acrocentric chromosomes (which have the centromere located near an end of the chromosome). Once effects of QTLs are identified, a fine mapping approach can be taken by looking at much smaller intervals, zooming in on the smallest interval that can be evaluated with available marker technology, currently around 2 cM. Whereas a genome-wide 2 cM marker map in a large sample is currently not practical, once a region is identified as containing QTLs using wider intervals, more markers in a specific interval can be obtained. A dense marker map can also be used to double-check a positive finding by shifting the interval frame slightly to a different pair of markers.

There is no simple solution to the multiple comparisons problem, but there is a tradeoff between statistical power and potential problems arising from multiple comparisons. The best resolution is replication of findings in independent samples. The outcome of exploratory investigation will, at a minimum, provide candidate regions which can be focused upon in future studies of more limited scope for the purpose of replication. Furthermore, positive findings are a non-problem if they lead to the identification of functional genes. Nevertheless, the optimal segment size between markers under various conditions is an area that needs further investigation.

An alternative approach which reduces the problem of multiple comparisons effectively is to focus on candidate loci or regions if they are known. Some quantitative phenotypes, such as lipids and lipoproteins, have been studied extensively over a long period, and a number of loci are known or hypothesized to affect quantitative variability of the phenotypes. The problem of multiple comparisons becomes much less acute when candidate regions are used.

There are a number of extensions to the QTL model which remain to be explored. If QTLs are located in several chromosomal regions, it is possible to incorporate a model of nonadditive genetic effects for loci in the different regions. At present, nothing is known about the power to detect nonadditive effects using our mapping approach.

It is frequently desirable to include covariate effects in this type of analysis. In a general sense, it is desirable to be able to model explicitly the effects of confounding covariates such as age. At a more specialized level, a comprehensive program to understand the genetics of complex quantitative phenotypes will yield measured genotype effects that can be incorporated into the model so that the search for additional QTLs involves the residual genetic component of variation.

Numerous analyses have demonstrated correlated genetic effects on a variety of phenotypes, and a vast literature has been built up on multivariate genetic analysis of twin data since the seminal paper of Martin and Eaves (1977). A multivariate extension of the methodology in this paper has the potential to test for pleiotropic effects of QTLs using the gene mapping methodology.

Some investigators (Carey & Williamson, 1991; Plomin, 1993) advocate the use of nonrandom samples to increase power to detect QTLs. Certain study designs, particularly those aimed at disease rather than normal quantitative variation, use selected samples more appropriately than random samples. A correction for ascertainment can be built into the model presented in this paper.

Years of work in genetic epidemiology have provided a wealth of information regarding quantitative phenotypes that have genetic influences that are not inherited in a simple Mendelian fashion, and are likely not monogenic. Gene mapping techniques such as the method proposed in this paper provide a link between complex models of familial resemblance for such traits and and molecular genetic technology. This methodology will serve as a powerful tool in understanding the relationship between gene products at the basic level and complex multifactorial phenotypes.

ACKNOWLEDGMENTS

This work was supported in part by National Institutes of Health grants NIGMS GM-28719 and MH-31302.

REFERENCES

American Heart Association, (1991). *1991 Heart and stroke facts*. Dallas, TX: American Heart Association.

Amos, C.I. & Elston, R.C. (1989). Robust methods for the detection of genetic linkage for quantitative data from pedigrees [published erratum appears in Genet Epidemiol 1989;6(6):727]. *Genet Epidemiol, 6*, 349-360.

Burns, T.L. & Lauer, R.M. (1986). Blood pressure in children. In M.E. Pierpont & J.H. Moler (Eds.). *Genetics of cardiovascular disease* (pp. 305-317). Boston: Martinus Nijhoff

Carey, G. & Williamson, J. (1991). Linkage analysis of quantitative traits: increased power by using selected samples. *Am J Hum Genet, 49*, 786-796.

Clifford, C.A., Hopper, J.L., Fulker, D., & Murray, R.M. (1984). A genetic and environmental analysis of a twin family study of alcohol use, anxiety, and depression. *Genet Epidemiol, 1*, 63-79.

Eaves, L.J. (1977). Inferring the causes of human variation. *Journal of the Royal Statistical Society, 140*, 324-355.

Fulker, D.W., Cardon, L.R., DeFries, J.C., Kimberling, W.J., Pennington, B.F., & Smith, D. (1991). Multiple regression analysis of sib pair data on reading to detect quantitative trait loci. *Reading and Writing: An Interdisciplinary Journal, 3*, 299-313.

Fulker, D.W. & Cardon, L.R. (1994). A sib-pair approach to interval mapping of quantitative trait loci [published erratum appears in Am J Hum Genet 1994 Aug;55(2): 419]. *Am J Hum Genet, 54*, 1092-1103.

George, V.T. & Elston, R.C. (1987). Testing the association between polymorphic markers and quantitative traits in pedigrees. *Genet Epidemiol, 4*, 193-201.

Goldgar, D.E. (1990). Multipoint analysis of human quantitative genetic variation. *Am J Hum Genet, 47*, 957-967.

Goldgar, D.E. & Oniki, R.S. (1992). Comparison of a multipoint identity-by-descent method with parametric multipoint linkage analysis for mapping quantitative traits. *Am J Hum Genet*, *50*, 598-606.

Gora-Maslak, G., McClearn, G.E., Crabbe, J.C., Phillips, T.J., Belknap, J.K., & Plomin, R. (1991). Use of recombinant inbred strains to identify quantitative trait loci in psychopharmacology. [Review]. *Psychopharmacology (Berl)*, *104*, 413-424.

Haley, C.S. & Knott, S.A. (1992). A simple regression method for mapping quantitative trait loci in line crosses using flanking markers. *Heredity*, *69*, 315-324.

Haseman, J.K. & Elkston, R.C. (1972). The investigation of linkage between a quantitative trait and a marker locus. *Behav Genet*, *2*, 3-19.

Hill, A.P. (1975). Quantitative linkage: a statistical procedure for its detection and estimation. *Ann Hum Genet*, *38*, 439-449.

Hopkins, P.N. & Williams, R.R. (1989). Human genetics and coronary heart disease: a public health perspective. [Review]. *Annu Rev Nutr*, *9*, 303-345.

Iselius, L. (1988). Genetic epidemiology of common diseases in humans. In B.S. Weir, E.J. Eisen, M.M. Goodman & G. Namkoong (Eds.). *Proceedings of the second international conference on quantitative genetics* (pp. 341-352). Sunderland, MA: Sinauer Associates, Inc.

Kaprio, J., Koskenvuo, M., & Rose, R.J. (1990). Change in cohabitation and intrapair similarity of monozygotic (MZ) cotwins for alcohol use, extraversion, and neuroticism. *Behav Genet*, *20*, 265-276.

Kendler, K.S., Neale, M.C., Kessler, R.C., Heath, A.C., & Eaves, L.J. (1993). A test of the equal-environment assumption in twin studies of psychiatric illness. *Behav Genet*, *23*, 21-27.

Lamon-Fava, S., Jimenez, D., Christian, J.C., Fabsitz, R.R., Reed, T., Carmelli, D., Castelli, W.P., Ordovas, J.M., Wilson, P.W., & Schaefer, E.J. (1991). The NHLBI Twin Study: heritability of apolipoprotein A-I, B, and low density lipoprotein subclasses and concordance for lipoprotein(a). *Atherosclerosis*, *91*, 97-106.

Lander, E.S. & Botstein, D. (1989). Mapping mendelian factors underlying quantitative traits using RFLP linkage maps [published erratum appears in Genetics 1994 Feb; 136(2):705]. *Genetics*, *121*, 185-199.

Loehlin, J.C. & Nichols, R.C. (1976). *Heredity, environment and personality: A study of 850 sets of twins*. Austin, TX: University of Texas Press.

Martin, N.G. & Eaves, L.J. (1977). The genetical analysis of covariance structure. *Heredity*, *38*, 79-95.

Matheny, A.P.,Jr., Wilson, R.S., & Dolan, A.B. (1976). Relations between twins' similarity of appearance and behavioral similarity: Testing an assumption. *Behav Genet*, *6*, 343-351.

Namboodiri, K.K., Green, P.P., & Kaplan, E.B. (1984). The collaborative lipid research clinics program family study. IV. Familial associations of plasma lipids and lipoproteins. *American Journal of Epidemiology*, *119*, 975-996.

Neale, M.C. & Cardon, L.R. (1992). *Methodology for genetic studies of twins and families*. Dordrecht: Kluwer Academic Publishers.

Olson, J.M. & Wijsman, E.M. (1993). Linkage between quantitative trait and marker loci: methods using all relative pairs. *Genet Epidemiol*, *10*, 87-102.

Paterson, A.H., Lander, E.S., Hewitt, J.D., Peterson, S., Lincoln, S.E., & Tanksley, S.D. (1988). Resolution of quantitative traits into Mendelian factors by using a complete linkage map of restriction fragment length polymorphisms. *Nature*, *335*, 721-726.

Paterson, A.H., DeVerna, J.W., Lanini, B., & Tanksley, S.D. (1990). Fine mapping of quantitative trait loci using selected overlapping recombinant chromosomes, in an interspecies cross of tomato. *Genetics*, *124*, 735-742.

Penrose, L.S. (1983). Genetic linkage in graded human characters. *Annals of Eugenics*, *9*, 133-138.

Perusse, L., Despres, J.P., Tremblay, A., Leblanc, C., Talbot, J., Allard, C., & Bouchard, C. (1989). Genetic and environmental determinants of serum lipids and lipoproteins in French Canadian families. *Arteriosclerosis*, *9*, 308-318.

Plomin, R. (1993). Molecular genetic investigation of low and high cognitive ability in children. *Behav Genet*,

Plomin, R. & McClearn, G.E. (1993). Quantitative trait loci (QTL) analyses and alcohol-related behaviors. [Review]. *Behav Genet*, *23*, 197-211.

Plomin, R., Willerman, L., & Loehlin, J.C. (1976). Resemblance in appearance and the equal environments assumption in twin studies of personality traits. *Behav Genet*, *6*, 43-52.

Rao, D.C., Wette, R., & Ewens, W.J. (1988). Multifactorial analysis of family data ascertained through truncation: a comparative evaluation of two methods of statistical inference. *Am J Hum Genet*, *42*, 506-515.

Rice, T., Vogler, G.P., Perusse, L., Bouchard, C., & Rao, D.C. (1989). Cardiovascular risk factors in a French Canadian population: resolution of genetic and familial environmental effects on blood pressure using twins, adoptees, and extensive information on environmental correlates. *Genet Epidemiol*, *6*, 571-588.

Schork, N.J. (1993). Extended multipoint identity-by-descent analysis of human quantitative traits: efficiency, power, and modeling considerations. *Am J Hum Genet*, *53*, 1306-1319.

Tambs, K., Moum, T., Holmen, J., Eaves, L.J., Neale, M.C., Lund-Larsen, G., & Naess, S. (1992). Genetic and environmental effects on blood pressure in a Norwegian sample. *Genet Epidemiol*, *9*, 11-26.

Vogler, G.P., Rice, T., Perry, T.S., Laskarzewski, P.M., & Rao, D.C. (1991). Familial aggregation of lipids and lipoproteins in families identified through random and nonrandom probands in the Oklahoma Lipid Research Clinic Family Study. *Coronary Artery Disease*, *2*, 167-174.

Wilson, A.F., Elston, R.C., Sellers, T.A., Bailey-Wilson, J.E., Gersting, J.M., Deen, D.K., Sorant, A.J., Tran, L.D., Amos, C.I., & Siervogel, R.M. (1990). Stepwise oligogenic segregation and linkage analysis illustrated with dopamine-beta-hydroxylase activity. *Am J Med Genet*, *35*, 425-432.

SUBJECT INDEX

diet, 97, 102, 242
divorce rates, 103
DNA markers, 195, 196, 198, 199, 200, 201, 202, 204, 230, 231, 233
DZ twins, 7, 11, 12, 25, 31, 33, 34, 35, 36, 50, 56, 59, 60, 75, 78, 129, 131, 134, 141, 142, 149, 154, 186, 236, 238, 239

E

education, 92, 209
Endophenotypes, 202
environmental factors, 4, 5, 6, 9, 14, 15, 16, 17, 26, 27, 31, 37, 39, 54, 55, 57, 58, 59, 62, 63, 68, 69, 95, 97, 98, 99, 100, 101, 104, 105, 114, 118, 119, 120, 121, 122, 129, 130, 131, 132, 133, 134, 141, 144, 146, 147, 149, 150, 185, 186
environmental influences, 2, 6, 7, 8, 9, 12, 15, 16, 17, 18, 20, 21, 26, 42, 45, 47, 75, 80, 89, 90, 91, 97, 186
ethnic relations, 152
etiological factors, 4, 8, 14, 17
exercise, 3, 39, 48, 135, 138, 139, 149, 242
extended families, 210
extended family, 149

F

familial aggregation, 5, 155, 156, 157, 167, 168, 169, 170, 180, 181, 183
Familial Aggregation, 155
Familial Transmission, 127
family design, 4, 7, 243
family environment, 5, 79, 86, 127, 128, 129, 131, 132, 134, 141, 150, 185
family experiences, 15
family planning, 154
Family Studies, 127, 128
fibrosis, 185, 206
Finland, 73
FORTRAN, 219
fragile X, 4, 3, 14, 20
France, 129, 151

G

gene expression, 4, 23, 26, 28, 31, 32, 33, 34, 38, 39, 47, 50
gene frequency, 175, 197
gene regulation, 35, 39
genetic diseases, 198, 204
genetic epidemiology, 89, 167, 235, 244
genetic etiology, 19, 21, 185, 186
genetic factors, 4, 13, 14, 15, 17, 24, 25, 27, 59, 63, 95, 97, 98, 101, 116, 118, 120, 122, 168, 173, 238, 242
Genetic Heterogeneity, 196
genetic loci, 31, 48, 187, 188
genetic markers, 199, 204, 213, 235
genetic modeling, 3
genetic recombination, 187
genetic relationship, 24, 166, 168, 185, 238
Genetic Simplex Model, 58, 60
genetic susceptibility, 168
genetic variance, 24, 26, 28, 31, 32, 33, 36, 38, 47, 48, 63, 76, 79, 80, 86, 129, 131, 146, 238, 239, 240
genetic vulnerability, 5
genetically complex diseases, 197
Genome Screening, 209
genotype, 4, 21, 54, 56, 57, 63, 65, 66, 67, 68, 69, 84, 86, 96, 97, 120, 121, 132, 141, 169, 173, 175, 180, 187, 189, 191, 192, 199, 203, 209, 216, 244
genotype-environment interaction, 4, 54, 58, 63, 65, 66, 69
Genotype-Environment Interaction, 53, 55
genotyping, 5, 211, 231
Germany, 4, 72
group familiality, 9, 10, 11

H

heart disease, 209, 246
heritability, 6, 7, 9, 11, 12, 28, 29, 33, 35, 36, 46, 47, 75, 76, 79, 83, 88, 129, 130, 146, 148, 149, 153, 238, 239, 240, 242, 243, 246
human genome, 4, 95, 120, 208, 210, 230
Huntington's Disease, 14, 185, 206
hyperactivity, 3, 6, 11, 19, 21, 204
hypertension, 209, 211, 234

I

identical twins, 9, 15, 16, 72, 75, 87, 88, 89, 92, 186
individual differences, 4, 2, 4, 5, 6, 7, 9, 11, 12, 13, 19, 20, 21, 23, 32, 34, 35, 39, 43, 53, 58, 75, 80, 82, 90, 124
Individual Differences, 4, 7, 49
insulin, 213
intelligence, 4, 75, 82, 83, 87, 88, 90, 96, 101
interaction factors, 57, 67

Iowa, 95, 97, 99, 100, 101, 102, 103, 104, 108, 110, 111, 114, 116, 117, 119
IQ, 10, 11, 26, 49, 69, 73, 75, 80, 81, 83, 86, 88, 89, 92, 93, 152, 153, 233

J

Japan, 73

K

Knowledge, 128, 159, 170

L

language and communication, 4
LDL, 242
lifestyle, 89, 139, 173
linear logistic regression, 159, 169
linkage analysis, 5, 13, 120, 187, 188, 189, 191, 194, 196, 199, 205, 206, 208, 210, 211, 212, 220, 224, 230, 231, 232, 233, 234, 235, 242, 243, 245, 247
Linkage analysis, 206, 230, 232, 235, 245
lipoproteins, 244, 246, 247
Lod Score, 188, 190, 191
longitudinal behavior genetic analysis, 4
longitudinal model, 55, 63

M

maladaptation, 1, 3, 13, 14, 18
maladaptive behavior, 2, 6, 7
maladaptive behaviors, 3, 4
Maladaptive Development, 7, 15
marital duration, 129, 133, 135, 136, 138, 139, 140, 149
Marker Loci, 235
marriage, 129, 138, 140, 151, 152, 153, 173
mental ability, 88
Mental Disorders, 121
mental retardation, 3, 10, 11, 20, 204
microsatellite markers, 235, 243
Modelling, 51, 171, 173, 175, 177, 181, 182
molecular genetics, 5, 95, 121, 235, 243
motor problems, 4
multiple regression, 11, 86, 135
muscular dystrophy, 185, 207
MZ twins, 12, 25, 33, 34, 35, 54, 60, 73, 78, 79, 82, 89, 130, 131, 134, 141, 142, 149, 186, 237, 238. 239. 241. 242

N

NATO, 70
neonatal period, 96
nervousness, 138, 139, 140
Netherlands, 53, 70
New Zealand, 4, 72
nonshared environmental effects, 238
nonshared environmental factors, 15, 16, 17
normal behavioral development, 3
normal development, 1, 3, 4, 8, 13, 14, 15, 17, 18, 20, 102
Normal Development, 13, 17
North Atlantic Treaty Organization, 70
Norway, 127, 133, 150, 151
nuclear familes, 180
nuclear families, 68, 127, 131, 140, 153, 232

P

panic disorder, 98, 118, 182, 204
path analysis, 74, 75, 105, 140, 146, 235
path models, 31, 40, 75
Pathological Development, 2
pedigrees, 120, 155, 160, 191, 196, 197, 199, 203, 205, 210, 211, 213, 220, 223, 230, 231, 236, 245
personality, 4, 5, 2, 3, 5, 21, 76, 80, 82, 83, 86, 87, 88, 90, 91, 92, 98, 103, 117, 128, 130, 135, 139, 149, 151, 152, 153, 199, 235, 246, 247
phenocopies, 97, 98, 120, 192, 193, 194, 196, 199, 209, 212, 224, 225, 226, 228
phenotype, 4, 13, 24, 31, 32, 35, 36, 37, 38, 39, 49, 54, 55, 58, 60, 63, 86, 96, 97, 98, 111, 129, 130, 132, 140, 141, 146, 151, 189, 191, 194, 195, 203, 209, 210, 211, 213, 237
phenotypic variance, 26, 33, 34, 36, 146, 238, 239, 242
phenylketonuria, 2, 97
prenatal care, 102
problem behaviors, 4, 2, 5, 7
psychiatric illness, 104, 110, 121, 151, 185, 246
psychological differences, 88
psychopathology, 1, 2, 3, 4, 5, 8, 14, 15, 17, 18, 19, 20, 21, 40, 103, 106, 108, 110, 111, 114, 116
pubertal stages, 39, 43, 45, 49
puberty, 28, 32, 33, 34, 39, 41, 43, 46, 47, 48, 49, 50

Q

QTL Model, 237
quality of life, 134
quantitative traits, 5, 50, 213, 230, 231, 235, 236, 242, 243, 245, 246, 247
Quantitative Traits, 235

R

reading disability, 4, 12, 19
recombination, 188, 189, 190, 191, 200, 210, 215, 216, 218, 219, 221, 226, 227, 229, 237, 238, 239, 240, 241
Regressive Logistic Model, 155, 169
Retardation, 3
risk factors, 8, 14, 127, 169, 180, 181, 183, 235, 242, 243, 247

S

schizophrenia, 5, 15, 16, 19, 96, 97, 98, 124, 125, 185, 186, 187, 194, 196, 197, 198, 199, 200, 202, 203, 204, 205, 206, 207, 208, 213
separation studies, 95
sex differences, 142, 144, 162
sexual orientation, 204, 231
Shared environmental, 237, 241
shared environmental effects, 12, 238, 239, 241
sib pairs, 197, 200, 214, 223, 224, 231, 236, 238, 243
siblings, 6, 7, 9, 10, 11, 12, 15, 16, 17, 19, 20, 26, 99, 103, 104, 129, 131, 132, 134, 140, 141, 142, 143, 144, 146, 147, 149, 150, 157, 170, 171, 172, 173, 177, 178, 180, 212, 231, 238
sickle-cell anemia, 14
sisters, 141, 144, 147, 148, 149, 207, 233
smoking, 5, 96, 102, 132, 156, 157, 160, 161, 162, 163, 165, 166, 167, 168, 172, 173, 175, 177, 178, 182
social homogamy, 129, 130, 134, 140, 141, 149
social interaction, 3
social security, 100, 102
socioeconomic status, 80, 81, 93, 114, 154
Socioeconomic status, 87, 90
Spouse Resemblance, 127, 138
Spouse selection, 151
stereotyped behavior, 3
Substance Abuse, 111
Sweden, 4, 72

T

Tourette's syndrome, 204
TRA method, 75, 76, 86
traits, 3, 5, 2, 5, 18, 23, 39, 72, 75, 76, 77, 79, 82, 83, 86, 88, 91, 128, 129, 132, 149, 152, 157, 164, 169, 181, 182, 187, 202, 203, 204, 209, 210, 211, 213, 229, 232, 233, 234, 235, 236, 241, 242, 245, 247
twin design, 4, 7, 55, 237, 243
Twin Model, 237
twin pairs, 4, 16, 21, 23, 27, 30, 34, 35, 39, 40, 44, 54, 56, 59, 63, 64, 66, 72, 128, 155, 173, 186, 237, 239, 240
twin study, 5, 6, 7, 19, 21, 23, 87, 90, 121, 134, 148, 149, 151, 205, 236

U

unique environment, 26, 43, 238
United Kingdom, 4, 72, 80
United States, 4, 72, 80
Utah, 185

W

whole-genome screening, 5, 210, 211, 224, 229